ECONOMIC GEOLOGY OF NATURAL GAS HYDRATE

Coastal Systems and Continental Margins

VOLUME 9

The titles published in this series are listed at the end of this volume.

Economic Geology of Natural Gas Hydrate

By

Michael D. Max
MDS Research & Hydrate Energy International,
St. Petersburg, FL, U.S.A.

Arthur H. Johnson
Hydrate Energy International,
Kenner, LA, U.S.A.

and

William P. Dillon
Geological Survey Emeritus and Hydrate Energy International,
Woods Hole, MA, U.S.A.

With contributions of

Sarah Holman, Michael Kowalski, George Moridis, John Osegovic,
Shelli Tatro and George Taft

Springer

A C.I.P. Catalogue record for this book is available from the Library of Congress.

ISBN-10 1-4020-3971-9 (HB)
ISBN-13 978-1-4020-3971-3 (HB)
ISBN-10 1-4020-3972-7 (e-book)
ISBN-13 978-1-4020-3972-0 (e-book)

Published by Springer,
P.O. Box 17, 3300 AA Dordrecht, The Netherlands.

www.springer.com

Cover Illustration
Flare image shown by Satoh, et al. (2003). Picture taken and supplied by T. Collett. Image enhancement by Rachel Max.

Printed on acid-free paper

Printed in the Netherlands.

DEDICATION

This book is dedicated to **Dr. Keith A. Kvenvolden**, a pioneer in the studies of gas hydrate and the broader issues of petroleum in the natural environment. Keith has been one of the most knowledgeable scientists in the field of gas hydrate geochemistry. Furthermore, he is a true gentleman who has encouraged others and has been a guiding force to his peers and younger scientists.

And also to:

Dr. Burton G. Hurdle, a well-known facilitator and scientist of the Acoustics Division of the Naval Research Laboratory for over half a century, whose personal support for younger scientists trying to do breakthrough research led directly to the passing into law of the Gas Hydrate Research Act (of 2000).

TABLE OF CONTENTS

Preface

This book is a companion to "Natural Gas Hydrate in Oceanic and Permafrost Environments" (Max, 2000, 2003), which is the first book on gas hydrate in this series. Although other gases can naturally form clathrate hydrates (referred to after as 'hydrate'), we are concerned here only with hydrocarbon gases that form hydrates. The most important of these natural gases is methane. Whereas the first book is a general introduction to the subject of natural gas hydrate, this book focuses on the geology and geochemical controls of gas hydrate development and on gas extraction from naturally occurring hydrocarbon hydrates. This is the first broad treatment of gas hydrate as a natural resource within an economic geological framework. This book is written mainly to stand alone for brevity and to minimize duplication. Information in Max (2000; 2003) should also be consulted for completeness.

Hydrate is a type of clathrate (Sloan, 1998) that is formed from a cage structure of water molecules in which gas molecules occupying void sites within the cages stabilize the structure through van der Waals or hydrogen bonding. Hydrate crystallizes naturally in permafrost cryosphere and marine sediment where water and sufficient gas molecules are present, and pressure and temperature conditions are suitable to support spontaneous nucleation and growth (Chapter 2). Hydrate is mainly composed of water and a hydrate forming gas (Fig. P1). When gas hydrate forms, it concentrates the gas in the hydrate crystal lattice. Where methane is the hydrate forming gas, about 164 m^3 of methane (at STP) can be contained within the solid crystalline hydrate at any pressure-depth. This element of concentration and the large volumes of hydrate known or projected, are the attributes that render hydrate a potential economic resource of combustible natural gas on a national or world scale.

Natural gas hydrate is stable in a zone of that extends downward from some depth below the Earth's surface in permafrost regions to a greater depth than water ice is stable (4.4.2; Fig. 4.9). In oceans and deep lakes, gas hydrate is stable from some depth in the water column down to some depth below the seafloor that is also determined by rising temperature (3.2.1; Fig 3.1). Natural gas hydrate that forms in the water, however, is positively buoyant and floats upwards and naturally dissociates. Hydrate that forms on the seafloor may be held in place by intergrowing with sediment. A region referred to as the gas hydrate stability zone (GHSZ) includes hyrate on the seafloor and hydrate formed in sediments beneath the seafloor, as well as the analogus zone of hydrate development in permafrost regions.

The general physical chemistry of gas hydrate, its formation in relation to its general location and the depth of GHSZs in which hydrate may occur, are relatively well known. Our concern is to apply those aspects of the physical chemistry of natural gas hydrate that are most important to identifying the best conditions for the formation of hydrate concentrations of economic proportions. Also, we identify and discuss difficulties that must be overcome in both finding and recovering the natural gas, as well as pointing out hydrate-specific opportunities in the developing field of commercial recovery of gas from natural gas hydrate deposits.

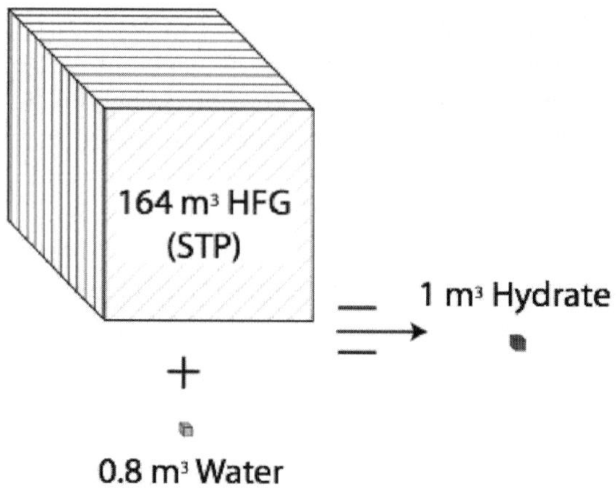

Figure P1. Proportional volumetric relationship between water, gas, and hydrate in a fully saturated hydrate. The compressional attribute of hydrate formation and its concentration of natural gas within the hydrate make concentrations of hydrate potential energy resources.

In some respects, a book on the topic of gas hydrate economic geology is premature because there are no proven economic deposits of gas hydrate, with the possible exception of the Messoyakha gas field of western Siberia. Even at Messoyakha, however, there is considerable uncertainty about extraction of the natural gas bound up in the permafrost hydrate. Although Makogan (1981) identified about 5 billion m^3 ($Bm^{3)}$ of gas from dissociated gas hydrate, Collette and Ginsberg (1997) suggested that hydrate has not substantially contributed to the volume of extracted gas.

Large volumes of natural gas hydrate, at least in its oceanic environment however, appears to occur widely (Kvenvolden & Lorensen, 2001; Soloviev, 2002a, 2002b) although Laherrere (2000) and Lerche (2000) draw attention to the uncertainty of estimates. For instance, over 65% of a 20,000 km^2 area of the seafloor off Taiwan in the northern sector of the South China Sea appears to have well developed BSR (Bottom Simulating Reflector on reflection seismic records) (3.3.2) in water depths from 700 m to 3,500 m, where hydrate was originally identified from poor reflection seismic records (McDonnell et al., 2000), (Liu et al., 2004, pers. com, 2004). The widespread BSRs off the U.S.

east coast are well known, and have been subject to ground truth studies by well constrained drilling (Dillon & Max, 2003; Goldberg et al. 2003). In addition, new discoveries and valuation of permafrost hydrate, which may be more amenable to near-term commercialization, provide an starting point for development.

There are no existing industry-standard practices for detailed delineation of economic hydrate deposits or for volumetric assessment of 'grade' or 'value' as there are for conventional hydrocarbon and mineral deposits. Nor is there any economically constrained extraction experience on which to base commercial valuation of other gas hydrate deposits. In fact, methods for identifying potential concentrations of hydrate that may have economic potential do not yet exist. The most ubiquitous indicator of the presence of gas hydrate is the first order identifier of BSR on seismic records (3.3.2), but a BSR actually identifies the top of a gas-rich zone beneath sediments whose porosity may be effectively sealed by hydrate. The existence of a BSR does not identify high-grade hydrate concentrations. In addition, drilling has proven little about geologic models that are proposed by us as controlling hydrate development and distribution (Chapters 2, 4, 6). There is presently no undisputed methodology for identifying the extent, reservoir character and strength, or the volume of in-situ hydrate development. Finally, it is not yet known whether hydrate actually constitutes a producible energy resource on the scale of its apparent volume of up to twice the combustible content of all conventional hydrocarbons on Earth or whether it constitutes any potential as an energy resource.

Until recently, gas hydrate has been regarded as a scientific curiosity. Lee et al. (1988) for instance, although identifying the replacement of petroleum by natural gas as the main source of the Earth's combustable energy, make no mention of natural gas hydrate as an econmic gas resource. Industrially, hydrate has been, and continues to be as an impediment to flow assurance in gas and petroleum pipeline systems while its potential for separation of materials has gone largely unresearched. Indeed, industry is still spending on the order of two million dollars a day on inhibiting and remediating unwanted gas hydrate (OTC, 2004). Attention is now turning to the potentially very large energy resource possibilities of naturally occurring hydrate (Kvenvolden and Lorenson, 2001). The equivalent of giant and super giant gas fields may occur in concentrated and economically exploitable hydrate deposits. Gas hydrates constitutes a new unconventional gas play, and may prove to be one of the major energy resources of the 21st century. For a number of countries, hydrate may be the only indigenous option for non-renewable, combustible energy resources.

This book examines broadly the economic geologic potential of gas hydrate in both permafrost and oceanic environments. We have developed geologic and paragenetic models for hydrate concentration and extraction that merge lessons learned from experiments nucleating and growing hydrate in natural seawater, which is similar in composition to the pore water of marine sediments in which natural gas hydrate occurs. Suggestions are made for exploration and extraction scenarios, especially of oceanic hydrate where geological structure may not have the same significance as it does in permafrost hydrate deposits.

Chapter 1 discusses hydrate as part of a spectrum of naturally occurring hydrocarbon resources. Specifically, different sources of conventional and unconventional gas deposits are discussed including coalbed methane, which may provide the most relevant commercial development model for bringing an unconventional gas resource into production and profitability. Because the energy needs and national economic and political parameters governing decision-making may vary considerably, the impetus driving development of hydrate is urgently felt in some countries while it is ignored in others. Superdemand for energy worldwide has also lifted hydrocarbon prices to a new plateau. Energy prices are likely to be maintained substantially above the levels of the inexpensive world energy paradigm that previously had been controlled by the United States.

Chapter 2 focuses on those elements of gas hydrate nucleation and growth that are important to the formation of hydrate concentrations. Growth media, including both gaseous and aqueous environments, are discussed, along with the natural mechanisms that are likely to yield high pore saturations through heterogeneous nucleation and slow growth as a result of naturally modulated supply of hydrate-forming reactants. The principals of both dissociation and dissolution are also described because these are important for recovery of natural gas from hydrate. Physical chemistry is used to illustrate growth models that have been tested through experimentation, and are constrained by thermodynamics, to produce solid hydrate. This section contains enough physical chemical information to allow a geologist or economic geologist who is not a specialist in physical chemistry and hydrate paragenesis to better understand the hydrate system and environmental constraints that may lead to the formation and recovery of natural gas from economic deposits of hydrate.

Chapter 3 characterizes hydrate as part of the geological environment. Gas hydrates comprise an unconventional diagenetic hydrocarbon mineral-like deposit that may be associated with conventional deposits of natural gas. One of the present difficulties hampering many petroleum geologists and geophysicists is that hydrate, and especially concentrations of hydrate, are not governed by the same rules as are conventional hydrocarbon deposits. Gas hydrate is unique among hydrocarbon resources in that it is a solid crystalline material in the natural state, which can rapidly transform to its constituent water and gas in response to changing pressure and temperature.

Porous sediments containing high concentrations of hydrate have been investigated through drilling in the Mackenzie Delta of Canada and in the deep continental shelf margin offshore Japan. Data derived from drilling is now providing the ground truth needed for realistic assessment of hydrate grades and values.

Chapter 4 summarizes known hydrate localities in both permafrost and oceanic areas from the point of view of their modes of formation. Clear distinctions between them are described, as are the similarities and distinctions between them and conventional gas deposits and the means available for recovery of the natural gas from hydrate deposits. Hydrate is identified as an economic mineral deposit and compared with other solid, crystalline stratabound mineral deposits of diagenetic origin, with which there are many similarities. A great deal is understood about both metallic and non-metallic mineral deposits with similar paragenesis. Application of that knowledge should aid in the identification of geological settings most appropriate for hosting significant concentrations of hydrate economic targets or 'sweetspots', so as to guide exploration.

Chapter 5. Natural gas derived from hydrate may be inherently more expensive to recover than most conventional gas deposits because the equivalent of secondary recovery methods must be applied from the beginning. Whereas conventional gas and petroleum deposits are artesian or only have to be pumped to the surface, the hydrate must first be converted to gas. The three main methods for converting the hydrate to recoverable gas (Depressurization, inhibitor stimulation, and thermal stimulation) require different infrastructure and materials and consequent costs. It is also likely that more than one method may be used together in some way, for instance, heating and depressurization. Existing exploration and assessments of hydrate developments are described. Nonetheless, because of the higher energy cost levels that are likely to persist, and because better economic cost and extraction models for hydrate are being developed, extracting natural gas from hydrate may become an attractive investment.

Chapter 6 takes forward the current knowledge of hydrate formation and models likely economic deposits of oceanic gas hydrate. Oceanic hydrate deposits are diagenetic and have grown in place in the sediment by incorporating pore water, under pressure and temperature conditions similar to those in which they now exist. Identification of BSR (3.3.2), is a first order technique for identifying the phase boundary between the hydrate stability zone and underlying gas (Chapters 3 and 5), but more refined techniques are required to identify economic deposits of hydrate. The different models for economic deposits of hydrate are assessed for their response to seismic and other exploration methods and some guidance is offered. Permafrost hydrate deposits appear to be fundamentally different from oceanic hydrate deposits. For the most part, they have been formed from existing conventional gas deposits that were converted to hydrate through the depression of the cryosphere across at least part of the gas zone during intensification of glacial periods, rather than being formed through the migration of hydrate-forming gas into the GHSZ.

Hydrate will not be mined, of course. It will be converted to its constituent natural gas and recovered as gas. But extraction models in some deposits in weaker geological strata may do well to follow mining practices for extraction where some of the ore is left behind in order to promote reservoir stabilization and prevent collapse. This chapter focuses on applying different hydrate nucleation and growth models from Chapter 2 in different natural groundwater systems and geological situations.

Chapter 7 presents one of the most recent thermodynamic models for In-Situ Conversion of Gas Hydrate to Natural Gas that is particularly applicable for high-grade oceanic hydrate deposits, which will probably be the initial source of natural gas from oceanic hydrate. Physical cases of hydrate occurrence that are likely to be encountered in potential economic deposits of gas hydrate are analyzed and shown to have significant economic potential for safely converting and extracting natural gas from hydrate.

Chapter 8 considers the offshore regulatory and permitting environment for gas hydrate, which is governed by UNCLOS, the Law of the Sea. The articles that pertain to the geographic position of likely hydrate deposits along continental margin, as well as examples for particular countries, are explained in detail. National exploration and extraction resource legislation is only briefly discussed. Existing and potential claims for areas beyond the present limit of 200 nmi are not discussed, because the framework for resolution of those claims and for competing claims to continental shelf areas claimed by more than one nation exist in the present UNCLOS documents and procedures.

Chapter 9 summarizes the main characteristics for both permafrost and oceanic natural gas hydrate and the timeliness of the recognition of natural gas hydrate as the likely next big gas play. Emphasis is placed upon the sufficiency of the present level of technology in deep water drilling and hydrocarbon exploration, extraction, and infrastructure for recovering natural gas from oceanic and permafrost hydrate deposits.

Glossary. References are mainly from mining, hydrocarbon exploration, physical chemistry, and geological sources.

References from the text are included in this selected list only where they are not already referenced in the first book for brevity and because the two books are intended to be complementary. There are also a number of references in this list that are not referenced in the text. This is particularly true in the case of some foreign references and where a large number of references have a common theme. In this case, one or only a few references are used in the text.

Miscellaneous. This short section includes the full contact information for authors, a comment on the fresh water sequestered in natural gas hydrates that may be of environmental significance, and a short discussion of the first known experiment that produced gas hydrate by Joseph Priestley during 'one frosty winter' in January, 1779 (1790).

In addition to references of publications, cross-references between sections are made between chapters using section numbers such as (2.3.1), figure numbers (which have unique numbers in each chapter) or references to other chapters and the Glossary.

The authors of this book agree strongly with the visionary development efforts to develop natural gas hydrate as an energy source that are being carried out in a number of countries. The greatest step-increment in progress, however, took place as part of an international effort.

"March 7th 2002, an extremely cold winter night in Arctic Canada: a flare from dissociating natural gas hydrate deep below a test well burned for the first time in oilfield (hydrocarbon exploration) history. This flare is one of the products of an international joint project, "Mallik 2002 Gas Hydrate Production Research Well Program", undertaken by a partnership of seven organizations from five countries: the Japan Oil and Metals National Corporation (renamed from: Japan National Oil Corporation, JNOC), the Geological Survey of Canada (GSC), the GeoForschungsZentrum Potsdam, Germany (GFZ), the United States Geological Survey (USGS), the Indian Ministry of Petroleum and Natural Gas (MOPNG), the BP-Chevron-Texaco Joint Venture Group and the United States Department of Energy (US DOE), with the support of the International Continental Scientific Drilling Program (ICDP)" (Tsuji & Emmermann, 2003).

The image of the gas flare shown on the cover of this book was the immediately visible result of an in-situ stimulation test of controlled changes in temperature and pressure in the Mallik 5L-38 hydrate well that was designed to convert solid gas hydrate in a permafrost hydrate reservoir into its constituent gas and water and to produce a sustained gas flow. The conversion produced pressurized gas that was vented, and flared, according to industry practice. This image has been shown in a lower resolution format a number of times before but it is included here in uniquely high resolution, as it may be the most important symbol of progress in the development of natural gas hydrate as an energy source. This moment may come to be regarded as Time-Zero in the practical development of economic exploitation of natural gas hydrate resources.

INTRODUCTION

NATIONAL PROGRAMS FOR HYDRATE RESEARCH

Energy potential of natural gas hydrate is now the primary motivating agent for hydrate research at the National level of the United States and most other countries that are making significant investments in hydrate research. This thrust follows a long period (from the 1930s) during which the primary research interest in hydrate was driven by the energy industry's concerns in the field of flow assurance, or mitigation and remediation of unwanted hydrate that formed in pipes carrying wet hydrocarbons. Drilling safety and flow assurance appear to be the main concerns of most energy companies, many of which are involved with the government-driven hydrate-related energy research, while the carbon cycle and global climate modeling appears to be the research driver in other countries, particularly those which do not have a likelihood of hydrate energy resources in their oceanic (or permafrost) areas.

The United States Department of Energy (DOE) initiated the first national gas hydrate program at government level in 1982. The Departmental program made extensive use of contractors and was based at the Morgantown West Virginia DOE laboratory that was the precursor to the present National Energy Technology Laboratory (NETL). The program was active until 1992, after $8 million had been well invested, but was terminated owing to the low price of conventional energy sources and internal DOE policies. The program was invaluable for transforming the field of hydrate science to a potential energy program and for establishing the framework for further development worldwide. The Japanese and the Indian governments built on the results of the U.S. program and initiated national hydrate programs in the mid-1990s. The United States established a formal national hydrate research program in 2000 with the passing of the Gas Hydrate Research and Development Act. Since then, a number of countries having energy or foreign currency issues have initiated hydrate research programs or at least have raised their level of awareness as to the existence of potential hydrate energy resources.

Countries with Developed National Hydrate Energy Interests

Canada: Early pioneering work in the early 1970s proved the existence of hydrate in permafrost terrane through drilling. Hydrate has been identified in over 250 wells in five areas: (1) the Cascadia margin of western Canada, (2) the Mackenzie Delta and (3) the northern shelf of the Arctic Islands bordering the Arctic Ocean, (4) the western margin of the Labrador Sea (indications of the presence of hydrate has been observed on reflection seismic lines of the corresponding Greenland shelf by M.D. Max), and (5) the Atlantic coast of Canada (Majorowicz and Osadetz, 2001; 2003). Relatively sophisticated

Michael D. Max et al. (eds.), Economic Geology of Natural Gas Hydrate, 1–16.

estimates of the volume of hydrate in these fields has been made (Mosher et al., 2005; Osadetz et al., 2005)

Perhaps of greatest boost to understanding the energy potential of gas hydrates is the research since 1997 centered on the Mallik gas hydrate research site in Canada's Mackenzie Delta. The Geological Survey of Canada (GSC) and the Japan National Oil Corporation (JNOC) have led this work. Among the participants are the GSC, JNOC, USGS, DOE, GeoForschungsZentrum Potsdam (GFZ), India Ministry of Petroleum and Natural Gas (MOPNG)/Gas Authority of India (GAIL), and the ChevronTexaco-BP-Burlington joint venture group. The project has also been accepted by the International Scientific Continental Drilling Program, which provided a broadening of the scientific research goal. At present, the Mallik deposit is the best-evaluated hydrate deposit in the world (Chapters 3 and 5) and the only one in which a natural gas production test from hydrate has been attempted.

In early 1998, the JAPEX/JNOC/GSC Mallik 2L-38 gas hydrate well was drilled to a depth of 1,150 m in the Mackenzie Delta. Gas-hydrate rich sandy to pebbly clastic strata were identified at depths between 890 to 1,110 m beneath 640 m of permafrost. Silt and clay-rich sediments such as silts and clays, which separated the main gas hydrate layers, were free of hydrate or contained little hydrate. Typically, hydrate-bearing strata were 10 cm to 1.5 m thick with an estimated porosity of 25 to 35%. Hydrate concentrations were up to 80% of pore saturation (Uchida et al., 2001). Other wells were drilled and in 2002, a brute-force production test in the 5L-38 well was capable of sustaining a large flare (Satoh et al., 2003). Although the hydrate conversion test consumed more energy than it produced from an area of hydrate-enriched sediment, continuous conversion of hydrate was demonstrated.

The GSC recently established a new gas hydrate research and development program as part of Natural Resources Canada (NRCan), which is a federal government department specializing in the sustainable development and use of natural resources, energy, minerals and metals, forests and earth sciences. The new science program consolidates GSC hydrate researchers. The focus is on gas hydrates as an environmentally friendly source of fuel for North America. University researchers are funded by a scientific funding agency similar to the U.S. National Science Foundation. Other government agencies appear to operate independently. The mechanism for the coordination of overall hydrate research in Canada is unclear.

A joint international research program that has been largely funded by the Japanese government succeeded in 2002 in carrying out a short production test (cover figure) at Mallik in the Mackenzie delta. This test showed that conversion of hydrate to recoverable gas was a physical possibility and substantiated thermodynamic recovery models. When the gas pipeline to the Mackenzie Delta from Alaska is completed (by 2007 or 2008?), it is likely that some natural gas from hydrate will be recovered along with the associated conventional gas, even without a hydrate-specific hydrate recovery program. The Mallik and nearby related fields could be developed for hydrate natural gas on a fast track if required.

Chile: More than 70% of Chile's natural gas is imported from Argentina. Chile's experience has been that during periods of social and economic upheaval in Argentina, their gas supplies are likely to be interrupted. During two of these periods in the recent past, when gas supplies were cut off for weeks, Chile was subject to considerable economic distress because, as with almost every other country, they have no fallbacks for sudden energy shortages. Southern Chile produces a small amount of gas, but most of the long Chilean margin has not been explored for either conventional gas or hydrate deposits using modern technology.

Gas hydrate investigations to date have been conducted by an international collaboration that includes the Pontificia Universidad Catolica de Valparaiso, the U.S. Naval Research Laboratory, the University of Hawaii, and the Universities of Kiel and Bremen, Germany. These investigations have included piston coring, heat flow measurements, and collection of both normal and deep-tow seismic data. Gas hydrate has been recovered from some of the shallow cores.

Researchers collected the first hydrate-relevant data from Chile and the Universities of Bremen and Kiel (GEOMAR) along the Chilean margin in 2003. In November 2004, the Chilean government approved an expanded program to investigate the national gas hydrate resource potential. The second of two hydrate research cruises in Chilean waters as part of an international consortium led by the Naval Research Laboratory and Pontificia Universidad Catolica de Valparaiso (Chile) took place in the summer of 2004 (Gardner et al., 2004). These cruises involved seafloor sampling, chemical analyses, and high-resolution seismic surveys. Subsequent phases of the program are scheduled to commence in the latter part of 2005.

China: In 2000, three national natural science foundations with an interest in different aspects of the gas hydrate system commissioned research focused on gas hydrate. This research built on earlier surveys to identify gas hydrate undertaken in 1999 by the Guangzhou Marine Geological Bureau. In May 2004, the Center for Hydrate and Natural Gas Research was established in the Guangzhou Institute of Energy Conversion (Chinese Academy of Sciences), which is heading multidisciplinary research among academic and company interests. The Second Institute of Oceanography of the State Oceanic Administration is involved with some gas hydrate research, but has no gas hydrate program. In 2001, a gas hydrate project was established (Second 863 Program), and the Geological Survey of China has initiated a number of marine research projects focusing on the identification of hydrate (Yang et al., 2003). In 2002 a national gas hydrate project was initiated with the equivalent of 100 million dollars allocated as start-up funding. The first Chinese scientific program meeting of this project was held in Beijing in November 2003, with mainly Chinese and Japanese scientists attending. First order assessments of sea areas adjacent to China have identified considerable hydrate shows (Huang, 2004; Jiang, 2004; Wu, 2004; Wu et al., 2005).

Recent seismic surveys and research, including seismic data processing, complex trace analysis, AVO analysis with full waveform inversion, show that indications of gas hydrate occur in the marine sediments of the South China Sea and East China Sea passive margin sediments. BSRs have been recognized in the northern margin in the Xisha Trough and Dongsha regions (Song et al., 2001a, b, 2002b, 2003a; Fu et al., 2001; Ma et al., 2002; Hu et al., 2002) and on the western slope of Okinawa trough and other areas (Yao, 1998; Song, 2000, 2001c; Meng et al., 2000, Fu et al., 2001; Zhang et al., 2002; Qian et al., 2002; Liu et al., 2002; Wu et al., 2005). The Xisha Trough and Dongsha regions and the western slope of the Okinawa Trough are the principal areas of national gas hydrate interest in China. The Guangzhou Marine Geological Survey is carrying out hydrate research with the Leibniz Institute of Marine Science (GEOMAR).

India: The Indian national gas hydrate research program has moved from an early phase of preliminary identification of gas hydrate resources in their offshore area (including along the eastern side of the Bay of Bengal sector of the northern Indian Ocean) to one of focused research (Das, 2004). The Indian Department of Ocean Development (DOD) has announced that large quantities of hydrate have been identified along India's 7,500 km coast. The Institutes of Oceanography and the Institute of Geophysics have identified the Kerala-Konkan offshore region as having significant hydrate shows. \

There has been a sharp increase in funding that the Indian government has allocated to hydrate research and development. This interest in India's marine resources may track a general recognition by the Indian government that they must improve their naval and marine research capabilities. The availability of excellent naval platforms for use in a disaster relief role following the late-December 2004 tsunami in the Indian Ocean is due to this existing focus by India on their huge maritime area. Increased funding is aimed at making India one of the leading hydrate research nations. A multibillion-rupee budget (currently estimated at Rs. 12.5 billion) for developing technologies to tap ocean power (OTEC) has also been announced in 2004 (the time period over which this funding will apply is unclear).

New resources will aid this effort, including a new research ship (at a cost of Rs. 1.55 billion) that is largely dedicated to gas hydrate research. The new research vessel is scheduled to be operational by the beginning of 2006 and is intended to deploy new technology. The vessel will have a 48 m^2 deck, from which equipment can be lowered to the seafloor. It is planned to use advanced engineering seafloor drilling equipment. Drilling of the thickly sedimented submarine fans in the Indian Ocean is being contemplated by the Integrated Ocean Drilling Program (Clift & Molnar, 2003). IODP will provide high-resolution climatic records along with data relevant to the presence of potential source beds for the production of natural gas. The Indian government is aggressively exploring their hydrate potential resources, and has licensed commercial exploration interests for hydrate as well as conventional gas and oil.

Following discussions with the Naval Research Laboratory (U.S.) in the late 1990s, the Indian DOD also allocated Rs. 800 million to initiate a collaborative gas hydrate exploration project. Discussions are underway to collaborate with Russia in joint research programs in Indian waters. The National Geophysical Research Institute in Hyderabad has identified at least nine potential hydrate resource areas where research interests will focus on exploration.

Japan: The Japanese government, through its Ministry of Economy, Trade and Industry (MITI), has commissioned the greater part of current hydrate research funding, which for the last five years has been greater than the rest of the world combined. Japan National Oil Corporation (JNOC) has integrated hydrate research and development of both basic research and field surveys with an aim of exploiting methane hydrate as a commercial energy resource. The budget for 2004 was originally $100 million, which included a drilling program in the Nankai Trough. The funding also supports research within Japan, where there are excellent established laboratory facilities. The research is mainly aimed at improving production rates, studying models of potential pressure regimes and gas migration paths within a reservoir during production, and assessing drilling and completion issues related to soft sediments. The Japanese are using seismic methods to optimize exploration techniques and locate hydrate-rich areas but have not carried out extensive modeling of the depositional system in which the hydrate resides, relying principally on the study of seismic data.

A research consortium for methane hydrate resources in Japan (also known as the MH21 Research Consortium) was established to undertake research in accordance with an R&D plan prepared by the Advisory Committee for National Methane Hydrate Exploration Program. There are currently about 250 people in 30 organizations working on the MH21 program. By the time phase 1 of MH21 wraps up in 2006 it is intended to select two sites off their coast for production tests. Phase 2 extends from 2007 to 2011 and includes two offshore production tests. Successive phases are intended to exploit hydrate.

The Japanese are also the only nation currently carrying out assessment drilling of potential hydrate deposits, although their field program is currently in a state of flux. The latest program was carried out based on planning for drilling and coring between 10 and 20 wells in the Nankai Trough off Japan's East Coast. Initial results indicate that their geologic model was incomplete. Produced gas did not behave as anticipated, resulting in an incomplete test program and results that were not completely satisfactory. This result is not an unusual occurrence in a program of testing resource deposits whose actual character and response cannot be known exactly. It is anticipated that the data will lead to improved understanding of the occurrence of gas hydrate in the reservoir. The Japanese program is thus currently going through a stage of reassessment that has set their program back from its planned milestones. This reassessment may have some impact on plans for a 3-6 month production test at either Mallik or the North Slope (with BP). We consider a physical model case similar to the Nankai hydrate deposits in Chapters (4 & 6).

Russia: Scientists in Russia were the first to recognize the energy potential of gas hydrate in its permafrost regions and the first to develop methodology for the in-situ conversion of natural gas hydrate to recoverable gas from permafrost hydrate deposits (5.7). Because Russia has such a large resource base of conventional natural gas, however, little emphasis has been placed by any national agency or energy company in Russia on the development of gas hydrate resources, although GASPROM, the State energy company, briefly investigated hydrate resources. Scientific research on hydrate has increased recently as part of individual initiatives and in step with the attention that hydrate is receiving worldwide. However, an integrated national program in hydrate research is apparently not being planned by Russian central or regional governments despite the availability of intellectual resources and experience and the clear evidence for large quantities of permafrost hydrate.

There are a number of established research groups that carry out hydrate research. A number of publications are in Russian only (RCM, 2003, S&E, 2004), but are often used as the basis for papers published in English. The Laboratory for Gas Hydrate Geology at VNIIOkeangeologia, St. Petersburg, is primarily responsible for the study of oceanic gas hydrate in the field. This is the main group that has sea-going capabilities. The Institute of Physical Chemistry in Novosibirsk primarily studies fundamental properties of gas hydrate. These studies include crystal structure and theoretical investigations of mainly high-pressure gas hydrate formation, such as would be found in oceanic hydrate. In Tyumen, the kinetics of gas hydrate formation and the influence of inhibitors are focused on developing methodology for controlling hydrate formation or remediating unwanted hydrate. VNIIGAS principally is concerned with gas hydrate concentrations in permafrost areas, particularly the complex thermodynamic aspects of relatively shallow hydrate. The Department of Permafrost (Cryogeology) at Moscow University has a close collaboration with VNIIGAS and conducts experimental studies of gas hydrate formation and decomposition and the composition and properties of hydrate-saturated sediments. The Department of Hydrocarbon Studies (Oil and Gas) at Moscow University also studies gas flow and gas hydrate accumulations in the field. In addition to these programs, there are a number of researchers working in the gas hydrate field, for instance, at the State University of Yakutsk. See Chuvillin, Ershov, Ginsburg, and Soloviev references, in particular.

Recent discoveries of hydrate in the northern slope of the Black Sea (Lüdmann et al., 2004; Naudts et al., 2005) suggest that the proto-delta of the Dnieper River, which forms a steep slope along the southern margin of the now submerged shelf region abutting a number of coastal States, provided organic rich sediments to the basin. The extent of hydrate mineralization and the source of the gas not yet been identified. There may be extensive deposits of hydrate, similar to the Caspian Sea to the east (Lerche, 2000).

United States: The Gas Hydrate Research and Development Act of 2000 (signed by President Clinton in May of that year) has been operated under the auspices of the U.S. Department of Energy (DOE). A Congressionally sponsored review of the research and development activities was undertaken in 2003 by the National Research Council (NRC, 2004) to review the progress made under the act and to provide advice on future research. Slightly over $29 million dollars was expended in funding hydrate research under the act since its inception up to the time of completion of the NRC report.

The United States Geological Survey (USGS) has maintained a continuous, broad research program in gas hydrate studies since 1990, and the USGS holds the greatest non-core repository of information on gas hydrate in the U.S. This work has included extensive field seismic studies on the Atlantic, Gulf of Mexico, Pacific, and Alaskan continental margins of the U.S., and also theoretical seismic analyses. Well logging, geochemistry and geotechnical studies have been carried out on USGS cruises and in cooperation with drilling offshore by the Ocean Drilling Program and onshore with the Geological Survey of Canada and the Japanese National Oil Corporation as well as other international cooperators. Laboratory geotechnical and petrophysics studies have complemented the field studies at both the Woods Hole and Menlo Park offices of the USGS. The Department of Energy (DOE) has provided partial support for field and some laboratory expenses to USGS in 1990-1993 and from 1997 to the present.

Since 2001, a Naval Research Laboratory (NRL) in-house methane hydrate research program developed a collaborative agreement with the University of Hawaii (Hawaii Natural Energy Institute) and NRL to form an international consortium for methane hydrate research. This collaboration has grown to include five nations (U.S., Canada, Chile, Germany, and Japan) dedicated to investigate the presence of methane hydrates off the coasts of the US, Canada, Japan and Chile. These collaborations were developed during the course of three workshops over the last three years (International Workshop on Methane Hydrate Research and Development), with up to 12 nations participating. Research goals of the collaboration focus on the basic NRL objectives to develop international efforts on methane hydrate exploration and Chilean goals, which are to locate and study hydrate distribution and composition along the Chilean coast, and to assess energy potential and geohazards.

In addition to DOE and individual agency funding, the National Science Foundation has continued to support research involving hydrate. DOE has also taken the lead in the establishment of a new Asia Pacific Economic Cooperation

(APEC) gas hydrate consortium that was proposed at the November 2003 gas hydrate workshop in Chile.

BP Exploration (Alaska) and the DOE also have undertaken a project to characterize, quantify, and determine the commercial viability of gas hydrates and associated free gas resources in the Prudhoe Bay, Kuparuk River and Milne Point field areas in northern Alaska. The University of Alaska in Fairbanks, the University of Arizona in Tucson, and the USGS also are participating in the Alaska BP project. Several Gulf of Mexico programs are currently under way. The most comprehensive study is a Joint Industry Project (JIP) led by ChevronTexaco, designed to further characterize gas hydrates in the Gulf of Mexico. Participants include ConocoPhillips, Total, Schlumberger, Halliburton Energy Services, U.S. Minerals Management Service (MMS), Japan National Oil Corp. and India's Reliance Industries. The primary concern of U.S.-based energy companies at present appears to be seafloor stability aspects of hydrate in near-seafloor sediments in order to mitigate drilling hazards.

Countries Showing Early Interest in Hydrate

Australia: Australia's ocean territory is about 16 million km^2, about twice as large as its land area. There are considerable thicknesses of continental slope and marginal basinal sediments in which gas hydrate can be expected to form, but exploration to date has focused on conventional hydrocarbon deposits. Australian is emerging as a major supplier of LNG and has recently completed a contract to supply China, amongst other countries.

Reflection seismics have been used to identify hydrate in a number of continental margin slopes and basins. For instance, a bottom-simulating reflector (BSR) has been identified in thick packages of Cretaceous and Tertiary sediment with numerous diapirs that fill the Southern Fairway Basin (SFB) on the Lord Howe Rise (LHR) of the Tasman Sea. Cores confirm the presence of hydrocarbon gases (Exxon et al., 1998; Dickens et al., 2001). Hydrate has also been inferred on the NW margin of Australia facing Indonesia. In addition to energy exploration issues, the Petroleum Exploration Society of Australia (PESA) hosted a workshop on seafloor stability aspects of gas hydrate and associated fluids and gases in seafloor sediments in October 2004. As in the U.S., energy companies are concerned about drilling safety and the impact of the hydrate systems on seafloor stability in the deeper water now being explored for hydrocarbon deposits. Australia presently has no national gas hydrate program although there is considerable activity among university earth scientists.

Belgium: Scientists at the Renard Centre of Marine Geology in Gent have been very active in marine hydrate research and have taken part in cruises and have organized and strongly participated in scientific meetings.

Brazil: Brazil has an extensive continental slope with thick marine sediments containing large amounts of organic carbon, a source for petroleum and gas deposits. The Amazon submarine fan bears a strong resemblance to the hydrocarbon-rich marine sediments of the Mississippi River delta, which is currently a focus of U.S. gas hydrate energy research. Indications of gas hydrate and subjacent gas deposits have been identified in the Amazon fan (Sad et al., 1998; Selva et al., 2000) in water depths between 600 and 2,800 m. Brazil is currently supporting considerable exploration and development of its abundant deep-water hydrocarbon resources. There is currently, however, no national gas hydrate research program.

European Union: with the notable exception of ***Ireland***, appears to be primarily interested in the hazard and the carbon cycle/global climate change aspects of hydrate, or for basic physical chemistry research. French, German, and Italian research vessels are maintaining aggressive marine research programs using state-of-the-art ships and technology in many ocean areas, most notably in Polar regions using icebreaker and ice-capable research vessels superior to anything the United States can field. Individual European universities and research centers, such as the Department of Geology and Geological Mapping, Institute of Geology and Mineral Exploration of Greece, Heriot-Watt University (The Hydrate Group, Institute of Petroleum Engineering) in Scotland, the School of Earth Science, University of Birmingham, and Geotec Ltd, Northants, UK, Geomar in Kiel, Technische Universität Berlin, and GeoForschungsZentrum Potsdam in Germany, the University of Aveiro, Portugal, the Istituto Nazionale di Oceanografia e di Geofisica Sperimentale (OGS) in Trieste, Italy, carry out laboratory and marine research hydrate studies. In southern Europe, in addition to hydrate in the deep Mediterranean Sea, there appears to be hydrate in the Gulf of Cadiz and possibly on the more sediment-poor continental slopes to the north. Northern European continental slopes display many indications of hydrate, especially along the Norwegian and Barents Sea coast.

The European Commission has sponsored and funded four research projects dealing with Gas Hydrate since 1997. The HYACE project (1998-2001) was targeted at developing and testing pressurized core apparatus. Two core-head pressure corers were developed to sample more consolidated sediment containing hydrate. Testing was carried out on and offshore. The HYACINTH project (2001-2004) was intended to put the HYACE system to operational use. The HYACE/HYACINTH system was first used on ODP leg 204 offshore Oregon in 2002. HYDRATECH (2001-2004) is a project that aims to develop techniques to identify acoustically and quantify methane hydrate and to establish relationships between varying amounts of hydrate and its seismic response in sediments. The purpose of the ANAXIMANDER (2002-2005) project is sampling of sediments containing hydrate and a methane-dependent biota in the Anaximander sea-mountains in the eastern Mediterranean Sea in the vicinity of mud volcanoes.

Indonesia: Scientists at the Center of Technology for Natural Resource Inventory in the Agency for the Assessment and Application of Technology are currently preparing a recommendation to the Indonesian government to carry out technical and economic feasibility to explore hydrate-gas occurrences in the offshore accretionary prism adjacent to Indonesia south of Java and Sumatra.

Ireland: In 1998, the Marine Institute of Ireland published a plan for the scientific and economic development of its large continental shelf and seafloor area. This document identified energy, amongst other issues and opportunities. A framework addressing these issues has been provided in the Productive Sector Operational Program of the National Development Plan (2000-2006) with an indicative budget of over fifty million Euro for marine research and technology developments over the period 2000-2006. These documents are available through the Marine Institute. Two research vessels have been acquired and appropriate scientific and technical resources based in Galway have been staffed. The marine work is coordinated by the Irish Government and includes a seabed survey, which is overseen by the Geological Survey of Ireland. The possibility of hydrate resources in the Irish seabed resulted in a preliminary, in-house assessment in 2003. International contractors providing expert oversight and technology transfer to the Irish resource base began an assessment of existing seismic data during the early part of 2005. Ireland has informally designated ocean areas that might contain hydrate well beyond the 200-mile limit of national interest identified by UNCLOS (see Chapter 8).

Mexico: Indigenous oil and gas production is at a turning point. Two thirds of the nation's oil production is coming from a single field complex (Cantarell) that will begin a sharp decline in 2006. At present Mexico is a net importer of natural gas. Mexico is now beginning the exploration of its deepwater Gulf of Mexico acreage. The geology of the Mexican deepwater east coast has many similarities to the U.S. Gulf of Mexico, including diapiric and alochthonous salt, although there is no sediment supply on the order of the Mississippi River. Natural oil seeps are present throughout the deepwater area. The Mexican government plans to do all the development themselves rather than open exploration to foreign oil companies.

A conference, which was officially called the "First Forum on Natural Gas Hydrates in Mexico", was organized in the summer of 2004 by PEMEX, the Mexican Ministry of Energy, and the National University. Also associated were the Mexican Association of Exploration Geologists (AMGE) and the Mexican College of Geophysics Engineers (CIGM). This was essentially the first national gas hydrate conference in Mexico. The meetings covered two days and were held in the University Geophysics Department. The Mexicans invited speakers from the US, India (DNS), and Chile. About 75 Mexicans attended. However, there does not appear to be a hydrate research program at this writing, and petroleum remains the primary Mexican exploration objective.

Norway: Although there is no formal National Hydrate Program, has strongly supported research through STATOIL, which has carried out considerable research into the energy potential of hydrate both offshore Norway and Nigeria. In particular, the first 3-D seismic survey conducted specifically to assess slope stability and hydrate/gas in marine sediments (Bünz et al., 2003; Hjelstuen et al., 2004) was carried out in the vicinity of the uppermost Storegga Slide. This slide is one of the largest known mass flows whose generation is thought to be associated with hydrate dissociation. Researchers from the Unversities of Bergen and the University of Tromso, the Geotechnical Institute in Oslo, and the Geological Survey of Norway participate in hydrate research.

New Zealand: The presence of gas hydrates on the Hikurangi Margin east of northern New Zealand was first inferred from BSRs in 1981 (Katz, 1981). BSRs have also been detected on the Fiordland Margin to the southwest of New Zealand (Townend, 1997; Fohrmann et al., 2004). The New Zealand Foundation of Science, Research, and Technology has provided funding for a small gas hydrates project since 1997. This project has so far focused on an analysis of existing seismic data for the presence of BSRs to obtain first estimates of the amount of natural gas that may be stored in New Zealand's gas hydrate deposits. Gas hydrates surveys are planned on the Hikurangi Margin in collaboration with international partners.

South Africa: Widespread BSRs have been identified on multichannel seismic profiles in the upper continental slope in the southern periphery of the Orange River delta off South Africa. Although no hydrate has been drilled or found on the seafloor in the region, the presence of large quantities of gas hydrate is inferred (Ben-Avraham et al., 2002). The seafloor in the region appears to have many pockmarks and mud volcanoes indicating upwelling of gas-rich fluids.

South America: Only a few scientists other than those from Chile and Brazil appear to be taking part in hydrate research.

South Korea: Initiated preliminary hydrate research programs in the 1990s in conjunction with the U.S. Naval Research Laboratory (Gardner et al., 1998) and is now carrying out independent hydrate research through its universities and government research agencies. In March 2005, the Korea Gas Corporation issued a press release on the progress of several years of gas hydrate exploration that identified gas hydrate potential in the Uleung Basin, which lies in the constricted sea area between Korea and Japan. EEZ issues (Chapter 8) in the area are presently focused on the ownership of isolated islands that are about half way between Korea and Japan. Large enough deposits of hydrate are reported to have been identified to relieve Korea of the need to import substantial LNG for next 30 years, although the results are preliminary and the economic potential cannot be truly known at this time. Korea is presently the world's second largest importer of LNG.

The Korea National Oil Corporation and Woodside Petroleum Ltd., which is 34 percent owned by Royal Dutch/Shell Group and is Australia's second-largest oil and gas company, signed an agreement to explore part of the Uleung Basin area. The Korean government has allocated about $22 million per year for the next 10 years for hydrate research.

Taiwan: McDonnell et al. (2000) and Liu et al. (2004) have recognized BSRs and blanking in the northern sector of the South China Sea in the submarine Luzon accretionary wedge off Taiwan. In 2004, the Central Geological Survey of Taiwan funded a 4-year preliminary gas hydrate research program, which also involves university researchers. It is likely that following the confirmation of very large areas of BSRs so early in the preliminary program, considerable hydrate and subjacent gas is present and that further research and development will follow.

Turkey: Turkish scientists are attending hydrate research meetings and are reported as having initiated at least preliminary hydrate assessment programs.

Ukraine: Scientists known to the authors would like to have a gas hydrate assessment program as hydrate has been identified in the Black Sea. Discussions have taken place about the possibility of the U.S. Department of State funding or partially funding joint U.S.-Ukrainian hydrate research.

West Africa: Gas hydrate has been inferred from reflection seismic records along the southwest African continental margin off the Congo River in originally relatively homogeneous pelagic sediments. These shows of shallow hydrate are associated with pockmarks, high fluid flow from the seafloor, seafloor hydrates and carbonates, and thermal anomalies. There are similarities with seafloor venting of natural gas-rich fluids in the northern Gulf of Mexico (Sassen et al., 1999; Sassen, 2000; Sassen et al., 2001; Hagen et al., 2004, Wood et al., 2004)

Gas hydrate, in some form, is probably ubiquitous on most continental margins of the world. New identifications and inferences of gas hydrate are now being made with regularity as the spreading knowledge of hydrate means that more researchers are looking for hydrate indicators.

TERMINOLOGY OF HYDRATE AND ITS PROCESSES

Throughout this book, exploration, valuation, mining, and processing terms that are commonly used in economic geology of metallic and non-metallic mineral deposits are used for hydrate with no special qualification. This usage is adopted because hydrate is a solid crystalline material that forms diagenetically in sediment and rock hosts in a manner similar to those more familiar as conventional mineral deposits. Hydrate deposits also can be described using terminology of conventional gas and petroleum deposits, to which they are closely related.

A number of terms are used interchangeably for various aspects of the natural gas hydrate system. Most prominent of these is that both the terms 'hydrate' and 'hydrates' are applied to naturally occurring natural gas hydrate. In the strict sense, the term hydrate should apply where a single species, such as methane hydrate, occurs. Because there are often small amounts of other hydrate forming gases present, principally ethane, but often also propane and butane especially where there is a thermogenic gas component, the plural is appropriate where individual hydrate occurrences are discussed. Hydrate-forming gas (HFG) can be used broadly to refer to any gas or mixture of gases that forms hydrate, but in this book it refers to hydrocarbon gases.

Natural gas hydrate containing either nearly pure methane or methane-based, mixed hydrocarbon gases is referred to in this book simply as 'hydrate', which is widespread in both permafrost and marine environments. Where enough hydrate occurs in sufficiently high concentrations to allow the natural gas to be recovered commercially, it will comprise a type of economic deposit. Hydrate in smaller concentrations has been formed in a similar manner and is distributed in Polar regions and marine sediments worldwide. Local geology determines definition of some sub-types of hydrate deposit, but the physical chemical conditions necessary for hydrate to form are the critical factors, particularly in oceanic environments. Following the terminology of the conventional gas industry where the singular term 'gas' is applied to all hydrocarbon deposits that are gaseous in form, and even despite the fact that some liquid condensates may be carried in the gas derived from conventional gas deposits, the term 'hydrate' is used here rather than hydrates. Thus, we regard 'exploring for gas hydrate' or 'hydrate deposits' for instance, as being a more correct usage than 'exploring for gas hydrates'.

As in the first book, the word 'hydrate' is used throughout for simplicity and consistency to refer to all types of natural gas hydrate (clathrate) deposits. Further, and mainly for convenience and brevity, and because it is understood that in the case of hydrate, the composite hydrate forming material is mainly hydrocarbon gas, the word gas is not used here as a modifier for the terms 'hydrate' or 'hydrate deposit'. The plural is 'hydrate deposits'.

Hydrate almost always occurs as diagenetic mineralization in the pore space of marine sediments. We do not regard large solid masses of hydrate resting on the seafloor and growing from seawater as having economic significance. Where hydrate is present, the sediments are hydrate-bearing in the same sense that disseminated stratabound metaliferous ores that occur in sediments are ore-bearing. Terms such as, 'hydrated sediments' are regarded by us as being less proper than 'hydrate-bearing sediments'.

The energy that is produced or consumed during the process of a chemical transformation at constant pressure and temperature can be enumerated and quantified as heat. This energy is known as the reaction enthalpy or heat of reaction. In describing the transformations of the hydrate system, a number of terminologies are presently used. For instance, hydrate can form or crystallize (combining the processes of nucleation and growth) and the two words can be used interchangeably. Heat of formation, crystallization, or heat of fusion, which is less appropriate but is also used, are appropriate to describe the heat produced when hydrate is formed. Formation of hydrate adds heat to its environment.

The transformation of solid hydrate to water and gas is often referred to as 'melting' although it is less appropriate than the term 'dissociation'. Melting describes the process wherein a solid material is altered to a liquid state usually by the application of heat alone. Water ice, in contrast, is almost isobaric under normal circumstances, although there are slight pressure effects on the water to ice transformation. Dissociation, on the other hand, describes the process by which a chemical combination, such as hydrate, becomes unstable and breaks up into its component constituents through the application of either pressure or temperature changes, or both. The term dissociation is thus more appropriate than 'melting' to describe the transformation of hydrate to its components when it is removed from its field of stability. 'Heat for dissociation' is approximately the same quantity for a particular volume of hydrate as the heat of formation, but with a negative sign. Dissociation consumes heat energy from its environment. Normally, when dissociation of natural gas hydrate takes place, liquid water and gas are produced.

A special case of dissociation of gas hydrate occurs when the temperature at which the dissociation takes place is below the freezing point of water. In this case, the hydrate forming gas can exit the hydrate by a solid-state diffusion process. Informal reports suggest that even where methane hydrate has been stored in liquid nitrogen, but at near atmospheric pressures, after a period of time the methane has been found to have diffused from the hydrate, leaving behind a crystalline structure that is slowly reorganizing itself from the cubic structure of SI hydrate (Sloan, 1998) to hexagonal water ice. In nature, this may take place in a permafrost terrane if pressure were to be lowered on hydrate that resides within the water-ice cryosphere (Chapter 3) where no heat is applied (Chapter 10).

A further special case of dissociation of gas hydrate can occur when the temperature of a stable hydrate is raised so that it becomes unstable, in the presence of pressure that is high enough so that the hydrate forming material directly forms liquid rather than gas. Both chlorine and carbon dioxide, amongst common hydrate forming gases, may often dissociate under these conditions, but this process is unlikely to occur in nature.

Finally, hydrate is **converted** to its constituent gas and water through dissociation that is artificially caused by any method or combination of methods. It is **recovered** from the geological strata by collector apparatus based on drilled holes within the geological strata. The gas is **produced** when it is recovered to the surface for use.

FROM RESOURCE TO RESERVES

Conventional gas deposits consist of pressurized gas held in porous geological reservoir traps. Much of the gas in conventional traps has been in its particular trap for considerable periods of geological time (Selly, 1998) where it has been isolated from the biosphere. Recovery of gas from conventional deposits takes place spontaneously where the natural pressure of the gas reservoir forces gas (and often associated fluids such as water, condensates, and petroleum) to the surface once a conduit is provided by drilling. Secondary recovery techniques that increase permeability of the reservoir may also be used, but the aim is to stimulate the flow of existing natural gas. Unconventional gas deposits, with the exception of hydrate (Chapter 1), are also in gaseous form and require special techniques for exploitation but are not dependent on a change of state.

Natural gas hydrate comprises unique, unconventional, diagenetic, nonmetallic mineral deposits that bear striking similarities in paragenesis and form to other mineral deposits, especially strata-bound mineral deposits (Chapter 4), many of which are metaliferous. Natural gas hydrate constitutes a very unusual gas play. Gas hydrate is a crystalline solid. The economic material is methane, along with minor amounts of other natural gases. The methane in hydrate is in solid form, and in most respects (e.g., physical form, paragenesis, valuation, method of exploitation) may be described as an 'ore' of natural gas (within the general group of non-metallic mineral deposits) in that it is a continuous, well-defined mass of material of sufficient concentration to make extraction economically feasible. Unlike most other mineral deposits, which once formed are relatively stable and remain fixed in their geological host, natural gas hydrate is an intimate part of the biosphere and is very responsive to natural changes in pressure and temperature that cause it to either form or dissociate, sequestering or releasing methane (Haq, 2003). Hydrate deposits often may not occur in a mechanically strong geological situation, which will have an impact upon the development of safe and productive extraction techniques.

The boundary between petroleum geology (which includes gas deposits) and economic geology, which is normally concerned with metal and other solid ore deposits that must be mined and subsequently processed to extract the valuable components, converges in extraction models for gas hydrate. There are certain exceptions to this generalization, such as using hot water to dissolve economic minerals such as sulfur and soluble evaporites such as halite and other evaporitic deposits. Recovery of natural gas will be a matter of converting the hydrate *in-situ* to gaseous form (Chapter 7) and collecting it from the geological strata. The gas produced from the hydrate will form a temporary conventional gas deposit in close proximity to the hydrate from which it has been derived. The gas will then be recovered by means that will be similar to those used for conventional gas. Ideally, there will be existing gas deposits in association with dissociating hydrate. However, the creation of confined temporary gas deposits where none previously exists, and which will keep dissociated gas from escaping, will be one of the keys to the efficient recovery of natural gas from gas hydrate.

Recent evaluation of existing geophysical information and scientific study of natural gas hydrate in both permafrost and oceanic environments is substantiating the concept that very large amounts of gas hydrate exist (Kvenvolden, 1988; Kvenvolden & Lorenson, 2001). However, except for certain permafrost hydrate deposits and one oceanic deposit being studied in Japanese waters, significant concentrations of hydrate that may approach the level of economic deposits have not yet been recognized. Thus, the potential of hydrate as an energy resource is of a speculative nature at this time (Fig. IN 2).

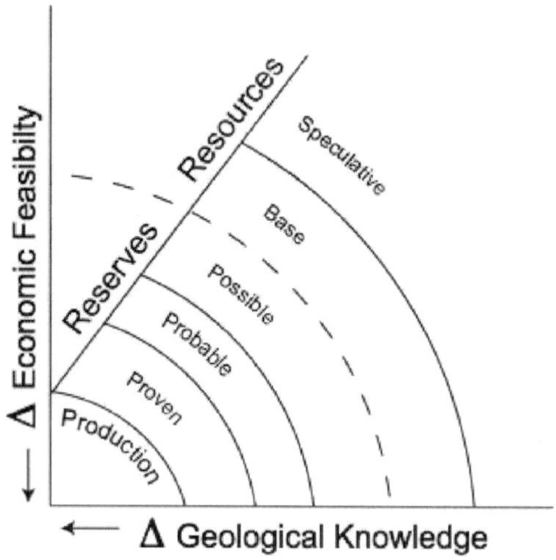

Figure IN 1. Diagram of development spectrum for natural gas hydrate. After Ion (1979). Courtesy of HEI.

Gas hydrate was established as a speculative resource (Fig. 2) in the late 1960s and early 1970s when it was identified in both oceanic and permafrost regions. Improvements in geological knowledge, including theoretical and laboratory studies that supported remote sensing and direct observation and sampling in the natural environment, brought the realization that gas hydrate widely occurs in nature and is a huge store of methane. The resource base was established when estimates of the abundance of hydrate became realistic (Kvenvolden, 1988). With the establishment of the first gas hydrate program by the United States in the mid 1980s, commercial and economic aspects of the hydrate resource base also began to be assessed seriously. At present, the Nankai deposits of the Japanese continental margin probably can be assigned a status of 'probable reserve' while well known sites in the Mackenzie delta of Arctic Canada can be assigned as 'proven - probable' status.

Not only must certain concepts normally applied to conventional mineral deposits be applied to hydrate to describe the modes of formation of different types of hydrate deposits, but certain practices common within the mining industry may also have to be applied. This idea is particularly important for oceanic hydrate because these deposits occur in relatively unconsolidated, mechanically weak marine sediments in near-seafloor situations (Chapters 3, 5), and the hydrate reservoir must be stabilized as the hydrate is converted, which will almost certainly weaken the sediment. Mineral deposit models must be established for hydrate that will guide both exploration and extraction. Despite all the apparent difficulties, however, we believe (1) that contiguous, well-defined masses of hydrate of sufficient concentration will be found, (2) that reserves will be defined, and (3) that extraction techniques can be perfected so that natural gas hydrate can be commercially extracted.

Chapter 1

Why Gas Hydrate?

1.1. INTRODUCTION

The pursuit of unconventional natural gas resources, such as gas hydrate, is often viewed as being completely unnecessary given the world's vast proven reserves of conventional gas. Any consideration of gas hydrate as a resource must therefore take place in the broader context of natural gas supply and demand. The proven reserves of conventional gas worldwide are enormous – in excess of 6,000 TCF – and have doubled over the past 20 years (Fig. 1.1), even as the annual production of natural gas has increased worldwide by over 67% (Fig. 1.2). The current proven reserves represent 67 years of supply at current rates of consumption (Fig. 1.3). In addition, large new conventional discoveries are still being made. This leads to the question: Why Gas Hydrate?

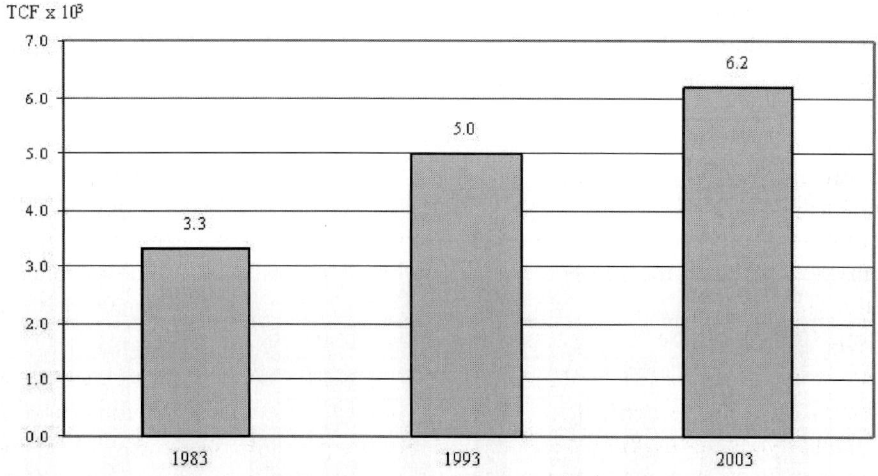

Figure 1.1. Current worldwide proven reserves of conventional natural gas exceed 6,000 TCF (from BP, 2004). In addition, the conventional reserve base has been growing during the past two decades. This conventional resource poses challenges for the development of unconventional gas resources.

Michael D. Max et al. (eds.), Economic Geology of Natural Gas Hydrate, 17–44.

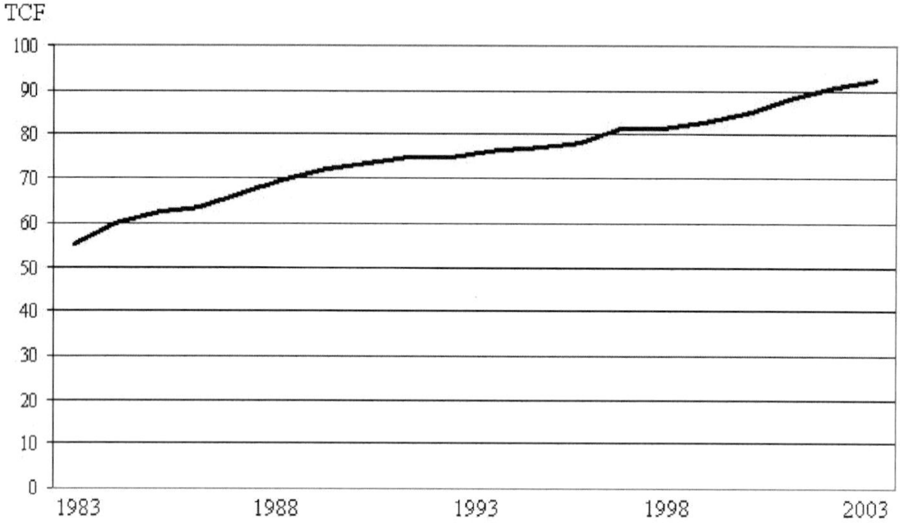

Figure 1.2. Global production of natural gas has increased by over 67% over the past two decades (from EIS, 2004).

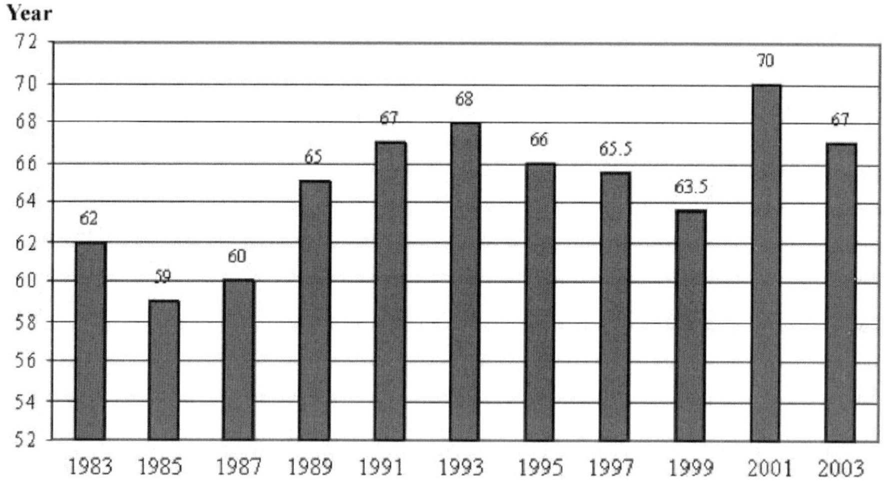

Figure 1.3. Historical natural gas P/P ratio. The ratio of reserves (R) to production (P) shows how long proven reserves would last at the production rates of the given year. While the R/P ratio varies from year to year, the overall trend for the past two decades shows an increase (from BP, 2004).

1.2. RESERVES VERSUS MARKETS

In regions with large reserves and few consumers, natural gas is essentially a worthless byproduct of petroleum operations. Throughout the twentieth century large volumes of gas were flared or vented for lack of a viable market. According to Prindle, 1981, as much as 90-95% of the gas produced from large fields was vented or flared during the mid-1900s. While this practice is decreasing, primarily due to environmental considerations, nearly 10 billion ft^3 (BCF) of gas continues to be flared or vented worldwide each day (Gerner, et al., 2004). This volume is equivalent to the combined gas consumption of Central and South America.

In contrast to the locations where natural gas has very little value, it is in short supply and relatively expensive in other markets, such as Japan. These disparities in price exist largely due to transportation issues. On a BTU basis, gas is more expensive to transport than oil, even through pipelines. Thus, even where pipelines exist, gas is typically priced lower than oil for the equivalent amount of energy in BTUs.

Where there are no pipelines connecting supply to demand, the economics of natural gas are far different from those of oil. Without pipelines, oil is still very easy to transport by ship. Transporting natural gas by ship requires its liquefaction. To liquefy natural gas the temperature is reduced to minus 260°F, turning 600 cubic feet of natural gas into one cubic foot of Liquefied Natural Gas (LNG). LNG is transported at near-atmospheric pressure in specially designed, double-hulled ships with insulated cargo tanks. At the receiving installation, the LNG is warmed and converted back to a gaseous state.

LNG facilities are among the world's most expensive energy projects, although costs vary considerably depending on the size of the facility and its location. According to the Gas Technology Institute a liquefaction plant with an annual output of 390 BCF (8.2 million tons) will cost between $1.5 and 2.0 billion (EIA, 2003). The largest LNG tankers currently being built hold 145,000 cubic meters of LNG (equivalent to 3 BCF of natural gas at STP). Larger ships are being studied having capacities of 200,000 to 240,000 cubic meters (4 to 5 BCF at STP).

Although the Gas Technology Institute has reported that LNG processing and transport costs have decreased 35 to 50 percent over the past 10 years (EIA 2004), the liquefaction, transport, and regasification involved in the LNG process remain expensive. Despite these costs, many in industry see LNG as the solution to the natural gas needs of the industrialized world. In the United States, this view is often echoed by government leaders (e.g. Greenspan, 2003).

The expansion of LNG markets challenges the viability of unconventional gas resources. While LNG facilities are expensive, the technology behind LNG is well known and the economic risks, mainly related to natural gas price, are manageable. The technology required for unconventional gas resources is still evolving and will require investments with unknown returns.

For gas hydrate, in particular, there are many uncertainties regarding reserve estimation, production rates, and operating expenses that the LNG industry does not face. For these reasons plans for 40 new LNG terminals and/or expansions have been announced and are currently under review (Fig. 1.4).

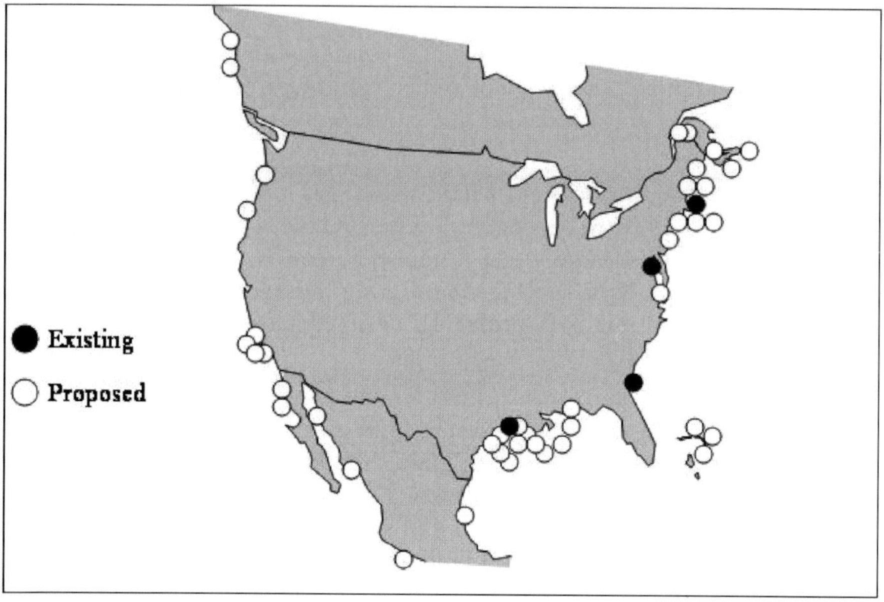

Figure 1.4. Plans for 40 new LNG terminals have been announced for mainland North America, in addition to the four that are currently in operation (from FERC, 2004).

Unconventional gas resource development is also impacted by the potential expansion of conventional gas infrastructure. New pipelines are being considered for the Alaskan and Canadian Arctic that would make large conventional gas reserves accessible to North American markets (Figure 1.5), decreasing the need for unconventional gas. New field development and pipelines in the Rocky Mountains will further reduce the incentive for pursuing unconventional gas.

Large new discoveries of conventional gas are also being made in both the Gulf of Mexico shelf and its adjacent deepwater areas. Yet, the pipeline capacity for natural gas in the deepwater Gulf of Mexico is very constrained. The National Petroleum Council has reported that there will be no capacity for transporting gas from hydrate for 20 years (NPC, 2003). These issues represent a challenge for unconventional gas resources, and have led the major players in the North American natural gas market to reject significant investment in gas hydrate as a resource.

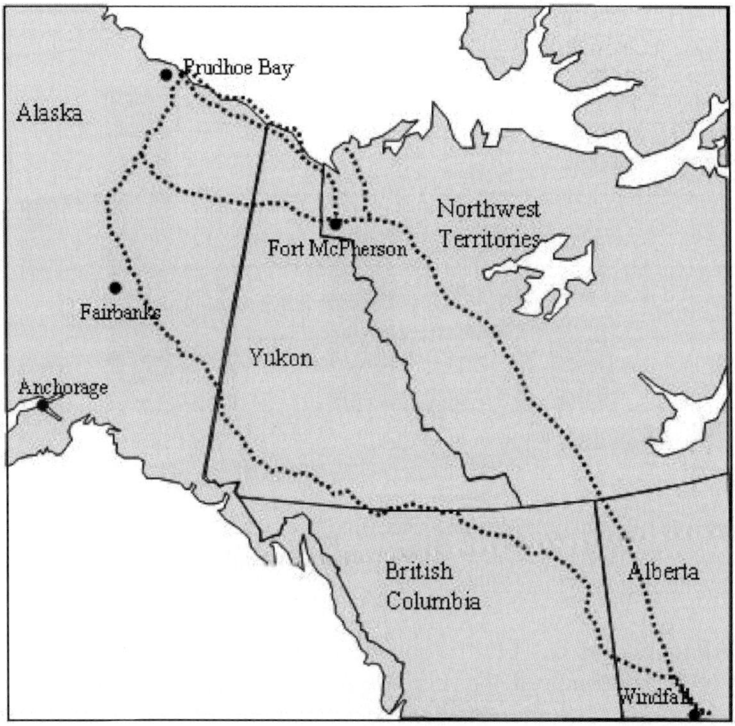

Figure 1.5. Several pipeline routes have been proposed to bring stranded gas from the Canadian and Alaskan Arctic to North American markets (adapted from BP, 2004).

1.3. THE CASE FOR UNCONVENTIONAL GAS

The global energy environment, as seen from the viewpoint of the well known field of conventional gas, at first might appear to make the near-term development of any unconventional gas resources highly unlikely as viable commercial enterprises. At the initiation of developing any new venture, risk is greatest and the likelihood of success can only be estimated better as greater knowledge of the new play is gained. Yet, a realistic case can be made for unconventional gas in general. Two different unconventional gas sources, coalbed methane, and shale gas already have been successfully developed and are important components of the North American gas supply. Their contribution has grown dramatically in a little over a decade (Fig. 1.6). Development of each of these unconventional gas sources in turn went through the same sequence of being initially regarded as being a fringe issue that was being promoted by a vociferous small group of enthusiasts, to an early grudging acceptance of the economic potential, to significant economic development. We regard hydrate as being on the same developmental path.

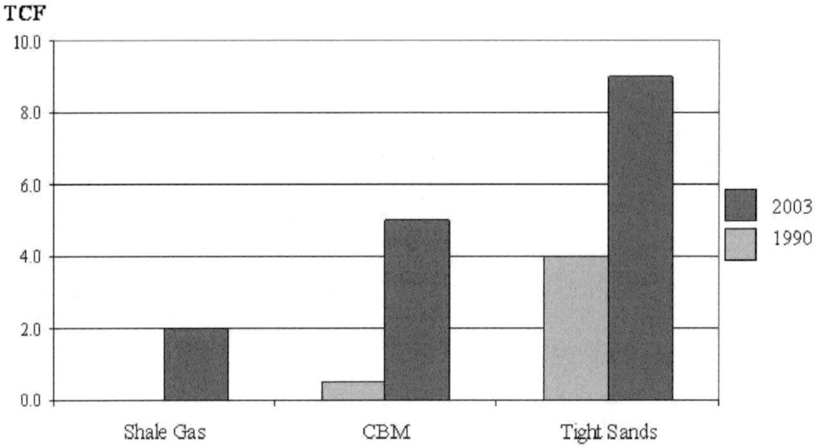

Figure 1.6. Production of Shale Gas, Coalbed Methane, and Tight Sands has increased significantly since 1990, demonstrating the commerciality of unconventional gas (EIA, 2004).

Although gas has been produced from shale in the Appalachian Basin for over 60 years, much of the increase in U.S. shale gas production has been achieved through the more recent development of the Barnett Shale in Northeastern Texas. The commercialization of the Barnett Shale was made possible by the development of improved technology, specifically the application of an improved fracturing method ("gel frac") by Mitchell Energy in the late 1980s (Gonzalez, 2004).

Tight gas sands are defined as sandstone formations with less than 0.1 millidarcy permeability (Surdam, 1995). Developing these reservoirs uses much of the same technology as with conventional gas sands. The primary difference is that the recovery-per-well is substantially lower so that more wells are required to develop a field. Tight sand development began, on a limited basis, in the San Juan Basin in response to improvements in fracture technology in the 1960s. The rate of development increased in the 1970s due to improving technology and significant increases in gas prices, with tight sand development expanding to several areas of the U.S. (Hugman, et al. 1993). Development increased further during the 1980s and 1990s in response to tax credits and additional improvements in technology that have reduced exploration costs, improved drilling results, and increased production rates. Since the 1980s drilling for gas in shale and tight sand in the U.S. has essentially tracked the natural gas price (Perry, et al., 2004).

Coalbed Methane (CBM) is a natural gas produced from coal seams. It occurs as adsorbed gas in natural fractures (cleats) and within the coal matrix and as free gas. Methane in coal seams was traditionally viewed as a hazard rather than as an energy resource because it contains only 1-2% of the energy capacity of the coal itself. While gas has been commercially produced from coal for over 200 years, until the late 20th century most gas derived from coal was manufactured by heating coal at an industrial plant. CBM is produced by reducing the natural formation pressure within the coal seam to allow the gas to be released from the coal. Initial development of CBM began in 1951 in the San Juan Basin of the Southwestern U.S. and accelerated in the mid-1980s in response to improved technology, gas prices, and tax credits. By the early 1990s CBM development had spread to the Warrior and Central Appalachian Basins; and by the late 1990s development included the Uinta, Raton, and Powder River Basins. At present, most coal basins are under review for CBM potential (Lombardi and Lambert, 2001).

While shale gas, tight sands, and coalbed methane comprised 17% of U.S. gas production in 1990, by 2003 they represented 32% of the U.S. total (Fig. 1.7) (EIA 2004). According to EIA estimates, by 2010 unconventional sources will supply 36% of U.S. production, and 40% by 2015.

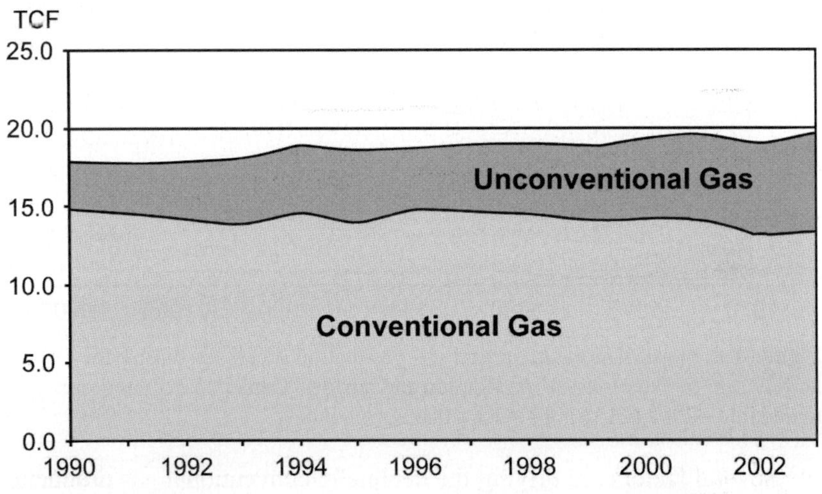

Figure 1.7. Unconventional gas (shale gas, tight sands, and coalbed methane) represents an increasing proportion of U.S. gas production (EIA, 2004).

The increases in production in shale gas, tight sands and CBM share a number of common elements. First are improvements in technology that have made the commercial development of each of them viable. Second is the relation between development and gas price. While each appeared to be an unlikely candidate for commercialization early in its history, the potential of a solid return on investment encouraged a few people in industry to find the means of exploiting each resource. Third, government incentives in various forms accelerated the commercial development of these unconventional resources.

These lessons provide valuable insights that should be taken seriously by anyone interested in the development of gas hydrate as a resource.

1.4. MEETING FUTURE DEMAND

Is there a need for an additional North American source of natural gas besides conventional production, the unconventional source described above, and LNG imports? As may be seen in Figure 1.7, the growing contribution of unconventional gas is only serving to keep U.S. production basically flat, as conventional production declines. The combined natural gas production for North America (Fig. 1.8) shows a decline in Canadian production after increases during the 1990s (Canadian Gas Association, 2004).

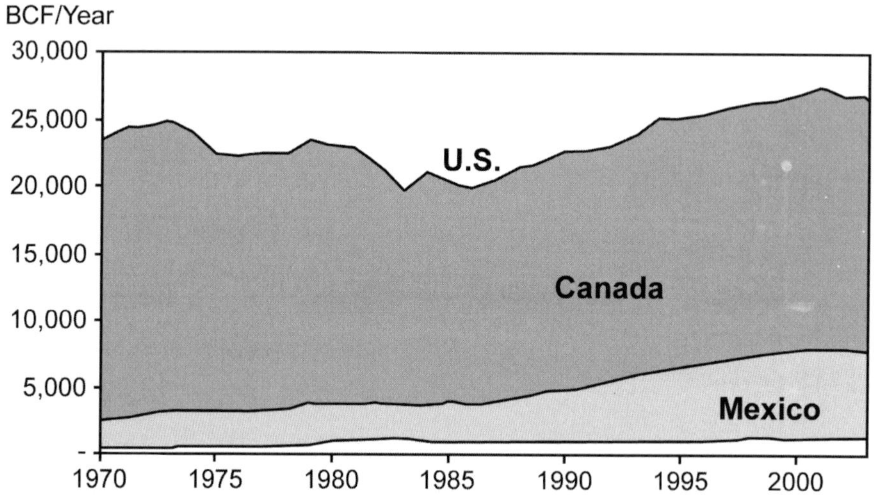

Figure 1.8. North American natural gas production increased during the 1990s, largely due to expansion of production in Canada. Canadian production began to decline ~ 2001 (EIA, 2004, CG, 2004 & SC, 2003).

Several factors are driving the decline in conventional gas production in North America. Exploration and production are banned or heavily restricted in many prospective areas. The National Petroleum Council (2003) estimates 148 TCF of gas is being kept from North American markets due to development bans or limitations in the Rocky Mountains, the Pacific Coast, Atlantic Coast and Eastern Gulf of Mexico. Conversely, other areas in North America have been developed for decades, making significant new conventional discovery increasingly rare. Improvements in production technology have made higher production rates possible, while exploitation of smaller reservoirs has also led to higher initial decline rates for gas wells. As a result, the volume of gas produced per well has declined in the U.S., even as the number of wells brought on stream has increased (Fig. 1.9). As noted by the NPC (2003), producers are "running harder to stay even". Without additional drilling, the NPC notes that U.S. gas production would decline by 25-30% per year. Wells in Canada show a similar decline (Canadian Gas Association, 2004). It will be a major challenge to maintain, let alone increase, North American natural gas production.

Yet as the outlook for increasing North American natural gas production grows more questionable, the forecast demand is projected to increase. Estimating either future supply or demand is difficult at best, and published figures can vary considerably. Many factors affect supply and demand. Supply is affected by government policy, weather (such as Gulf Coast hurricanes that can damage facilities), natural gas prices, and improvements in E&P technology.

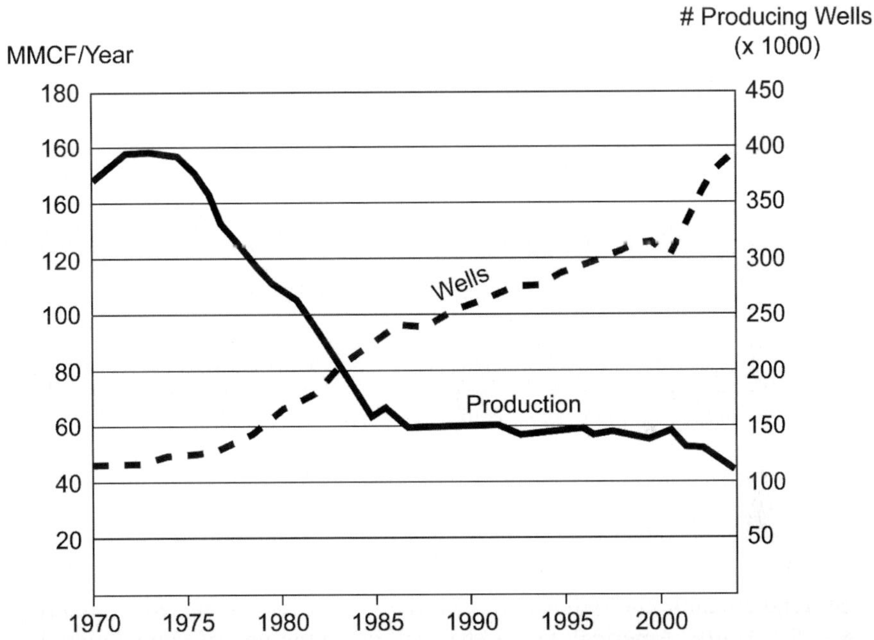

Figure 1.9. The rate of natural gas production per well in the U.S. has declined significantly during the past 30 years. Total production has been maintained by a steep increase in the number of producing wells (EIA, 2004).

Demand for natural gas is affected by improvements in utilization technology, availability of alternative fuels (fuel switching), economic cycles, and the weather. If the price of natural gas becomes too high, industrial customers have the option of moving their operations to other regions of the world, a strategy that has been increasingly pursued since 2000. While residential gas customers have fewer options, higher prices often lead to conservation measures.

Despite the uncertainties involved, the forecasts made by the National Petroleum Council (2003) utilize sound assumptions and are often cited. The NPC forecast (Fig. 1.10) projects North American natural gas demand to increase by as much as 48% by 2025 to a reach a total of 30-34 TCF/year. This forecast does not include the potential use of natural gas for the large-scale manufacture of hydrogen as part of the Bush administration's "Hydrogen Fuel Initiative". The NPC further estimates that traditional, regional producing areas will provide only 75% of North America's natural gas needs for the foreseeable future. This leaves a shortfall of 25%, which, if not remedied, will have severe economic consequences for the U.S. and Canada (Smil, 2003; Roberts, 2004). In the U.S., natural gas accounts for 23% of all energy consumed, with half of all homes being heated by natural gas (MMS, 2004a).

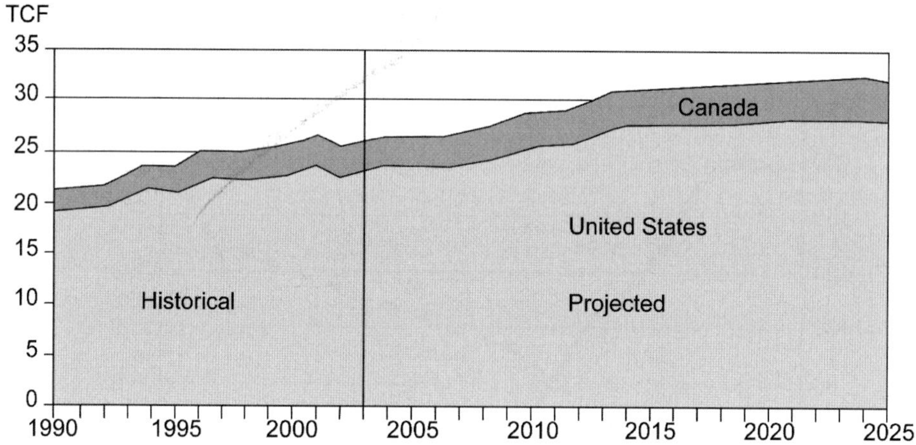

Figure 1.10. Natural gas demand is projected to increase significantly during the next two decades (NPC, 2003).

1.5. OPTIONS FOR INCREASING NORTH AMERICAN GAS SUPPLY

Three fundamental alternatives are available for increasing the natural gas supply for North America: (1) additional development of conventional gas resources, (2) increased imports of LNG, and (3) further exploitation of unconventional gas resources, including gas hydrates. Each of these options has inherent difficulties but all need to be pursued to ensure that gas supplies are adequate to meet growing demand.

1.5.1. Increased Conventional Gas Development

Since the beginnings of the petroleum industry, improvements in exploration and production technology have resulted in new discoveries and increased recovery. Continuous application of technology will play a critical role in efforts to maintain current rates of production. Government initiatives that provide royalty relief have also resulted in increased gas production by making development of marginal fields commercially viable. It is unlikely, however, that technological improvements and government incentives will be able to do more than slow the rate of production decline. Additional sources of natural gas are needed.

The Rocky Mountains are an important gas producing area at present and hold significant potential for increased production. While development continues in the Rockies, the regulatory process is restraining much of the region's potential. Leasing restrictions are preventing development in some basins; however, more serious impediments arise from restrictions on development and on regulatory delays after leases have been secured. As a result, 29% of the Rocky Mountains resource base is effectively put off limits (NPC, 2003). In areas of the Rocky Mountains where development is allowed, production is expected to increase substantially each year through 2025, partially offsetting declines elsewhere.

The natural gas potential of the U.S. continental shelf and margin is available for development only in the Western and Central Planning Areas of the Gulf of Mexico, and portions of Alaska. Even so, the offshore supplied 25% of U.S. gas production, with deepwater gas production in the Gulf of Mexico up 407% since 1996 (MMS, 2004b). Leasing moratoria on the remainder of the U.S. offshore has prohibited access to those areas. The U.S. Congress imposed the first offshore moratoria in 1982 on acreage off northern and central California (MMS, 1997). Additional moratoria have been progressively imposed on offshore acreage, with the Eastern Gulf of Mexico being withdrawn from leasing in 1998 and except for the western Gulf of Mexico; virtually the whole of the U.S. coast is now under moratoria (Manheim, 2004). The reserve potential of the inaccessible offshore areas is difficult to determine owing to the lack of data. The NPC (2003) estimates recoverable reserves of ~70 TCF.

Reducing restrictions on petroleum operations has been proposed as a means of increasing gas production (National Energy Policy Development Group, 2001) however no serious efforts are being made to initiate leasing in areas currently under moratorium. Anti-development sentiments run high despite concerns over natural gas supply and price, and it is unlikely that any moratoria or development restrictions will be lifted in the near term.

Even if moratoria were lifted from offshore areas, at least a decade would pass before additional gas would reach consumers. Before production could commence, industry would have to conduct surveys and studies to assess the value of specific tracts, and the U.S. Minerals Management Service would have to conduct lease sales. These activities would be followed by the permitting and drilling of exploratory wells, the creation and approval of development plans, and the design, construction and installation of platforms and pipelines. Each of these steps is complex and typically time-consuming, and each would be subject to delays by legal action brought about by those opposed to development. Opening up additional areas to exploration and production is unlikely to significantly address near-term gas shortages.

The Canadian Arctic and North Slope of Alaska have proven gas reserves of approximately 40 TCF, and estimated undiscovered reserves of over 140 TCF (USGS, 2004, National Energy Board, 2004). The lack of pipelines has kept these reserves from contributing to North American gas supply, except for local use. Several pipeline routes from the North Slope have been considered during the past decade, with large capital investments required for each option. Economic and political considerations have kept North Slope pipelines projects from moving forward. In late 2004, the U.S. Senate and House of Representatives passed loan guarantee provisions and tax incentives that will decrease the economic risks associated with the estimated $18 billion cost of a pipeline from the North Slope to Alberta (World Oil, 2004). There is no timetable as yet for construction, but the NPC (2003) estimates that 4 BCF/day (1.5 TCF/year) may be added to North American markets by 2013 or 2014.

Plans for a $7 billion (Canadian) pipeline from the Mackenzie Delta in the Canadian Arctic to Alberta are moving forward and construction may be complete by 2010. Initial production is estimated to be 1.2 BCF/day (0.4 TCF/year) (Stevenson, 2004).

While the two new pipelines will add nearly 2 TCF/year to North American markets, the availability of those volumes in Alberta may encourage the expansion of tar sand operations in Western Canada. Tar sand extraction uses 1000 cubic feet of natural gas to convert 1 barrel of bitumen light crude oil for refineries (Romero, 2004). Increased tar sand development would decrease the amount of Arctic gas available for other North American markets. Thus, the construction of new Arctic pipelines will not necessarily make a substantial impact on future natural gas shortages.

1.5.2. Increased LNG Imports

The U.S. Energy Information Administration (2004) forecasts increases in domestic natural gas production that will account for 57% of the additional supply needed, with imports accounting for the remaining 43%. In contrast, the Canadian Gas Association (2004), reports that North American gas production peaked in 2001. Even if the lower numbers are overly pessimistic, it is doubtful that existing gas resources can expand as projected by EIA. There is a clear need for dramatically higher imports of natural gas into North American markets.

The international trade in liquefied natural gas began in 1959 with the commissioning of the *Methane Pioneer*, a converted liberty ship (American Bureau of Shipping, 2004). Its first shipment was 5000 cubic meters of LNG from Lake Charles, Louisiana to the U.K. (IELE, 2004). The LNG market was slow during the 1960s, with shipments from Algeria to the U.K. forming the primary market. Japan began importing LNG from Alaska in 1969 and quickly became the world's largest importer of LNG (Curt, 2004). Global LNG shipments increased significantly in the early 1970s as natural gas supplies tightened. Four LNG import terminals were built in the United States between 1971 and 1980. They are in Lake Charles, Louisiana, Everett, Massachusetts, Elba Island, Georgia, and Cove Point, Maryland. Import volumes peak in 1979 at 253 BCF (1.3% of U.S. gas demand) (IELE, 2004) and then declined as a gas surplus developed in North America. The Elba Island and Cove Point receiving terminals were mothballed in 1980 and utilization at the Lake Charles and Everett terminals was minimal. North American interest in LNG was renewed in the late 1990s as natural gas demand and prices increased. A large LNG liquefaction plant in Trinidad came on production in 1999 and its proximity to U.S. markets supported greater use of LNG. The Elba Island and Cove Point Terminals were reactivated in 2001 and 2003, respectively (IELE, 2004). U.S. imports of LNG totaled 0.5 TCF in 2003 (Neil, 2004).

In 2003, thirteen nations had a total of 42 receiving terminals for importing LNG (EIA 2003) and imported 5.8 TCF (Reuters, 2004). The combined regasification capacity of these terminals is 15.1 TCF per year. Japan remains the largest customer of LNG, importing 2.6 TCF through its 23 terminals in 2002. However, the Japanese share of the LNG trade fell from 66% in 1990 to 48% in 2002 as other nations increased imports. Japan, South Korea, and Taiwan accounted for 68% of Global LNG imports in 2002, European nations accounted for 28%, and the U.S. accounted for just 4% (EIA, 2003).

The global trade in LNG has been supported by the expansion of liquefaction facilities. The world's annual liquefaction capacity stood at 6.6 TCF per year in late 2003. New facilities under construction will increase global capacity to 9.4 TCF per year by 2007. Iran, Yemen, Angola, Venezuela, and several other nations that do not currently have liquefaction facilities are considering such construction as a way of monetizing their natural gas reserves (EIA, 2003). The total liquefaction capacity of the existing and proposed facilities is 9.9 TCF per year (Ziff, 2004).

It is important to reiterate that 40 new LNG plants and/or expansions are being planned or proposed for North America. The cost of new LNG receiving terminals is approximately \$50 million per 100 million ft^3/day (MMCFD) of capacity (Pike, 2004). The average projected daily capacity of the proposed facilities is 1 BCF, yielding an average cost of \$500 million per terminal. To meet anticipated demand, the LNG tanker fleet will have to double in size by 2015; an increase of 212 vessels at an estimated cost of \$180 million per vessel, including the replacement of retiring ships (Pike, 2004). While the cost per MCF of LNG imports to North America has averaged \$3-4 per MCF in recent years (EIA, 2003), the planned expansion of LNG capacity will require extremely large capital expenditures.

Legal challenges and environmental concerns may impede the rapid increase in LNG offloading facilities. Local opposition to new LNG offloading installations is locally vociferous because of perceived dangers. In response to these concerns, the LNG industry is contemplating installing LNG offloading facilities offshore, where vaporization of the LNG may take place using the latent heat in seawater rather than persisting with the present practice of combusting some (about 2%) of the LNG to provide the necessary heat. Existing offshore pipelines may be utilized where they are available, such as offshore Louisiana, or relatively inexpensive new floating unloading points may be used.

1.5.3. Concerns for LNG

For North American markets LNG will play an important role in filling the gap between supply and demand. But an over-reliance on LNG imports raises serious questions that should be of great concern to North American consumers and policy makers.

- Vulnerability of Supply
- Vulnerability of Facilities
- Import Capacity
- Competition for Supply

The potential for gas supplies to be disrupted over the next 20 years is quite high. Most of the large and growing conventional gas reserves available for export are in unstable regions of the world – the Middle East, the former Soviet Union, Southeast Asia, and West Africa (Fig. 1.11).

A second concern is the potential for supply disruptions either due to accidents or terrorist attacks. The transportation of LNG has had an excellent safety record for over 45 years, and LNG facilities are designed to minimize the impact of operational error. The most serious incident involving LNG (a leak and explosion in Cleveland that resulted in 128 fatalities) occurred over 60 years ago (U.S. Bureau of Mines, 1946), and technological improvements have successfully prevented its recurrence. Yet, the ability of LNG liquefaction plants, tankers, and regasification terminals to withstand a terrorist attacks remains an open question. Advocates contend that there is virtually no possibility of a catastrophic explosion involving LNG (IELE, 2003), while others portray potential disaster (Clarke, 2004). A worst-case scenario at an LNG facility could involve an extremely large zone of destruction. In addition, there will be relatively few of LNG regasification terminals so the loss of even one will have a significant impact on supply to customers.

The unresolved issues of safety and public misconceptions have led to significant opposition from the public against the construction of LNG facilities near populated areas. Of the 40 new plants and expansions already noted, many are unlikely to be built, and delays from site restrictions, permitting issues, and legal action will add many years to the completion of others (Poruban, 2004). Thus, there may be insufficient LNG import facilities to fill the gap between North American production and demand.

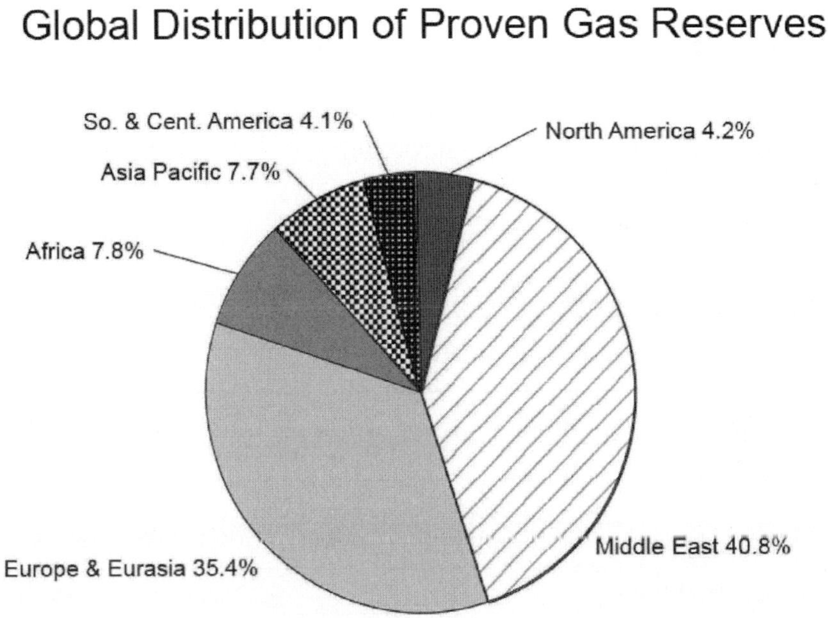

Figure 1.11. Most of the world's proven conventional gas reserves are in the Middle East and other unstable regions. This fact should raise concerns in North America about relying on LNG imports. (Data from BP, 2004).

Most of the public discussion regarding LNG has centered on the availability of regasification terminals when imports are needed to satisfy natural gas demand. The availability of LNG should also be discussed. Demand for LNG during the past decade has centered on Japan, South Korea, and Taiwan, with North America gradually increasing LNG imports. New construction of LNG liquefaction plants seems adequate to meet the needs of these markets. This ignores other nations that have LNG import facilities coming on line or being planned. New plants include India's Hazira facility (on stream in 2004) and China's Guangdong facility (on stream by 2007). Nations in the planning stages for new or expanded LNG import facilities include Spain, the Netherlands, France, Italy, Turkey, Greece, New Zealand, Indonesia, and Jamaica (EIA, 2003). During the early years of LNG trading, LNG was sold under long-term contracts at fixed prices, and with long-term contracts for LNG tankers. The current trend is for increasing volumes of LNG to be purchased as spot cargoes, making price and availability less certain (Neil, 2004).

1.6. LOOKING TO THE FUTURE

A strong upward trend exists globally for consumption of all energy sources as people throughout the world strive for higher living standards. The link between a nation's per capita energy consumption and its per capita wealth is very clear (Fig. 1.12). Demand for energy generates and promotes wealth (Economides &

Oligney, 2000). This is most evident in the large increases in oil and gas consumption observed for China and India in recent years. China's oil consumption is expected to grow at an annual rate of 7.5% and India's at 5.5%, as compared to 1% growth for the industrialized countries (IAGS, 2004). The increasing demand for oil will keep prices high, precluding a significant switch from oil to natural gas as gas prices increase.

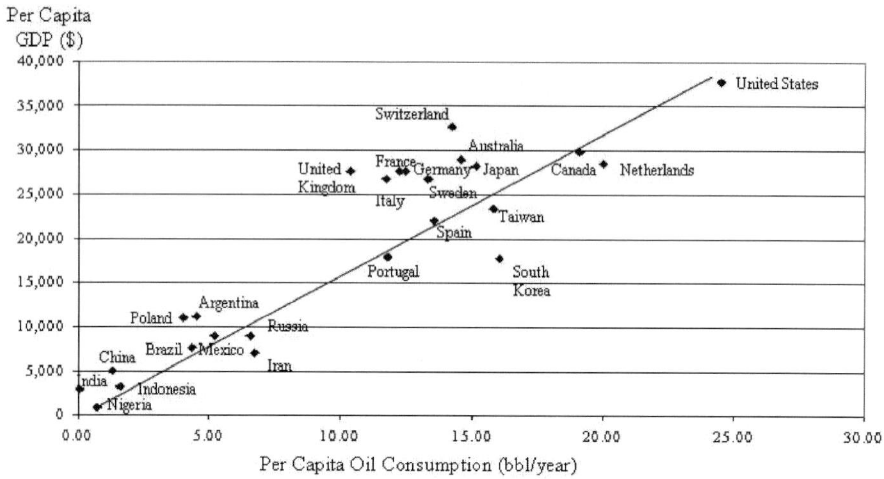

Figure 1.12. A linear trend is apparent between per capita oil consumption and per capita Gross Domestic Product. An increase in a nation's wealth is driven by an increase in energy use. (Adapted from Economides & Oligney, 2000; data from CIA, 2004).

A critical shortage of gas will impact residential customers through higher costs and possible supply interruptions (Campbell, 1997; Deffeyes, 2003). Most new electricity generation capacity is expected to be fueled by natural gas (EIA, 2004) due to advantages over coal-fired generators (capital expenses, fuel efficiencies, emissions, and permitting issues). Thus, shortages of natural gas will have serious consequences for all North American electric customers. The North American chemical industry uses natural gas as both a fuel and a feedstock. High natural gas prices have already led to closures of ammonia, melamine, and urea plants in several states (USGS, 2000, Associated Press, 2004) with severe impacts on local economies. Shortages in upcoming decades will accelerate this trend.

From the early 1980s until 2000 North American natural gas markets have experienced frequent periods of oversupply, commonly referred to as the "gas bubble' or "supply bubble". Many operators were forced to shut in large amounts of natural gas for lack of a market. That era has clearly passed and a new paradigm is in force (Fig. 1.13). While there will occasionally be periods of oversupply that will force prices down, these periods will be infrequent and short-lived.

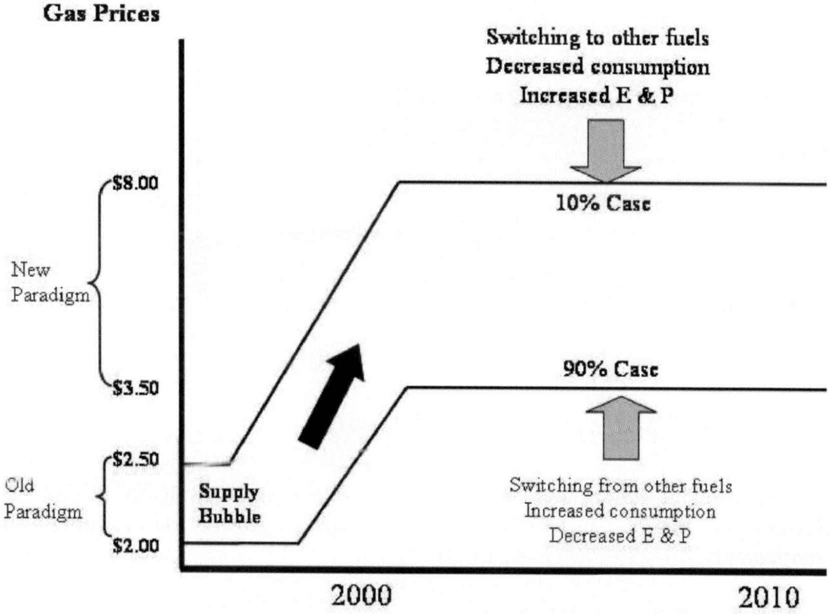

Figure 1.13. The "supply bubble" that characterized North American natural gas markets during the 1980s and 1990s no longer exists. Instead, there is a new paradigm in which low gas prices will be a rare and temporary exception.

Demand for natural gas is likely to increase further as concerns regarding greenhouse gas emissions continue to grow. Natural gas has an advantage over other fossil fuels (Fig. 1.14) because its combustion yields 44% less CO_2 than coal for the same energy release, and 29% less than oil (EIA, 1998). Increased use of natural gas as a replacement for coal may be required by law in many locations in the future.

By the year 2025, the United States will be consuming 29.1 to 34.2 TCF natural gas per year, an increase of 29-51% over 2002 (EIA, 2004). Conventional drilling, drilling for current unconventional sources, and imports from Canada will not come close to meeting that demand. LNG imports will fill a significant portion of the gap but are unlikely to be adequate.

Sufficient natural gas supplies are essential for the economies of North America. The portfolio of gas resources to meet the forecast demand will require additional components. Gas Hydrate has the potential to play a substantial role in contributing to North American gas supply.

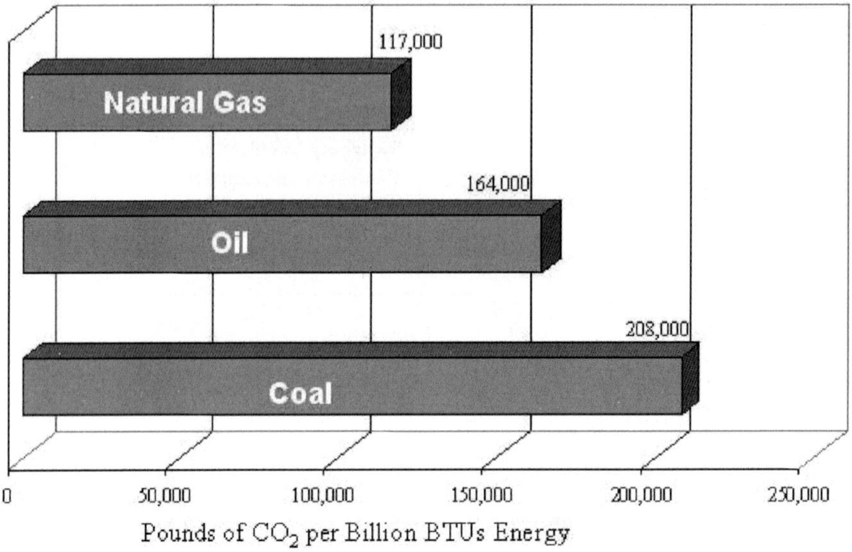

Figure 1.14. The lower carbon dioxide emissions that result from the combustion of natural gas rather than coal make gas or oil a more environmentally friendly fuel (EIA, 1998).

1.7. THE CASE FOR GAS HYDRATE

Gas hydrate is a crystalline substance that forms when gases (such as methane) combine with water under conditions of high pressure and low temperature. Since the 1960s gas hydrate has been recovered in cores from marine and Arctic environments in many parts of the world. Mounds of gas hydrate have been mapped and sampled from the seafloor of the Gulf of Mexico and the Pacific margin of North America. In addition, BSRs have been identified on continental margins throughout the world. These occurrences and indicators are described in more detail in Chapters 3 through 6.

Gas hydrate concentrates natural gas in a "cage" of water molecules. The concentration factor is such that the dissociation of a single cubic foot of saturated gas hydrate yields approximately 164 cubic feet of gas at standard temperature and pressure (MHAC, 2002). The concentration of natural gas by the gas hydrate structure significantly increases its resource potential. The energy density of gas hydrate (Table 1.1) is two orders of magnitude higher than that of natural gas at STP. As a result, relatively small accumulations of gas hydrate may have resource potential.

The total amount of gas hydrate in the world is not known with any certainty and estimates range over three orders of magnitude (Collett, 2004a). At the high end, the energy potential of gas hydrate exceeds that of all of the coal, oil, and other gas resources. More significantly, the low end estimates of

106,000 to 176,000 TCF of natural gas in gas hydrate (Milkov, 2003) are still several orders of magnitude higher than the estimates of recoverable discovered and undiscovered conventional gas.

Fuel and Form	**Density g/cc**	**Energy Density Btu/ft^3**
Hydrogen (Liquid)	0.071	229,000
CH_4 Gas (STP)	6.66x10-4	1,012
CH_4 Liquid (LNG)	0.42	570,000
CH_4 Solid (Hydrate	0.91	165,968
Reference liquid fuels		
Gasoline (Petrol)	0.74	876,000
JP-4 (Jet Fuel)	0.78	910,000
#2 Diesel fuel (U.S. Navy)	0.78	995,000

Table 1.1. Energy density of different forms of fuels. The increase in energy density of gas hydrate compared to methane gas is especially striking. (from Hoess and Stahman, 1969; Max, et al., 1997). *This was originally referred to as Figure 1.15 because it was captured as an image but it is now modified and updated.*

From a resource standpoint, however, the total volume of gas hydrate in the world is of little importance. What matters is the volume of natural gas that can be commercially produced from gas hydrate deposits; and for now, estimates of that volume are entirely speculative.

1.8. CURRENT KNOWLEDGE OF GAS HYDRATE OCCURRENCE

Much has been learned about the occurrence of gas hydrate in the natural environment during the past decade from the investigations carried out by universities and international consortia including the Ocean Drilling Program. These investigations have recovered and analyzed gas hydrate samples from a variety of deep marine and Arctic locations. While many prospective areas have not been sampled and much remains to be learned, the results to date reveal several fundamental concepts that may be used to guide gas hydrate exploration and development.

First, gas hydrates are widespread along continental margins and in Arctic regions where the appropriate temperature and pressure conditions occur along with an adequate source of hydrate-forming gas. These physical conditions occur at water depths greater than 500 meters at mid to low latitudes and greater than 150 to 200 meters at high latitudes (MHAC, 2002). At these water depths, gas hydrate may occur within a zone that extends into the sediment to depths of tens to hundreds of meters beneath the seafloor. In Arctic sediments, gas hydrate may occur within and beneath permafrost zones. While the pressure and temperature conditions are ubiquitous in these environments, the occurrence of gas hydrate is further controlled by the availability of hydrate-forming gas. In areas where the rate of gas flux is low, little or no hydrate may be present.

Second, the lithology of the host sediment has a significant influence on the distribution of gas hydrate in the sediment. Shales within the gas hydrate stability zone (HSZ, Chapter 3) typically contain low concentrations of hydrate, consisting of small dispersed crystals and occasional nodules and fracture fillings. The larger concentrations of gas hydrate in shale appear massive when cored but it is doubtful that they have significant lateral continuity. In contrast, the concentration of gas hydrate in the pore space of sands and gravels may exceed 80% (Winters, et al., 2002). The potential for such high concentrations of gas hydrate makes coarse clastic sediments the logical initial exploration target.

The only other large concentrations of gas hydrate are in hydrate mounds located on the seafloor in the Gulf of Mexico. The hydrate mounds occur at vents associated with hydrocarbon migration along faults. Large reserve potential has been attributed to hydrate mounds (Milkov & Sassen, 2002); however, there are several serious problems with producing natural gas from seafloor mounds. The mounds support unique and diverse biological communities and are protected by law. The base of the hydrate mounds is not well defined. Seismic data from some locations indicates that the temperature and salinity of migrating fluids has narrowed the hydrate stability zone so that the amount of gas hydrate present is far less than initially believed. It also appears that subsurface gas hydrate may be lining fault zones beneath the mounds. Dissociation of gas hydrates in fault zones has the potential to cause large slumps. These issues, along with the severe economic challenges resulting from the expenditure of large amounts of energy to recover natural gas, make the production of natural gas from seafloor hydrate mounds unrealistic.

The third important result from recent studies is the recognition that BSRs do not actually identify gas hydrate (3.3.2). The BSR, where present, is a reasonably good indicator of the phase boundary between hydrate-bearing and free gas-bearing sediments. Unfortunately a BSR yields little useful information about the concentration of gas hydrate in a volume of sediment. Well-defined BSRs are common in fine-grained sediments, especially in relatively stable basins with homogeneous stratigraphy. The location of the phase boundary in tectonically active basins will often be extremely variable rather than "simulating the bottom". A BSR may also be overwhelmed by seismic signature of the sediments in the basin. Thus, it is not surprising that BSRs may be absent where gas hydrate is known to occur (Pecher & Holbrook, 2000).

1.9. EXPLORATION FOR COMMERCIAL GAS HYDRATE PROSPECTS

Prospecting for gas hydrate accumulations requires far more than merely the mapping of BSRs in a basin. The successful development of commercial gas hydrate prospects involves a petroleum systems approach that is comparable to the development of conventional oil and gas prospects. This fact has not been widely appreciated by many of the government and corporate investigators throughout the world.

Gas hydrate prospects require reservoir rock with appropriate porosity and permeability. As difficult as it is to produce natural gas from tight formations, producing natural gas from gas hydrates in impermeable reservoirs is far more difficult since dissociation of the hydrate is also required. Reservoirs will also need to have sufficient areal extent to meet economic hurdles. Thus an understanding of the depositional history of the prospective formation is necessary.

The presence of high-quality reservoir rocks alone is no more a guarantee for hydrate prospects than for conventional oil and gas prospects. Exploring for commercial gas hydrate prospects also requires a suitable source of hydrate-forming gas, with appropriate migration paths into the reservoir at a time when a trap was present. With gas hydrate prospects the trap may form as a response to the hydrate-free gas phase boundary, provided the phase boundary was present at the time of hydrocarbon migration. Similarly, an up-dip seal is required for both hydrate and conventional prospects. With a gas hydrate prospect, hydrate crystals in the formation may act as a top seal.

The need for reservoir rock, trap, seal, source, timing, and migration are common to both gas hydrate and conventional prospects. Gas hydrate prospects also require mapping of temperature and pressure conditions to define the hydrate stability zone. The temperatures and pressures that define the phase boundary vary with changes in the chemistry of the hydrate-forming gas. Thus it is important to determine the expected gas composition of any basin as part of the exploration process.

1.9.1. Overview of Deepwater Production

The offshore oil and gas industry has a long history of applying new technology in innovative ways in order to pursue commercial opportunities. Any thought that gas hydrates will be too difficult to commercially exploit ignores the creativity that the industry brings to bear on technical hurdles.

The first "offshore" well in North America was drilled by Gulf Oil at Caddo Lake in northern Louisiana in 1907 (Schempf, 2004). The technology developed at Caddo Lake was applied in shallow-water developments elsewhere, including the Gulf Coast and Venezuela's Lake Maracaibo. The expectation that oil fields extended into the Gulf of Mexico led to several attempts at offshore drilling in the 1930s (Schempf, 2004). Pure Oil and Superior Oil made the first offshore discovery in 1937, at Creole Field in fourteen feet of water one mile from the Louisiana coast. The exposed location required new approaches for the design and construction of facilities. The first platform constructed beyond the sight of land was installed in the Ship Shoal Area by Kerr McGee in 1947 in eighteen feet of water, eleven miles from shore. A key innovation was the use of a tender so that the platform size could be reduced by 75%. This reduced the cost of offshore operations and essentially established the offshore oil industry (Schempf, 2004).

The opportunity for new discoveries and increased profitability led to a succession of new tools and technologies. The first jackup drilling unit was employed in 1954, followed by the first drill ship in 1956, and the first semisubmersible drilling unit in 1962 (Schempf, 2004). Exploration and evaluation technology progressed as well, with advances in geophysical tools and logging methods. These were combined with many less obvious innovations such as improved hoists, drill bits, and pipe-laying equipment. Subsea completions were initiated in 1961 (Schempf, 2004).

With the new technology, the industry pushed into progressively deeper water. A platform was set in 100 feet of water in 1955, 200 feet of water in 1962, and 1000 feet of water in 1979. The installation of Shell's "Bullwinkle" platform in 1615 feet of water in 1988 marked a turning point for offshore structures. With increasing water depth the installation of progressively larger fixed production platforms became uneconomical. This led to a succession of new designs including compliant towers, Sea Stars, Floating Production Systems, Tension Leg Platforms, SPARs, and a new generation of subsea completions (Fig. 1.16).

Twenty years ago, the concept of producing oil and gas in several thousand feet of water was viewed with skepticism, as was the concept of pursuing gas hydrates. The second wave of hydrocarbon exploration (the first wave was exploration on land) on the relatively shallow continental shelves has now given way to the third wave, which is deepwater exploration (Sorkhabi, 2004). As recently as 1991, energy industry representatives were of the opinion that hydrate would never be an economic play because the investment required for what they perceived to be a completely speculative energy source would never be made (Max et al., 1991). Now, the major part of the deepwater tools and technology required to explore and extract hydrate natural gas have been developed for conventional petroleum. Adapting that technology for producing hydrate natural gas fits well in the tradition of the petroleum industry, which has met one challenge after the other for over a century. The energy industry has already reached water depths where hydrate deposits can be expected and the depths of both startups and discoveries are increasing. In 2004, for instance, 14 new deepwater startups in the Gulf of Mexico averaged 4,596 ft (1,393 m) water depth while 12 deepwater discoveries averaged 6,479 ft (1,963 m) water depth (ON&T, 2005).

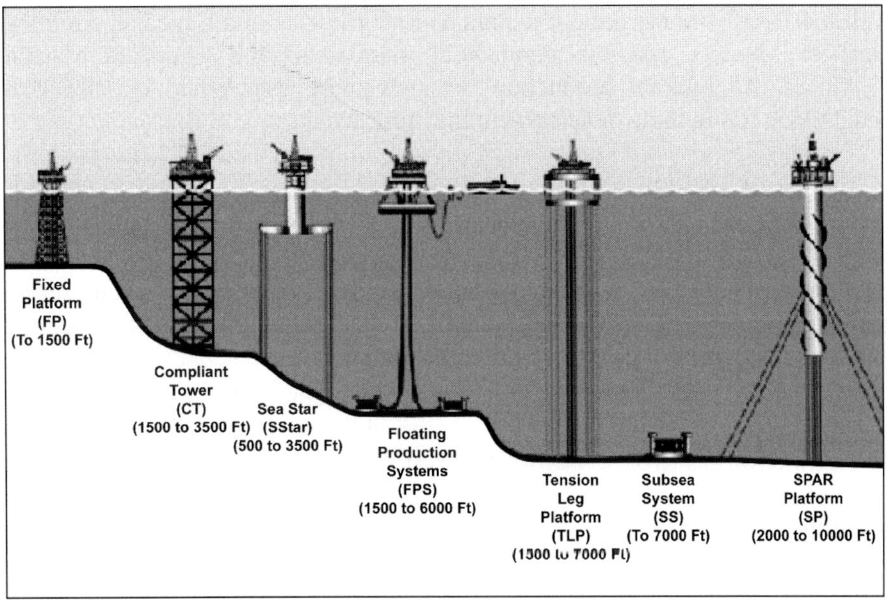

Figure 1.15. The commercial limitations of fixed production platforms with increasing water depth have led to a succession of new designs. Each design addresses the specific criteria of water depth and field size (MMS, 2004c). Hydrate is stable from 500m (~1500 ft) depth and deeper.

1.9.2. Models for Recovery

Gas hydrate contains no oil, or even condensate, a fact which should make hydrate development viable even in locations where conventional oil and gas development are banned. In the unlikely event of a spill, the only release would be of methane gas. While large releases of methane are dangerous, the environmental impacts of a methane release are likely to be far less than those of an oil spill (Patin, 1999). Nonetheless, during the initial stages of commercial gas hydrate development, the best economic results will be obtained from leveraging existing conventional oil and gas infrastructure such as pipelines and offshore platforms. Gas hydrate development will thus initially follow conventional development. By leveraging conventional facilities, gas hydrate production will avoid the enormous capital costs associated with deepwater and Arctic development that would be incurred as a stand-alone project.

Exploration and development of gas hydrate resources will involve the same activities of seismic acquisition and drilling that are currently banned from many prospective areas of North America. Gas hydrate exploration will thus be feasible only in areas that are open to conventional exploration. In addition, because leveraging existing infrastructure is essential for commercial gas hydrate development; the same issues of access to acreage that impact conventional oil and gas development will impact gas hydrate development as well. For North America these factors point to the U.S. and Canadian Arctic, the Gulf of Mexico, and the continental margin off the Canadian Maritime Provinces. As hydrate production becomes more established in later years, stand-alone gas hydrate development may proceed.

Many of the same technologies utilized for conventional gas development will be used in gas hydrate development. Yet the fact that dissociation will have to take place prior to extraction means that additional technology will have to be developed. Gas hydrate dissociation will occur when hydrate is sufficiently warmed or depressurized, or when a solvent such as methanol or glycol is introduced. Finding cost effective methods of dissociation is a critical element in gas hydrate commercialization.

Depressurization is the least expensive method of dissociation yet may not yield high enough flow rates for commercial viability. Depressurization may be feasible for gas hydrate-bearing sands if the sand has a sufficiently large free gas leg down-dip. The free gas could be produced by conventional means, resulting in a decrease in reservoir pressure that would cause up-dip gas hydrates to dissociate (Fig. 1.17), thus yielding additional free gas to the well bore.

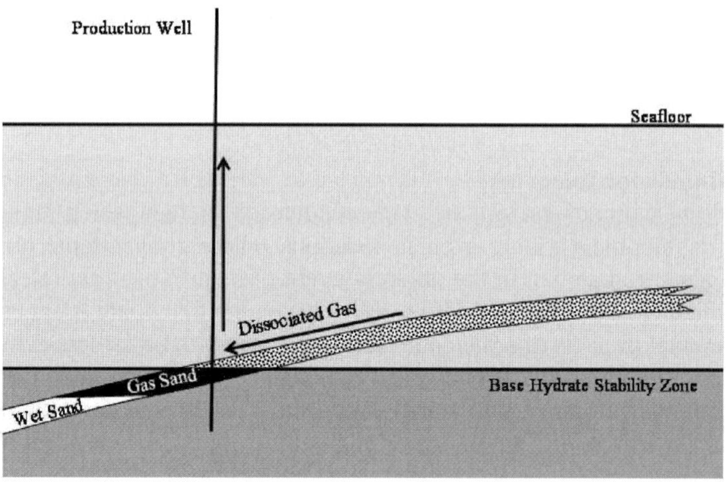

Figure 1.16. The least expensive approach for producing gas hydrate from a reservoir is to produce the free-gas that is continuous with the hydrate. This approach uses conventional technology. Courtesy of HEI.

Heating gas hydrate reservoirs will increase flow rates, but with added costs. The highest rates appear to be achieved by a combination of heating and depressurization (Moridis & Collett, 2003). Several sources of heat may be considered, including waste heat from the produced water of conventional oil

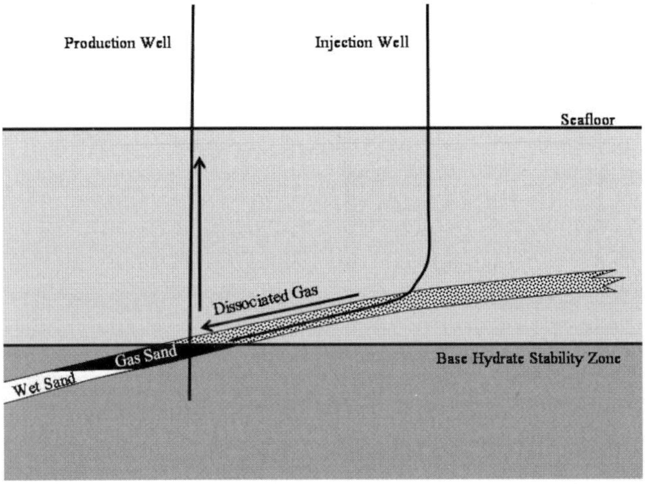

Figure 1.17. If the production rate is insufficient with depressurization alone, heat can be applied. In this example, a down-hole heating device such as a catalytic burner is used. Courtesy of HEI.

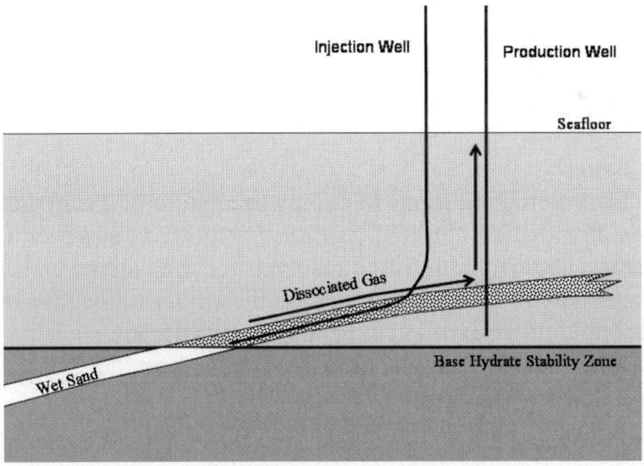

Figure 1.18. Where there is no free gas, heating is necessary, with the production well up-dip of the dissociation. Courtesy of HEI.

and gas operations, hot brines produced specifically for hydrate production, boilers at the land or sea surface, and down-hole heating units such as catalytic burners (Fig. 1.18). If the down-dip section of the gas hydrate-bearing sand is entirely wet, with no free-gas leg (Fig. 1.19), some form of heating will be needed to yield economical flow rates. For many potential prospects, gas hydrate-bearing sands may not cross the phase boundary (Fig. 1.20). Heating will be required in such settings.

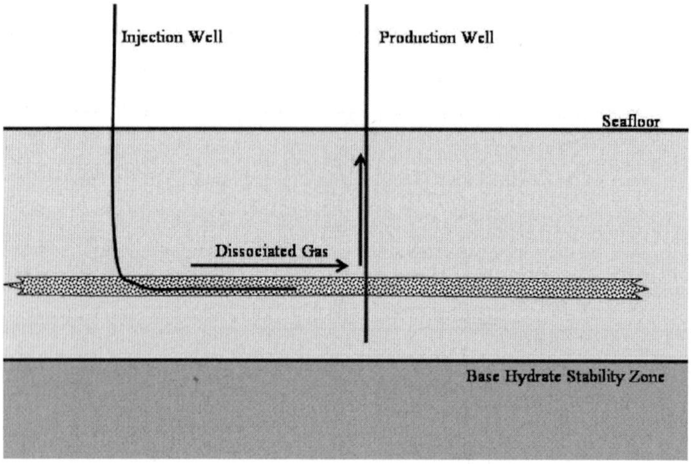

Figure 1.19. For gas hydrate-bearing reservoirs that do not cross the phase boundary, heating is required. Courtesy of HEI.

The nature of the occurrence of gas hydrates, combined with the production technology available, will determine which gas hydrate deposits may be commercially developed. The fundamental equation is that the gas must be produced at a high enough rate and at a low enough cost, whatever those may be for any particular geographic production/market area.

1.9.3. Business Issues

There are many unanswered questions regarding the nature of gas hydrate in the subsurface and its dissociation. From a business perspective, however, there are relatively few critical issues that must be resolved for the commercialization of gas hydrates as a resource. First among these is the ability to identify prospects using remote technology, in particular geophysical methods such as seismic interpretation. The required methodology would provide adequate risk assessment prior to leasing and drilling.

The second critical issue is the total volume of natural gas recoverable from wells drilled into a gas hydrate reservoir, and the rate of gas production. Recovery factors are no less important for gas hydrate plays than they are for conventional oil and gas plays. Estimates of the amount of gas in place for a gas hydrate prospect may yield large numbers, but those numbers become irrelevant if the per-well recovery is low. This is especially true in the high cost deepwater and Arctic locations where gas hydrate resources would be developed.

Along with the ultimate volume of gas recovered, the rate of production is a critical factor and is poorly known at present. Commercial gas hydrate development will require flow rates sufficiently high to offset operating expense and yield an acceptable profit. Resolving the questions of ultimate recovery and flow rates will require a better understanding of the occurrence of gas hydrate in geological materials including volumetrics, hydrate/sediment interactions, reservoir continuity, and the overall issue of gas hydrate as part of a petroleum system.

A third issue involves the safety of gas hydrate production, especially in the marine environment. The dissociation of gas hydrate will alter the physical properties of the host sediment, and could result in geological hazards including seafloor collapse and slumping (Fohrman et al., 2004; Pecher et al., 2004). While these effects could be minimal in some settings, their impact could be very serious in others, particularly where highly pressured pore fluids can blow out on the outside of drill casings, which can cause collapse of seafloor and sub-seafloor apparatus. Of particular concern are locations where the seafloor has a substantial dip and the prospective sand is within a few hundred meters of the seafloor.

A fourth issue is production technology. Much of the technology required for producing gas hydrate prospects either already exists or may be easily adapted from existing technology. Some new methods will also have to be developed, including not only efficient means of dissociating gas hydrate but also keeping hydrate from reforming in flowlines. Resolving production issues will require a better understanding of the occurrence of gas hydrate in a reservoir.

The final issue here is the cost of finding, developing, and producing gas hydrates. Gas hydrates occur in deepwater and Arctic environments where operations are expensive. There are, however, aspects of gas hydrate development that could result in lower costs than otherwise expected. Locations where gas hydrate developments are most likely to be pursued tend to have extensive seismic coverage, eliminating or reducing the need to acquire additional data. In addition, prospective gas hydrate reservoirs are relatively shallow so smaller, less expensive drilling rigs could be utilized. Developing and adapting technology for gas hydrate operations should lead to acceptable operating expenses. Also, in deepwater hydrocarbon provinces, it should be possible to piggy-back on conventional deepwater production facilities. There are over 100 deepwater discoveries either on production or with platforms being designed, built, or installed in the northwestern Gulf of Mexico. This offers a convenient base of operations for hydrate production.

Addressing the issues of recovery, safety, technology, and costs is essential if gas hydrates are to become a commercial resource. Current research efforts being conducted by industry, government agencies, and academic institutions in many parts of the world are making substantial progress on these issues.

1.10. THE GAS ECONOMY: ENHANCED EFFICIENCY AND SECURITY

Whether hydrate is developed as a new source of natural gas or not, the United States is moving from an oil to a gas-based energy economy. Considerable additional pipeline infrastructure will be required. Because new pipelines will be necessary in any case, we suggest that a distributed electricity production system be established wherein natural gas is distributed and smaller power stations are built to serve local areas. This distributed power generation system has two advantages. The transfer of energy by the flow of gas in a pipeline can be done with much less loss than by the flow of electricity in a wire. Thus a distributed electrical generation system using piped gas can be operated more efficiently that one employing centralized electrical generation and long electrical transmission lines. Pipelines are harder targets for terrorists as they can be securely buried and pumping and handling stations can be more easily protected than long high-tension electrical transmission lines.

1.11. CONCLUSIONS

The growth in energy demand in China, India, and elsewhere (Skislock, 2005) is affecting consumers throughout the world. The shift from regional to global natural gas markets has created a "superdemand" in which the United States no longer dominates energy consumption. In the face of global superdemand, the United States needs to increase energy efficiency and develop additional energy resources, and not just rely on increased imports.

During the past decade, gas hydrates have had a poor reputation as an energy resource within western energy companies, in part because they are very different from conventional natural gas and are essentially still in a basic research phase where all development paths have not yet been properly defined. As little as 10 years ago, gas hydrate development was viewed as being far into the future, with no near-term return on capital. There was a perception that hydrate development would be prohibitively expensive and require technology so completely different from that currently in use that hydrate development would not easily fit into an energy company's core business. Recent progress in understanding the physical chemistry of the natural gas hydrate system and progress in recovering gas from hydrate are changing that perception.

Much remains to be learned before gas hydrate begins to make a contribution toward meeting the demand for natural gas. Yet gas hydrate investigations are following the same path that led to the commercialization of coalbed methane, tight gas sands, and shale gas. These resources were also once widely viewed as insignificant and highly speculative, and impractical and uneconomical to develop. The demand for natural gas in North America and elsewhere will increase. The opportunity for gas hydrate to become an important component of North American supply is clear. Thus, the answer to, 'why gas hydrate', is not only 'why not gas hydrate' but rather an imperative to resolve the remaining technical issues involved and develop gas hydrate as an economic energy resource.

Chapter 2

Physical Chemical Characteristics of Natural Gas Hydrate

John P. Osegovic, Shelli R. Tatro, and Sarah A. Holman

2.1. INTRODUCTION

Natural gas hydrates (hydrate) are ice-like crystalline compounds of gas (mainly methane and other hydrocarbon gases) and water, which naturally occur both at very low temperatures in permafrost regions, and in the low temperature - high pressure regimes present in the deep oceans (Fig. 2.1). Hydrate is very different from water ice, however, in that under pressure it is stable at temperatures considerably higher than water ice. The hydrocarbon gases are thermodynamically stabilized within gas hydrates by interactions with a crystalline lattice of water molecules that are hydrogen bonded (Kvenvolden, 1993). In appearance, hydrates take the form of inter-grown, transparent/translucent-to-opaque, white-to-grey and yellow crystals, which have a poorly defined crystal form. Hydrates may "cement" the sediments in which they occur, and may impart considerable mechanical strength to them, but they may also occur in nodules and in the pore space matrix unattached to sediment grains.

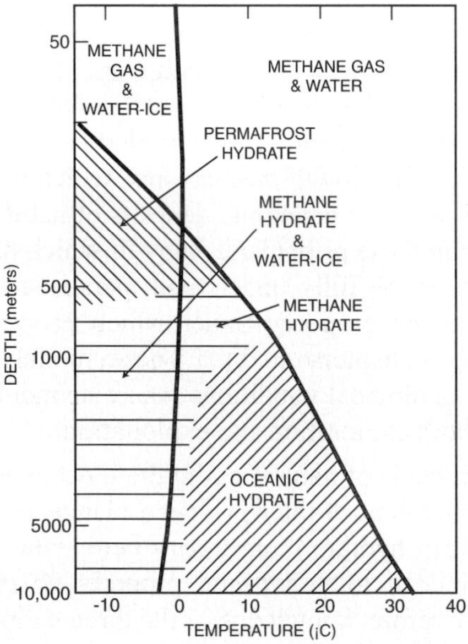

Figure 2.1. Diagrammatic methane hydrate pressure-temperature phase diagram. After Kvenvolden (1988) and Pellenbarg and Max (2000, 2003, Fig. 2). Permafrost and oceanic hydrate P-T fields delineated.

Michael D. Max et al. (eds.), Economic Geology of Natural Gas Hydrate, 45–104.

Hydrate formation forces methane molecules into closely packed guest lattice sites, which has the effect of concentrating the methane. Methane hydrates are non-stoichiometric, in that the water molecule crystal structure of the hydrate can be established without all the methane lattice sites being occupied. One cubic meter of methane hydrate contains about 164 m^3 of methane (at STP) and about 0.8 m^3 of water (Kvenvolden, 1993). At least three polymorphs of natural gas hydrate are known (Sloan, 1997)

Hydrate is a dynamic material that is very sensitive to its environment. Changes in pressure, temperature, salinity, and the degree of saturation of the hydrate-forming gas (HFG) in sediment pore water, for instance, can cause hydrate to nucleate, grow, dissociate, or dissolve. Hydrate appears to be very common in both oceanic and permafrost regions (Chapter 3; Max 2003) because the relative cold of Polar regions at the surface even on shallower seafloors allow for hydrate to be stable to considerable depths. In other geological times, for instance during the Mesozoic, when there was little or no permafrost and seafloor temperatures were warmer, the amount of hydrate present would have been substantially less (Max et al., 1999). More recently, relatively small changes in pressure and temperature have caused hydrate to form, and, more dramatically, to cause sediment redistribution through dissociation in sediments (Kennett et al., 2002). This particular sensitivity of its ambient environment may render natural gas hydrate unique amongst crystalline materials in the biosphere.

Even within pressure-temperature regimes where hydrate is potentially stable, the presence of large quantities of gas hydrate is not ensured. Like any diagenetic material or ore deposit in sediments, the provision of chemical components of the crystalline material must be delivered to a location in a manner suitable for nucleation and growth so that large quantities may form. It is these mass transport and growth mechanisms, which are the critical factors to the development and potential economic deposits of natural gas hydrate, that are considered here. Using these principals, many of which have only recently been understood or are yet to be fully understood, paragenetic models for different natural conditions can be proposed under which economic concentrations of hydrate will develop (Chapters 4, 7). These models should assist in the delineation of likely geological conditions where economic deposits of hydrate may occur and help both exploration and exploitation.

Descriptive models of natural gas hydrate occurrences must be based on chemical principles that control their presence. Understanding how to find and extract natural gas from hydrate is necessary before the gas can be used as an energy source, or, indeed to understand properly its effects on climate and seafloor stability. Therefore, knowledge of the three main chemical processes of nucleation, growth, and dissociation/dissolution is vital to understand how these processes may operate in the sediments where hydrate deposits may be found.

This chapter introduces these concepts in a manner that will highlight the significance of each component of each process, while striving to avoid becoming entangled in details of hydrate physical chemistry that are not directly related to the formation and exploitation of potential economic deposits of hydrate. Several models for hydrate deposition and concentration are presented, the consequences of which for exploration and potential development are discussed in Chapters 4, 6, 7 & 10. This chapter provides a working chemical description of the different manners of formation and breakdown of gas hydrate. It is not intended to cover all aspects of the physical chemistry, nor the in-depth thermodynamics of natural gas hydrates. For a thorough description of the physical chemistry of gas hydrate, Sloan (1998) should be consulted along other with appropriate references, such as Ripmeester (2000) and the considerable body of new scientific literature.

2.2. CRYSTALLINE GAS HYDRATE

Natural gas hydrates are part of a larger family of compounds known as clathrates, which are inorganic container compounds (Fig. 2.2). They are known in some fields as inclusion compounds (Kirchner 2004) because they are viewed as a complex water lattice that includes Hydrate Forming Gas (HFG) molecules. There are many types of container compounds (Pellenbarg & Max, 2003) but only natural gas hydrate is considered here. Natural gas hydrate is a methane-based mixed hydrate that may include ethane, propane, and butane. In the case of natural gas hydrate the host is a latticework of water molecules and the guests can be any of a number of small molecules including carbon dioxide, nitrogen, hydrogen sulfide, and, more particularly here, any of the hydrocarbon gases (Franks, 1973; Sloan, 1998). The water molecules surround and prevent or greatly reduce the mobility of the guest molecules. The presence of guest molecules in a high enough degree of saturation in a fluid or gas surrounding the hydrate is critical to both the formation and persistence of the hydrate. Without the interactions between host and guest molecules, the host molecules would not organize themselves to such a degree and the guest molecules would remain in either a gas or solvated form. The precise manner by which the presence of a guest molecule causes a host cage to form or partially form so that the repeating crystalline structure can initially develop and then grow is not well understood.

The interaction or degree of bonding between individual water molecules and the guests is very weak, but the overall interaction of the guests with the host structure can be quite strong (Durham et al., 2003). The strength of this interaction is why gas hydrates form and remain stable. Water molecules alone will not assume crystalline form above the freezing point of water, although water itself contains a considerable percentage of structured water molecules (Smith, 2004). The stability fields of all hydrocarbon hydrates extend considerably above the freezing point of water. Although complete occupation of every guest site by an HFG molecule is not required for hydrate to form and persist, it is necessary for a considerable majority of the guest sites to be filled.

It is not known to what extent unfilled guest sites may cause defects in the water latticework.

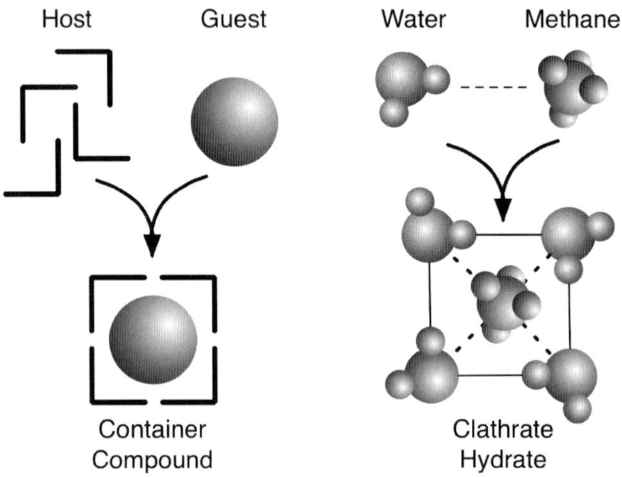

Figure 2.2. Natural gas hydrates are container compounds. Left: host molecules and guest. Right: In clathrate hydrates, water molecules form the host structure that accommodates the guest, in this case methane. The interaction between liquid water and hydrocarbon is weak (dashed line, top) while the interaction between the hydrate host structure and the hydrocarbon is strong. Once the compound is formed, it is difficult to remove the guest without breaking the structure. This diagram is not representative of the actual crystal structure of hydrate. For crystallographic models see Sloan (1998) and Max, (2003).

Hydrate is a stable crystalline form because the interactions or weak bonding between the molecules in the hydrate structure are persistent and fully formed whereas the interactions in an HFG/water solution are only partially formed and are thus constantly changing in a 'time averaged' manner (Fig. 2.3). Why gas hydrate is only stable at elevated pressures and low temperatures is beyond the scope of this text but relates to the ordering that occurs upon formation of hydrate (Sloan, 1998), among other things.

Gas hydrate can have several molecular structures or ways in which the water molecules can form the host network. Crystallographic data are well summarized by Kirchner (2004). Of the known structures the predominant structures among natural gas hydrate will be either sI (structure 1) or sII (structure 2) (Table 2.1). Natural gas will typically form sII hydrate unless it is essentially pure methane. It will then form sI. Mixed natural gas hydrates form under less lower pressure/higher temperature conditions than 100% methane hydrate. A third structure type, sH, forms with larger hydrocarbon gases but only if methane is present as well.

HFG -- Liquid Water Solution

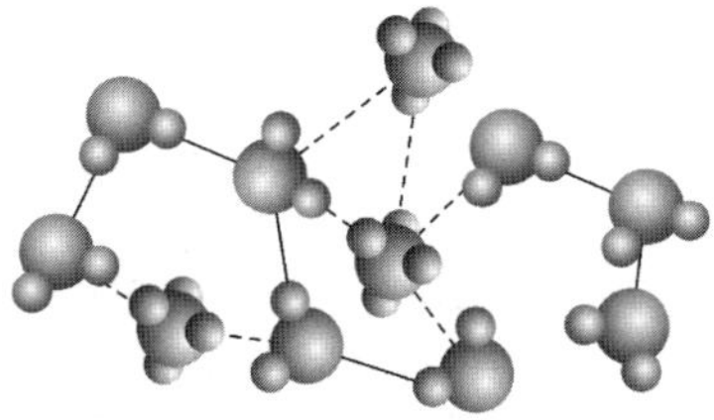

Interaction	Strength	Number
Water-Water	Medium	6
Water-HFG	Weak	6
HFG-HFG	Very Weak	1

Natural Gas Hydrate

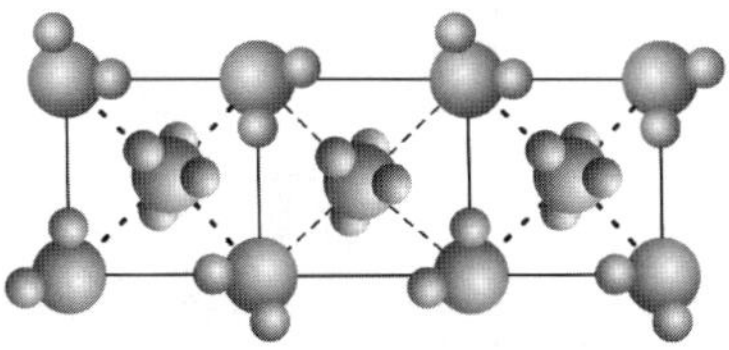

Interaction	Strength	Number
Water-Water	Medium	10
Water-HFG	Weak	12
HFG-HFG	Very Weak	0

Figure 2.3. Two dimensional schematic that emphasizes the differences in interactions in solution (top) and gas hydrate (bottom). The number of interactions, and thus the stability, increases from the solution to hydrate. The numbers demonstrate a trend and are not to scale. In pure liquid water a large fraction of the water molecules are interacting with four other water molecules (Smith, 2004). Additionally, the water to methane ratio in a room temperature solution at 10 MPa (equivalent to 1000 m below surface) is 420 to 1. In methane hydrate there are ideally 5.75 (Sloan, 1998) water molecules per methane molecule.

How a given guest or guests interact with the host to determine the hydrate structure formed is complicated (c.f., Schicks & Ripmeester, 2004). It is clear that a variety of structures may initially be present, but evolution of the mixture toward the pure and expected hydrate structure occurs over time. The process of forming a series of solid forms, with each step in the series being more stable than the last, is known as Ostwald Progression or following the Ostwald Series. The least stable material is the first to form. This material forms the next most stable solid after a given amount of time until the most stable material is reached.

Structure Type	Primary Component	Secondary component
sI	Essentially pure methane or ethane	Trace other HFGs
sII	Methane	Ethane, propane, hydrogen sulfide, carbon dioxide
sH	Small HFG must be present	Hydrocarbons with more than 4 carbon atoms

Table 2.1. Structure and HFG compounds found in natural gas.

2.3. FORMATION OF GAS HYDRATE

Gas hydrate crystals can form where the reactants, water and HFG, exist within the hydrate stability conditions of pressure and temperature. In particular, the methane hydrate stability field (Fig. 2.1), is most pertinent here because methane dominates the other hydrocarbon gases in both oceanic and permafrost hydrate deposits. See Sloan (1998) for an in-depth discussion of the effects of different dissolved solids and other chemicals, and where mixtures of HFG's occur and lead to the formation of compound hydrate. In these conditions the pressure-temperature field of hydrate stability alters, generally to become more stable at higher pressures and lower temperatures).

2.3.1. The Growth Dynamic

The nucleation and growth of gas hydrate requires a driving force. This driving force is called supersaturation. Kaschiev and Firoozabadi (2002) have defined supersaturation in an aqueous solution as the difference in chemical potential between the aqueous solution and the hydrate crystal. Chemical potential is a quantity that indicates the stability of a compound. The more negative the chemical potential, the more stable the compound. For example, the chemical potential of gas hydrate is less than the chemical potential of an HFG/water solution when the solution is supersaturated with respect to gas hydrate. A system always reacts to reduce chemical potential, thus the HFG/water solution described will react to form gas hydrate, assuming pressure-temperature conditions conducive to hydrate growth.

A supersaturated solution is formed when enough water and HFG are brought under conditions where hydrate is stable (Table 2.2). This solution can be more complex than a simple, homogeneous solution of natural gas dissolved in water or vice versa. The solution can contain gas bubbles, dirt, or other contaminates.

Where HFG-enriched pore water is in the presence of hydrate, the excess HFG dissolved in the solution reacts to generate more hydrate. The relative supersaturation of the HFG/water solution with respect to the hydrate mineral drives this growth. Dissolved gas molecules diffuse from the bulk water into the lower-concentration interfacial water. A gradient of HFG concentration should then be observed from the HFG source toward the HFG sink (the hydrate). Growth of the hydrate domain will bring these regions closer together as long as the pressure and temperature are sufficient to support the hydrate phase.

The three variables that determine supersaturation are not all independent of each other. The concentration is often a function of both pressure and temperature. Henry's Law is one way to describe the amount of HFG that can be dissolved in water as a function of pressure and temperature. Raoult's law describes the amount of water vapor that can be present in the gas phase as an indirect function of pressure and temperature.

Pressure	Temperature	HFG/H_2O Concentration	Descriptor
Just right	Just right	More than needed	Superconcentrated
Just right	Too cold	Just right	Supercooled
Too high	Just right	Just right	Superbaric
Too low	Too hot	Not enough	No hydrate

Table 2.2. Conditions and terminology for gas hydrate supersaturation.

For hydrate to remain stable and not dissolve or dissociate, solution parameters (Table 2.2) must be met. For hydrate to grow all of the three parameters must be 'super.' In any P-T-χ (χ is a notation for concentration or composition) space, the temperature, pressure, and concentration will always define a point (Fig. 2.4). Hydrate is stable if the point is on or above the top surface (labeled Hydrate Stable). For a driving force to be present, which is required for nucleation and growth, the system conditions must be above the surface. A supersaturated solution of HFG/water is always superconcentrated, supercooled, and superbaric.

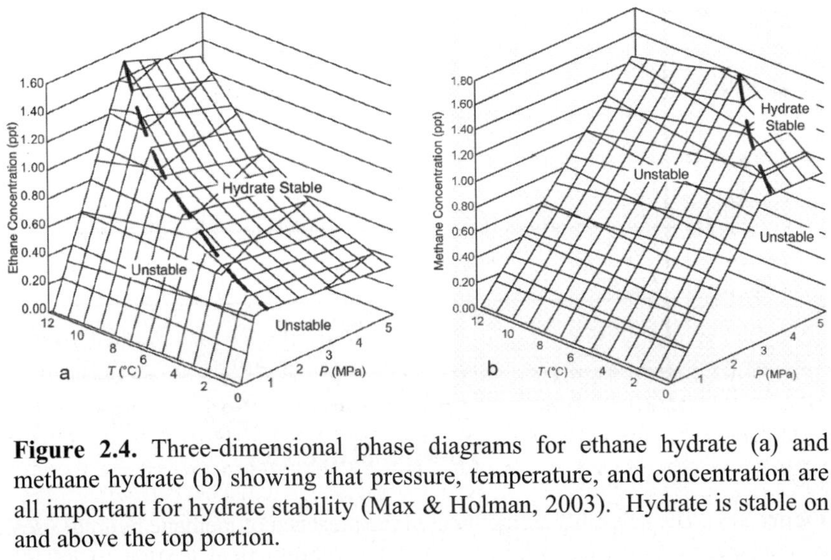

Figure 2.4. Three-dimensional phase diagrams for ethane hydrate (a) and methane hydrate (b) showing that pressure, temperature, and concentration are all important for hydrate stability (Max & Holman, 2003). Hydrate is stable on and above the top portion.

Henry's Law describes the amount of gas that can dissolve in water to form a supersaturated solution, which is most accurate for dilute systems such as natural gas dissolved in water. Henry's Law states that the amount of gas that will dissolve in a solvent is determined by the partial pressure of the gas at a given temperature. Increasing the pressure increases the amount of gas dissolved. Decreasing the temperature also increases the amount of gas dissolved; it is not entirely incorrect to think of this as the gas becoming more liquid-like as it cools. Once hydrate is present both general trends change. As the temperature decreases the solubility of HFG decreases (Fig. 2.5). This is because hydrate is a sparingly soluble mineral, and minerals tend to decrease in solubility with decreasing temperature (Table 2.3). Increasing pressure seems to have a small effect on the solubility of hydrate (Servio and Englezos, 2001, 2002) and (Davie, et al 2004). The effect of increasing pressure is to slightly lower the equilibrium methane concentration. The equilibrium partial pressures of methane hydrate at 275 K is 8.4 mM lower at 30 MPa than at 10 MPa (Davie, 2004). The difference increases with temperature: at 285 K the difference increases to 14 mM.

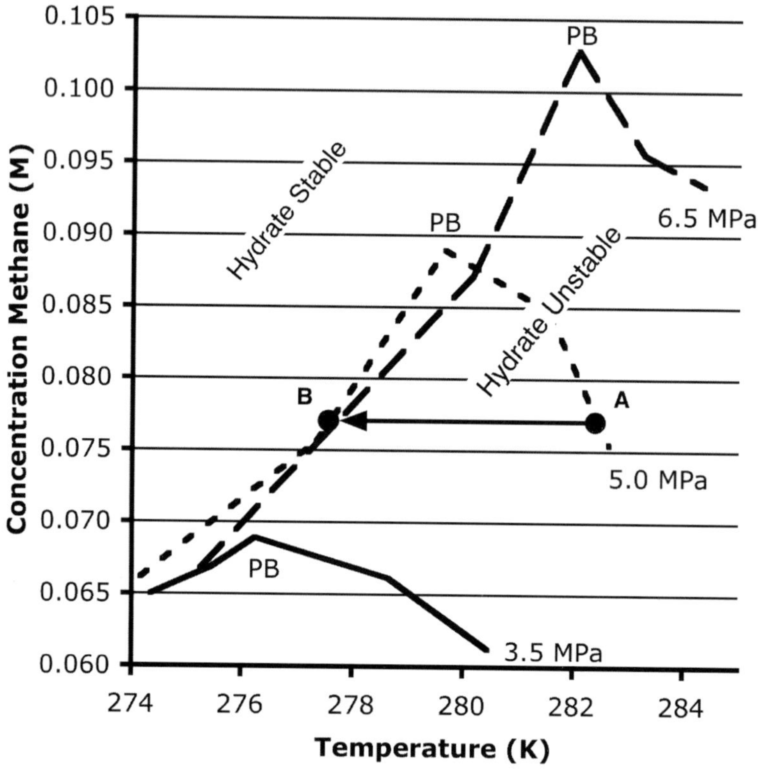

Figure 2.5. The solubility of methane in the presence of methane hydrate rises as temperature increases within the hydrate stability field (region to left of points PB) in fresh water. The solubility of methane decreases with temperature outside of the stability field (region to right of points PB). Point B represents the point where a solution saturated at Point A may begin to form gas hydrate. Data used for calculations (Servio and Englezos, 2002). Density of water at atmospheric pressure is a function of temperature (Lide, 2002). Equation to calculate water density as a function of pressure at a given temperature (Floriano and Nascimento, 2004).

	HFG solubility	Mineral Solubility	Hydrate Solubility
Temperature Decreases	Increases	Decreases[1]	Decreases
Temperature Increases	Decreases	Increases[1]	Increases
Pressure Decreases	Decreases	Decreases Slowly	Decreases Slowly[2]
Pressure Increases	Increases	Increases Slowly	Increases Slowly[2]

Table 2.3. Solubility trends with temperature. [1] General trend. Some solids show reverse behavior but natural gas hydrate is apparently not such a material. [2].There is no clear evidence, but Servio and Englezos (2001, 2002) note that pressure appears to have little effect on hydrate solubility.

The concentration of HFG in equilibrium with hydrate is lower than Henry's Law would predict (Fig. 2.6). Thus, solutions with HFG concentrations in the amount predicted by Henry's Law are supersaturated related to hydrate, and contain adequate driving force for hydrate formation. This driving force increases as the system pressure increases and the concentration of HFG increases.

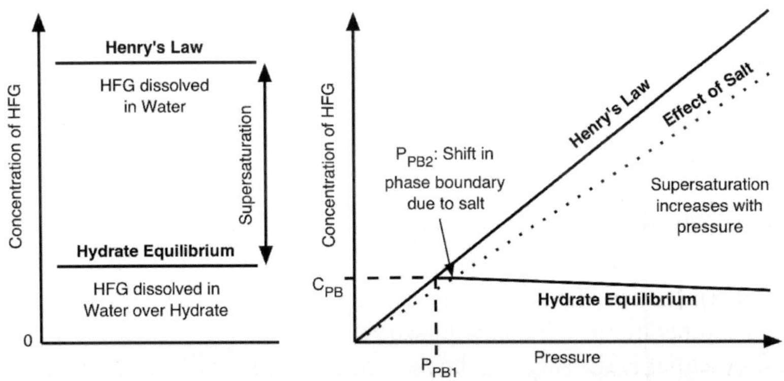

Figure 2.6. Left: the concentration of HFG that can be dissolved in water is much greater than the concentration that is in equilibrium with natural gas hydrate. The difference between these two is approximately the maximum supersaturation that can be easily achieved in liquid solution. Right: the supersaturation that can be generated with this method increases with pressure. The effect of salt on the concentration of HFG is shown. The hydrate phase boundary shifts from P_{PB1} to P_{PB2} in the presence of salts.

The relationship between pressure, temperature, concentration, and the hydrate heat of reaction results in a dynamic balance. If the pressure of a system is dropped slightly below the phase boundary, hydrate begins to dissociate. The dissociation of gas hydrate lowers the system temperature, which pulls the conditions back toward the phase boundary. Likewise, if the concentration of

HFG in equilibrium with hydrate is lowered, the hydrate dissociates until the equilibrium concentration level is reached. These behaviors, where a stress to the system is partially counteracted by chemical processes, are examples of Le'Chatlier's Principle: a system responds in a fashion to minimize a stress placed upon it.

2.3.2. Hydrate Growth Inhibition

In the natural system, only the ionic or other materials in the residual solutions formed by hydrate formation are concentrated. Thus, much of the chemical research on the development of artificial chemical inhibitors produced during flow assurance research is not relevant to this discussion of hydrate growth inhibition.

The presence of dissolved salts in water reduces the number of HFG molecules that can be dissolved in water (Fig, 2.6). This is referred to as the "salting-out" effect. As water molecules surround salt ions, less are available to dissolve HFG molecules, decreasing the amount of HFG available. Additionally, the chemical potential of the liquid also decreases, making it more stable; however, the hydrate chemical potential remains the same. This leads to 'inhibition' of hydrate formation.

Salts or other impurities dissolved in water do not have an effect on the hydrate equilibrium gas concentration level because the chemical potential of hydrate is stable regardless of changes in the solution. However, dissolved salts do cause a shift of the hydrate phase boundary because they affect the chemical potential of an HFG/water solution making it more negative and stable (Appendix A1). Because the edge of the hydrate stability zone occurs at conditions where the chemical potential of hydrate, water, and HFG are equal (the molecules have no preference for a particular phase), lowering the chemical potential of an HFG/water solution shifts the phase boundary to higher pressures and lower temperatures. The new phase boundary will occur where the hydrate chemical potential is the same as the new salt-modified chemical potential of the HFG/water solution. At a fixed temperature, a higher pressure is then required to reach the hydrate equilibrium gas concentration level (Fig 2.6). Dholabhai, et al. (1991), and Tohidi et al. (1994), present theoretical and experimental data on hydrate equilibrium conditions in saltwater solutions.

The phase boundary of gas hydrate can also be shifted by the addition of a component of HFG or a change in the percentage of the components of HFG. For example, the phase boundary of methane hydrate is shifted (Fig 2.7) with the addition of heavier hydrocarbons such as ethane, propane, and butane, which are commonly found in natural gas. For theoretical predictions of phase equilibria, the reader is referred to the program *CSMHYD* that accompanies Sloan (1998).

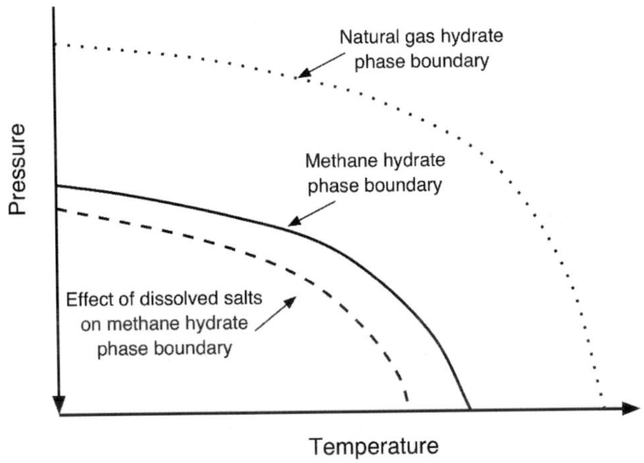

Figure 2.7. The effect of additional hydrocarbon components of HFG and dissolved salts on the phase boundary of methane hydrate. Natural gas hydrate includes methane and other hydrocarbon gases.

Higher pressures or lower temperatures are required for hydrate formation in the presence of dissolved impurities. In this way, saltwater solutions, including natural seawater, inhibit the formation of hydrate. The presence of heavier hydrocarbons in natural gas has the opposite effect and increases the methane hydrate field of stability by forming mixed, or compound, hydrates. In this way, these heavier hydrocarbons are hydrate promoters.

2.4. NUCLEATION

The process of producing a solid material begins with nucleation of small particles. The process of nucleation, creating new particles, is distinct from growth, making existing particles larger. The two are separated by size and energy. Nucleation occurs at small sizes (even single molecules) and requires energy. Growth occurs at larger sizes (clusters of molecules) and yields energy. Growth does not necessarily occur immediately even if the temperature, pressure, and concentration are right for hydrate to form, while nucleation begins as soon as conditions are correct (Bishnoi and Natarajan, 1996). The individual molecules that will make up the solid must first rearrange themselves into small groups that resemble fragments of the eventual crystal. The critical nucleus resembles sII rather than the stable sI material for pure methane hydrate (Moon, 2003); however, the basic hydrate cages are still present.

The surface of a hydrate crystal is unstable. It is composed of incomplete cages that have neither the stability of the bulk hydrate crystal phase nor the stability of the solution. The result is that the surface is constantly being dissolved and reformed. The dissolution of a single cage can result in the breakdown of the entire particle (Fig. 2.8). Until the hydrate particle reaches a sufficient size so that the dissolution of a few molecules does not jeopardize the existence of the particle, the processes of nucleation hold. The increase in size of the particle is a matter of chance and can be described in several ways (Moon, 2003); however, each mechanism results in the formation of a critical nucleus. Once the nuclei reach a large enough size so that the rate of addition equals the rate of dissolution, it is given a new name: critical nucleus. A new process occurs from that point, the process of growth, which is the subject of the next section.

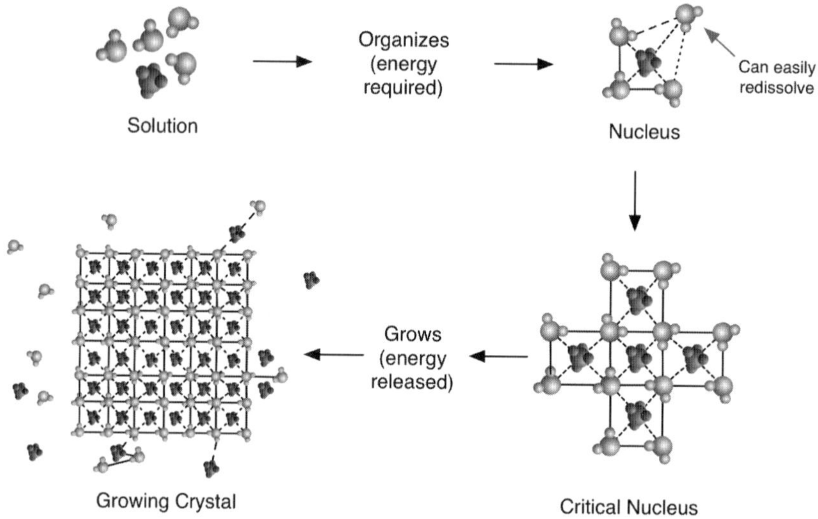

Figure 2.8. The process of making a crystal is broken into two parts. The first is the process of nucleation. The solution organizes into small units of the crystal. These small units are not stable: they can re-dissolve easily. Dissolution is no longer likely once the nucleus reaches a critical size. Growth of the crystal can then occur by addition of more structural units.

The manner in which a particle is formed can be divided into the subcategories of heterogeneous (HEN) or homogeneous (HON) nucleation (Fig. 2.9). HON requires the growth of a very small, isolated hydrate nucleus from the supporting solution and occurs at high supersaturation (Myerson, 2002; Kashchiev and Firoozabadi, 2003; Kashchiev and Rosmalen, 2003). The conditions for HON are very difficult to achieve and probably do not play a major role in natural hydrate formation except at extreme ocean depths. HEN requires the presence of a substrate, such as dirt or the walls of a container, for the nuclei to begin to form. The conditions for HEN are probably the typical conditions found in most hydrate reservoirs.

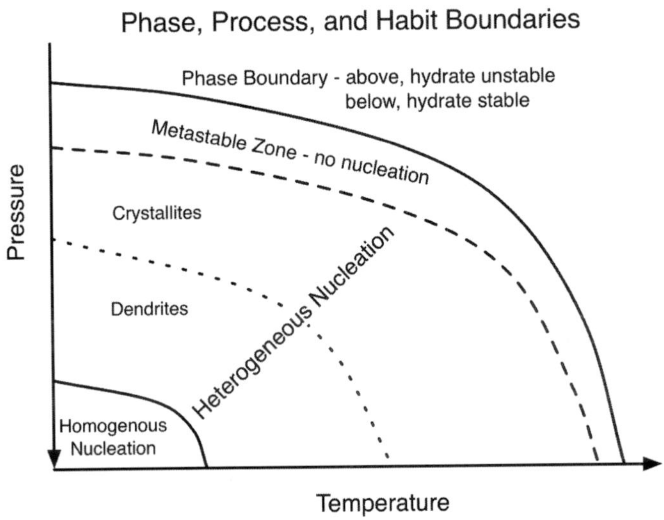

Figure 2.9. Phase diagram for gas hydrates indicating the location relative to the phase boundary of various chemical processes. Homogeneous nucleation occurs only under conditions of very high supersaturation. Heterogeneous nucleation occurs under less energetic conditions, but does not occur in the metastable zone (MSZ). The consequences of the metastable zone will be discussed more fully in the section on growth of gas hydrate as will the distinction between crystallites and dendrites.

Nucleation occurs because the components of the solution, the HFG and water molecules have excess energy. Forming hydrate can lower that energy. The energy is the driving force and is important in nucleation because very small particles of hydrate are unstable toward dissolution. Energy must be applied to form these particles, so that as they grow to larger particles, the overall system can achieve lower energy. Nucleation can be thought of as a barrier to formation of the hydrate phase. A certain amount of energy must initially be consumed to overcome the barrier. This energy is returned later, during the process of growth (next section). The energy source for overcoming the barrier comes from the driving force. The amount of driving force required to form the critical nucleus depends on the size of the nucleus required to start the growth process.

HON typically requires very clean solutions, as any impurity will lead to the initiation of HEN processes, which in turn lead to a lowering of the supersaturation. At the very high supersaturation possible in some natural locations HON is possible. The heating and the rapid consumption of reactants can lead to the predominant production of very small particles. This is related to, but not the same as, an annealing process wherein grain boundary area is minimized as the competing matrix of crystals seek their lowest energy state.

HEN occurs on surfaces or suspended particles. Production of critical nuclei is easier to achieve at lower supersaturation with HEN because the critical nucleus size is much smaller than for HON (Kashchiev and van Rosmalen, 2003). HON, being a complete particle with solution on all sides, requires a larger nucleus than HEN, which is only surrounded by solution in some directions and substrate in the others (Figure 2.10). The critical size is smaller because the interaction of the interface between the nucleus and the substrate is much more stable than the interface between the nucleus and the solution. The substrate acts to stabilize the nucleus by protecting part of it from dissolution.

As the supersaturation increases from zero, the rate of nucleation, generally considered the rate that detectable particles grow in number, increases. Initially, HEN, with little or no HON occurring, dominates nucleation. At relatively high supersaturation HON begins to occur. The rate of HON grows much faster than HEN because HEN is limited to surfaces but HON can occur anywhere in the solution. HON dominates HEN at these high supersaturations. Obtaining high enough supersaturation for HON to occur is very difficult. As the supersaturation increases, the rate of HEN increases, resulting in the consumption of reactant and lowering of the supersaturation. HON is only important where supersaturations can be raised to very high levels rapidly.

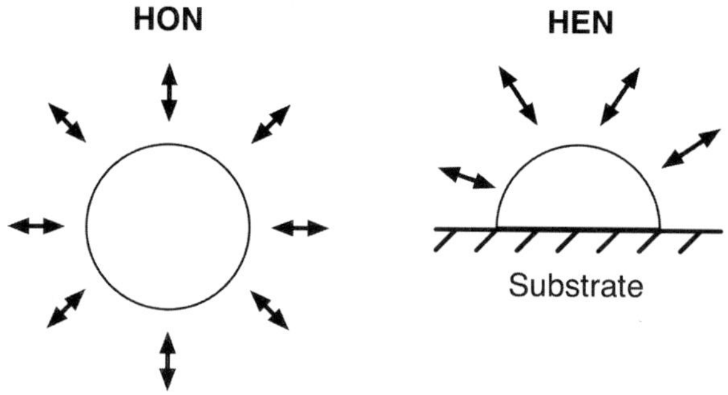

Figure 2.10. Nuclei produced by HON are more susceptible to dissolving than particles produced by HEN. The double-headed arrows indicate interactions with the surrounding solution.

The formation of crystals can be further categorized into primary and secondary nucleation. Primary nucleation occurs in the absence of pre-existing crystals and requires high supersaturation. Secondary nucleation, which arises from external sources in the presence of existing crystal faces, requires lower supersaturation because nucleation is autocatalytic. Secondary nucleation can be thought of as a subset of HEN. Whereas HEN requires a substrate be present for hydrate nucleation, secondary nucleation occurs when that substrate is specifically a hydrate crystal. Secondary nuclei can be formed directly from existing crystals when they are abraded macroscopically, commonly referred to as attrition, or when the crystal structure is broken (Myerson 2002).

Stimuli for secondary nucleation include turbulence and surfaces. Turbulence can cause the outer layer of crystals to be sheared off and it can also cause contact-type secondary nucleation, where two surfaces are involved. These surface interactions include collisions of crystals with other crystals, with impurities, with rotating mechanisms such as impellers, and with vessel or sedimentary walls. In a natural setting, the turbulence and surface interaction resulting from pressure waves, sediment mass flows, earthquakes, and the sudden resettling of sediment could cause secondary nucleation of gas hydrate crystals.

2.5. GROWTH

The process of growth can occur following the formation of critical nuclei. As with nucleation, growth rates depend upon the driving force, which is related to the availability of reactants and the ease with which they can be accommodated into the crystal lattice. Like nucleation rates, growth rates increase with driving force. Unlike nucleation, the end result of a growth process may involve different types, shapes, and sizes of crystals depending on the driving force (Myerson, 2002, Chapter 6). Where there is the greatest potential for growth outside of the region of homogeneous nucleation, within the dendritic and crystallite product zones (Fig. 2.9), growth may be very rapid. Imperfect crystal structures often result, in addition to overgrowth of other material, including growth media and its dissolved constituents.

Growth that is most likely to take place under natural conditions will lie in the pressure-temperature region of the metastable zone (MSZ) (Fig 2.9). This is a region where primary nucleation does not normally occur but growth (addition of mass to an already present particle) takes place. Unlike the normal phase boundary, which describes the location of a thermodynamic phase transition, the MSZ is a kinetic value. As time progresses the MSZ boundary approaches the thermodynamic phase boundary (discussed in section 2.6.5). At some point in time the MSZ and thermodynamic phase boundary may be coincident.

At the very low supersaturations found in the MSZ there is not sufficient energy to create critical nuclei. Again, this is a kinetic, not thermodynamic, restraint. The most stable phase, and the one that will ultimately develop, is the hydrate phase; however, it takes time for the molecules to rearrange to form the hydrate phase and to overcome any reaction barriers (such as nucleation). Thus, in the MSZ the extra dissolved HFG cannot form new crystals, but may join any hydrate crystal that is placed into the solution. In other words, a metastable solution will not produce solids because nucleation does not produce critical nuclei, but a crystal that is introduced to the solution, by mass transport of the crystal or solution, will grow. This phenomenon, known as seeding, is well known and is used to grow large crystals with high selectivity in industrial processes. Growing crystals at the low supersaturation typically found within the MSZ can have a second consequence related to the type of solid that is produced.

The heat released as the crystal grows controls the characteristics of the solid produced. At the phase boundary the amount of heat released is called the heat of formation. Within the stability field the heat produced is equal to the heat of formation at the phase boundary plus the driving force. As the supersaturation increases the driving force increases and more heat is released. The amount of heat and the rate at which it is produced and dissipated can affect what shapes a crystal can form. The growth of crystals is slow at the low driving forces found in the MSZ. The heat produced can be easily diffused into the surroundings, and the crystal can take an ideal shape. However, as the driving force increases the amount of heat generated increases. Because heat is transferred to the surroundings only through surface contact, the surface area relative to the volume of the crystal must be increased. This form of growth is called dendritic growth and may produce crystals with non-ideal shapes, such as needles (Subramanian and Sloan, 2002). At high enough supersaturations dendritic growth no longer produces crystals but results in dendrites. Dendrites are solids that grow in such a way that any given point of the growing material may serve as a nucleation site for new crystal growth. The pattern is essentially that of a river, which gives way upstream to smaller and smaller tributaries, each of which have their own sets of tributaries, and so on. Each of these dendritic crystals provides the roots for other crystals until the crystals may fill the entire space and overgrows and traps non-crystalline material. This growth mode will result in structures that have a very high surface area with respect to both the mass of hydrate and the volume in which the growth has taken place (Fig. 2.11).

Once growth processes have stopped and the system is once again at equilibrium, the hydrate mass may continue to evolve. An ensemble of hydrate crystals can participate in several processes. One of these processes, Ostwald ripening, will be discussed in the next section. The other process is grain boundary healing and occurs when the boundaries between grains fuse together. This process occurs rather quickly [observed in the Marine Desalination Systems (MDS) Laboratory to occur on a time scale of minutes with ethane hydrate], gaining mechanical strength and further rejecting non-participatory material. This process can lead, over time, to the growth of solid, polycrystalline masses of hydrate that internally will display a history of many smaller crystalline particles (also see 2.6.1).

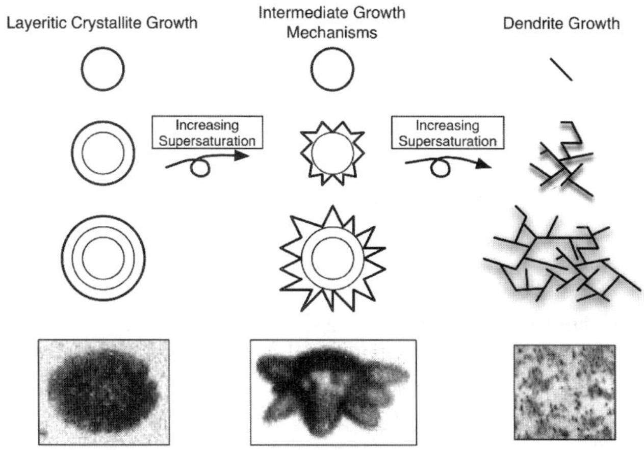

Figure 2.11. Examples of different growth patterns. Left: formation of a crystal at low supersaturations, perhaps within the metastable zone. The photograph bottom left is an example of layeritic growth on a bubble. Center: the growth is no longer regular and the surface area to volume ratio has increased. Crystals are formed at intermediate supersaturations. The photograph bottom center is an example of dendritic growth without dendrite production on the surface of a bubble. Right: at high supersaturations, the surface area is maximized to radiate heat by the growth of thin, long, interconnected dendrites. The photograph at the bottom right shows a dendrite mass. The dark spots are interspersed bubbles. Photographs of example morphologies with the permission of MDS.

2.5.1. Effects of Diffusion

The growth of a solid pulls reactant out of the surrounding solution. This will lead to a deficit of HFG near the crystal/nucleus, while the bulk solution will have virtually the same concentration. An HFG gradient is developed from the solid surface to the bulk solution. The gradient is affected by the viscosity, size of particle, proximity to other particles, driving force, and, in sediment, the tortuosity. These factors have a large effect on the size distribution of crystals produced.

If the diffusion gradient is very large compared to the size of the crystal, the growth rate of the crystals will decrease with increasing crystal size (Myerson, 2002, p. 146). The result will be a narrow distribution of relatively small particles. If the diffusion gradient is small compared to the size of the particles then the growth rate (change in radius) is independent of crystal size as a function of radius or mass per unit area. The product under these conditions will contain a wide distribution of particles sizes. However, if the overall supersaturation is low, as opposed to supersaturation lowered by formation of a gradient, then the growth rate increases with size. This is due to the Gibbs-Thompson effect (GTE) that states that the equilibrium concentration (above which nucleation and growth occur) decreases with crystal size. The GTE will cause large crystals to grow in solutions with low driving force or many small to intermediate sized crystals to grow in solutions with high driving force. GTE is also the theoretical reason behind Ostwald ripening. Smaller crystals effectively supersaturate larger crystals. Reactant is transferred from the smaller crystal to the large crystal. Overtime the small crystal may be completely consumed.

2.5.2. Growth from Mixtures of HFG

Natural gas is not composed of pure methane. Natural gas from the Gulf of Mexico, where hydrate is found naturally, can contain 12% ethane (Maekawa, 2001) and some regions produce gas that is less than 60% methane, although where the contaminants are CO_2 and SO_x, these contaminants are usually removed before the natural gas can be used. As described previously, mixtures of HFG gases generally cause the phase boundary to move to higher temperatures and lower pressures, forming compound hydrates. However, such mixtures result in a more complex hydrate growth process. The formation of hydrate from mixed natural gas is a two-stage process where heavier HFGs (ethane, propane, etc.) precipitate along with methane in one type of hydrate structure first (sII or sH). Once all of the heavier HFGs are consumed, the methane continues to precipitate but with a different host structure (sI) (Uchida, 2004). This process is called staged nucleation.

In staged nucleation, the driving force to form the compound hydrate is much higher than the driving force to form pure hydrate. The MSZs for the pure hydrate and compound hydrate are thus of different sizes, with the compound hydrate having a short induction time (time before critical nuclei are formed) compared to the pure hydrate. The compound hydrate begins precipitating, lowering the overall concentration of HFG available. In typical natural gas this mixture is predominantly methane (or carbon dioxide) with smaller amounts of other, heavier hydrate formers. The methane and heavier hydrate formers precipitate together to form the compound hydrate. Once all of the heavier hydrate former is consumed pure methane hydrate may form if there is still enough dissolved in solution. This formation will take place at a later time (Figure 2.12).

Stage nucleation can control the hydrate structure detected in a natural hydrate. At the base of hydrate found in a vertically rising HFG solution, it is likely that compound hydrates will form near the bottom, consuming nearly all of the heavier hydrocarbons and some of the methane, while higher up the hydrate region sI pure methane hydrate may be present. The initial hydrate production purifies the HFG as it moves toward the seabed, possibly giving rise to ethane and propane rich sII hydrate at the base of the hydrate and methane rich sI hydrate toward the top of the hydrate layer.

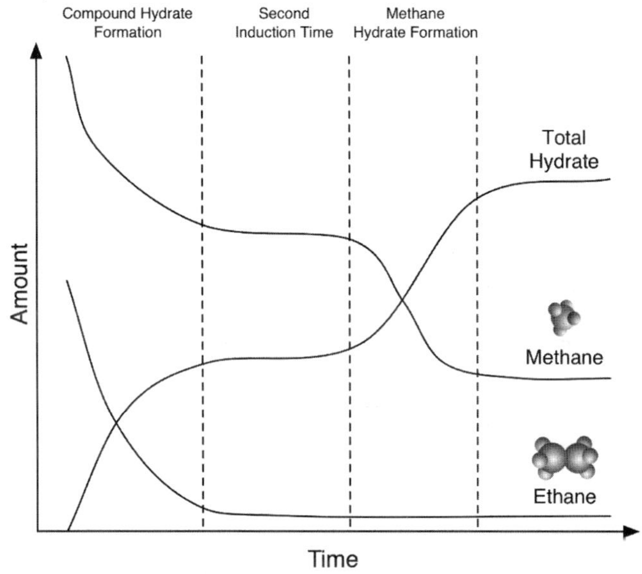

Figure 2.12. Two staged growth using a mixture of methane and ethane as an example. When multiple hydrate forming gases are present the first hydrate formed may be a compound hydrate, a hydrate that contains both gases. If one of the gases is completely consumed (in this case ethane), a second round of hydrate growth may occur to produce pure methane hydrate. The second induction time can be of variable length. Pure methane hydrate formation may possibly begin while compound hydrate formation is ongoing.

The ratio of methane to heavier hydrocarbon consumption during compound hydrate formation depends on the initial composition of the HFG, but at least one example (Uchida, 2004, Fig. 1) presents the decrease in partial pressure of methane and propane during the initial precipitation of compound hydrate with a ratio of 11 mol methane: 1 mol propane.

2.5.3. Hydrate Growth from Different Types of Solution

The discussion to this point has dealt with growth of gas hydrate from two different types of homogeneous solution. The first type occurs where a saturated liquid of HFG dissolved in water enters in the hydrate stability zone (HSZ). Such a system may begin super concentrated if it is coming from higher pressure and not too much higher temperature or from lower temperature and not too much lower pressure. Migration of this solution into the HSZ creates a metastable solution. The solution will migrate some distance into the HSZ before nucleation begins. Since nucleation is autocatalytic, nuclei and crystals generate more nuclei; hydrate will form above and below this initial hydrate formation site following the supersaturated solution until all of the excess HFG is consumed.

Because the amount of hydrate forming gas that can be dissolved in water is very small, a large amount of solution must be transported to the growth domain. Table 2.4 presents calculations that demonstrate this phenomenon. The data of Servio and Englezos (2001; 2002) deal with distilled water and has a reported accuracy of 5-10%. Each entry in the table is accurate to +/- 15% because of inaccuracies in Henry's Law and the possible non-linearity of solubility at the pressures measured. For comparison, data for carbon dioxide, a much more soluble constituent of natural gas, are presented in the last row of Table 2.4. Gases are generally increasingly less soluble in water as the salinity increases. The consequence of this is that the values reported in Table 2.4 will be shifted to lower temperatures and higher pressures with increasing salinity (the shift for normal seawater is about 2 K).

Conditions (MPa/K)[1]	Supersaturation (mM)[2]	Deposition Efficiency[3]	Cycles to Dryness[4]
3.5/274.35	14	18 %	675
5.0/277.35	35	32 %	274
6.5/280.15	48	36 %	201
CO_2: 2.0/275.95 [5]	*330* [5]	*25 %* [5]	*29* [5]

Table 2.4. Results of precipitation calculations for methane hydrate in pure water. [1]Represents approximately 2 K supercooling. [2]Calculated using Henry's Law corrected for temperature, the density of fresh water at (*P, T*), and the data from Servio and Englezos, 2002. [3]Represents the amount of HFG that will be deposited from a liquid solution of HFG in water that is not being recharged. [4]Calculation of how many times the solution must be recharged with gas to produce a solid mass of pure hydrate. [5]Data for carbon dioxide taken from Servio and Englezos (2001). Carbon dioxide is about 11 times more soluble in water than methane.

The second type of homogeneous solution is a saturated vapor. This phase will be largely HFG with water vapor 'dissolved' within it. Like the liquid solution, this solution can be super concentrated, but this time with water. If the gas is cooled at high-pressure (super cooled or super baric) hydrate may form, with the saturation calculated using Raoult's law. If the gas is in contact with a region that is super cooled at the right pressure then gas hydrate will form. Water vapor will be transported through the gas space up to the growing hydrate mass by diffusion. More on this concept will be discussed in the section on growth models (2.7).

The two systems described above are easy to understand and describe. They were chosen to introduce the concepts presented in this chapter to this point and represent homogeneous solutions. Homogeneous solutions are the same throughout. This means that every point in the bulk of the solution can be described in exactly the same way. Homogeneous solutions include liquid solutions with small, evenly distributed suspended particles. However, hydrate can form from heterogeneous solutions as well. Heterogeneous solutions have different macroscopic characteristics in different locations. Such systems include water droplets sprayed into a HFG atmosphere or gas bubbles in a water matrix. The interface between the gas and water can serve as a template for hydrate growth, and, as will be described, at the highest supersaturations normally possible.

The formation of hydrate on gas surrounded by water is perhaps the most common manner in which hydrate is produced (Ohmura et al., 2004). The interface between the gas and liquid is a skin very much like the top layer of water in a glass of water. The interface is not stable for many of the same reasons the interface between hydrate and the surrounding solution is not stable. The instability leads to surface tension. Surface tension is the interface energy. The system attempts to minimize surface tension. The only way to do this is to squeeze the bubble or droplet, which reduces the surface area but also increases the internal pressure. The smaller the bubble/droplet is, the larger this effect becomes. Because the pressure of the bubble/droplet determines the amount that will dissolve (Henry's or Raoult's Law) the squeezing raises the saturation limit and more gas dissolves, making the bubble/droplet smaller, which raises the pressure, causing more gas to dissolve. Thus, supersaturation outside a bubble or droplet is always higher than can normally be achieved in the solution; however, it is only with very small bubbles/droplets that this becomes truly important. The interface also serves as a potential location for nucleation.

Bubbles are segregated into two different classes, macro and micro, based on the magnitude of squeezing. Macro bubbles are large and experience little squeezing. The concentration around them is not enhanced very much, and they are considered stable. Microbubbles are so small that the concentration around them is very high. They are not stable, have high internal pressure, are surrounded by highly supersaturated solution and may collapse (Takahasi, 2003). The size that divides micro and macro bubbles is in the range of 1 μm.

The production of microbubbles in a laboratory situation in a short time can be difficult, but small gas bubbles may be more likely to develop naturally in a geological environment during rapid thermal changes. The production of mixed tetrahydrofuran/xenon hydrate (a sII hydrate not likely to be found in natural gas hydrate) using microbubbles occurred by raising the concentration around the microbubbles to such an extent that the metastable zone was avoided (Takahasi, 2003). The extent to which these results relate to natural hydrate formation has yet to be investigated.

In certain circumstances, mostly associated with shock waves, cavitation bubbles may form (Lohse, 2003). These bubbles are similar to microbubbles in that they are very small and unstable. Unlike micro bubbles, cavitation bubbles contain little material. They are essentially void spaces. Cavitation bubbles have very high internal energies and may interact with hydrate in two opposite manners. The first mechanism of interaction occurs when cavitation bubbles collapse. Very high pressures and temperatures are developed. The combination of high pressure and temperature is not conducive to hydrate formation. However, the pressure wave produced by the collapse can cause nucleation nearby the site of collapse (Devarakonda, 2003). Secondly, cavitation bubbles have been observed cutting through ice crystals (Chow, 2004) and can result in the breakup of large particles. The high energy released by cavitation bubble collapse can also melt small crystals and nuclei.

Hydrate formed on the surfaces of bubbles or droplets is subject to all the other rules for growth. The very high supersaturations possible near the surface of a bubble/droplet enhance the likelihood of dendritic growth mechanisms with the final material containing many overgrown inclusions. Homogeneous solutions can also produce dendritic growth, but the possibility of large crystal formation is more likely. Both systems will require recharging of reactants, to a lesser extent if more soluble HFG gases, such as carbon dioxide, hydrogen sulfide, or sulfur dioxide, are present in significant amounts.

2.5.4. Example of Hydrate Growth

Hydrate growth at the interface between an HFG atmosphere (in this case ethane) and an HFG/seawater solution is demonstrated in Figure 2.13. The morphology, or type of crystal produced, is a disk composed of concentric rings. The photographs in the figure follow the growth of a single nucleus. As time progresses the nucleus increases with size. The conditions are low driving force, but the growth takes place outside of the MSZ. The final pictures (Fig. 2.13) show an increase in additional nuclei near the original nucleation site as the nucleation rate increases owing to autocatalysis.

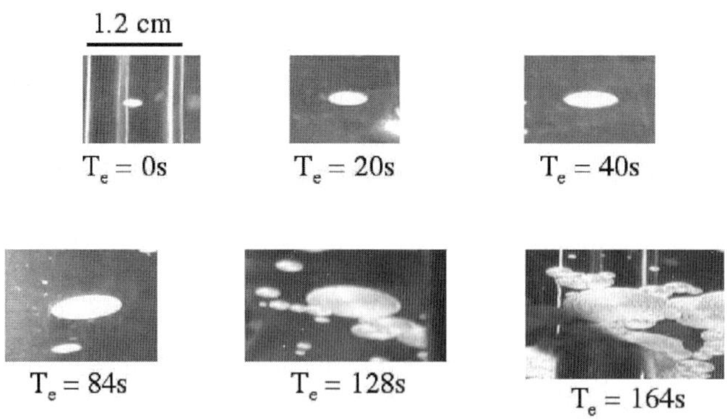

Figure 2.13. The growth of a hydrate disk-shaped crystal with time outside of the metastable zone. The number of nuclei increases with time due to primary nucleation. The time noted underneath each photograph is the elapsed time from the first picture.

This crystal morphology is representative of sI ethane hydrate formation at a large, flat ethane gas/seawater interface over a wide range of pressures and driving forces and shows concentric layeritic growth. However, nucleation rates increase rapidly with driving force causing a reduction of overall crystal size.

Nucleation is autocatalytic: the number of new nuclei produced increases with time (first order in nuclei). Eventually, the new nuclei dramatically outnumber the old, larger crystals, and the interface is sealed with solid hydrate. This is a typical result for precipitation outside of the MSZ. At high driving forces, the rate of nucleation can be so high that larger crystals do not have time to form before being completely surrounded by other crystals. The result is a seemingly homogenous crystal size distribution of small crystals.

2.6. HYDRATE DISSOCIATION AND DISSOLUTION

Dissolution and dissociation of gas hydrate both involve the breakdown of the water lattice and release of the gas. The breakdown of hydrate typically forms a liquid phase saturated with HFG. For sparingly soluble HFGs, such as the hydrocarbons in natural gas, a free gas phase saturated with water vapor also appears (Lekvam and Ruoff, 1997). However, other modes are available in which a liquid phase or a gas phase is not produced, and kinetics, as always, plays a role. (It is common to use the term decomposition to describe the breakdown of the hydrate lattice, either through dissolution or dissociation. With respect to terminology used in chemistry, this is an incorrect description. In chemical terms, when a substance decomposes, the process is irreversible and the reactants are converted to different chemical compounds. When the hydrate lattice breaks down, the components cease to be organized in the crystal structure. Both the water molecules, and the HFG remain intact, and the process is reversible through the reorganization of the water and HFG molecules.)

In the field of gas hydrate research, hydrate breakdown takes place in two principal manners. If the breakdown of the hydrate structure occurs outside of the hydrate pressure-temperature stability field, the general term dissociation is used. If the breakdown occurs under temperature-pressure conditions at which the hydrate is expected to be stable but where undersaturation of the surrounding solution is related to the diffusion of methane or other HFG from the voids, the term dissolution is used. Diffusion of the HFG from the hydrate cell structure appears to be related to, or causal, to the breakdown of the water molecule latticework structure (Appendix 6). There is limited information on the mechanism by which both of this processes occur, and the possibility exists that the two mechanisms may be similar.

$$\text{Dissociation: Methane Hydrate} \rightarrow \text{Water}_{\text{Liquid}} + \text{Methane}_{\text{Gas}}$$

$$\text{Dissolution: Methane Hydrate} \rightarrow \text{Water}_{\text{Liquid}} + \text{Methane}_{\text{Aqueous}}$$

Until recently, hydrate related research focused on exploring hydrate nucleation, growth, and dissociation (Sloan, 1998, and Rehder et al., 2004). Recent work has expanded the focus to include dissolution (Zhang, 2003;

Tsouris, 2004; Rehder et al., 2004), which, previously, has not been explored extensively.

2.6.1. Hydrate Dissociation

Hydrate dissociation has been proposed to be a two-step process in which the hydrate host lattice is destroyed followed by physical movement of gas molecules into the surrounding water (Kim et al., 1987). It occurs when an increase in temperature and/or decrease in pressure places hydrate into conditions at which it is no longer stable. As stated previously, hydrate formation produces heat energy. In contrast, dissociation requires an input of energy in order to break down the crystal structure of the hydrate and release the gas (Max and Holman, 2003). As hydrate dissociates, heat energy from the surroundings is consumed in order to complete the reaction causing the temperature of the surroundings to decrease. The energy required for dissociation is approximately equal but opposite to the heat released during formation, for example 54 kJ/mole for methane hydrate (Handa, 1986b).

A representative example of the temperature-pressure relationship for hydrate (Fig. 2.14) can be used as a generalization for all hydrocarbon hydrate. The shaded area to the left of the line represents conditions at which hydrate formation occurs. The area to the right of the line represents conditions where the hydrate is not stable, and will tend to dissociate. Along the line, or phase boundary, lie the temperature-pressure conditions at which the hydrate system is at equilibrium. If there is no change in pressure or temperature and there is no excess reactant, hydrate will form as fast as it dissociates. This results in no net change of the total amount of hydrate mass, although redistribution of hydrate can still occur. Additionally, this line represents the phase change conditions and if a stress is placed on a hydrate mass to either grow or dissociate, the system parameters will change in the direction that will bring the system to the phase boundary condition. If the hydrate exists at conditions where hydrate is not stable (Fig. 2.14A) and dissociation occurs, the surroundings will become cooler until the temperature of the phase boundary, which corresponds to the systems pressure, is attained (Fig. 14, Point B), or all of the hydrate is consumed. If the system conditions reach the phase boundary (B), the net dissociation will slow or possibly cease. The farther away from the phase boundary the decomposition conditions lie, the higher the driving force, the greater the degree of cooling, and the higher the rate of dissociation.

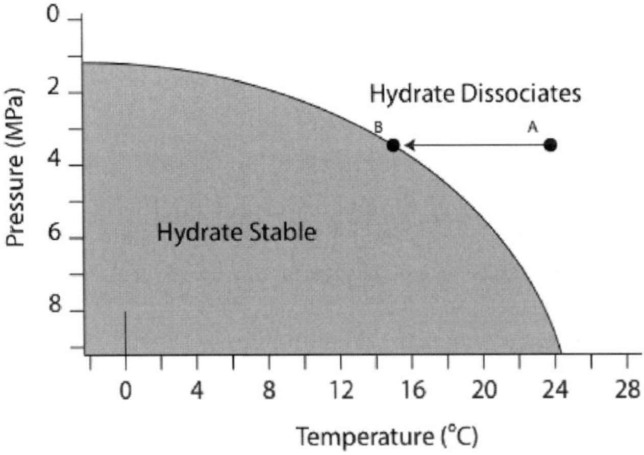

Figure 2.14. Representative diagram of the pressure-temperature stability relationship for hydrate. Point A: Temperature-pressure conditions at which hydrate is not stable. Point B: Equilibrium conditions at the same pressure found at point A.

A solid mass of sI ethane hydrate can be formed within a pressurized vessel filled with seawater. A resistance temperature detector is encased in the hydrate mass. Chemical stress is placed on the system by slowly lowering the pressure, and the temperature response of the hydrate mass is monitored. The hydrate temperature drops as the pressure is lowered. This process defined the phase boundary of ethane hydrate in seawater. There is an excellent correspondence to the theoretical phase boundary predicted by the program *CSMHYD* that accompanies Sloan, 1998 (Fig. 2.15). Phase boundaries can be mapped in this way because pressure and temperature are not independent along the boundary.

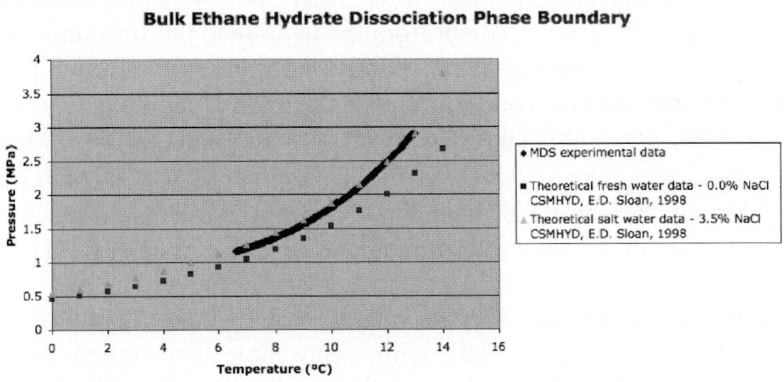

Figure 2.15. Experimental sI ethane hydrate dissociation experiment produces the ethane seawater phase boundary. Figure courtesy of MDS.

When hydrate is dissociated under the experimental conditions described above, two products are produced: hydrate-forming gas-saturated water and water vapor-saturated hydrate-forming gas. Both of these fluids have a high potential for reforming hydrate as all of the necessary components are contained within the mixture. Cooling or re-pressurizing may easily reform the hydrate. Reformation rates are also increased due to what some refer to as the "memory effect." For a reason that is not completely clear at this point, the induction time of a mixture that has previously contained hydrate or ice appears to be shorter than for a mixture that has not (Ohmura et al., 2003; Takeya et al., 2000; and Uchida, Ebinuma, and Narita, 2000). It has been suggested that the memory effect is due to hydrate precursors that remain in the fluid after hydrate ceases to be visually detected. If the reformation of hydrate must be avoided, heating the mixture may be beneficial to make the conditions less favorable for hydrate nucleation and may eliminate any hydrate precursors that exists within the fluid (Max and Holman, 2003).

Another phenomenon observed within hydrate masses is the fusion of grain boundaries within a hydrate mass. During this process, referred to as annealing, a very thin layer on the surface of the hydrate begins to dissociate due to a brief excursion from the stability field. The water and hydrate-forming gas released are already super-saturated when conditions return to the HSZ. This small amount of solution reforms hydrate and unites the grains. Through annealing, multiple grains of hydrate can be united to create a single solid mass.

2.6.2. Hydrate Dissolution

Temperature and pressure are not the only factors that control hydrate stability, as can be observed when hydrate that is exposed to free water or gas vanishes even though it is within the pressure-temperature stability field (Max and Holman, 2003). If the hydrate exists within an aqueous environment, the water must be saturated with hydrate-forming gas. If the hydrate exists within a gaseous environment, the gas must be saturated with water vapor and HFG. If there is a positive diffusion gradient from the hydrate to the surrounding water or gas, the hydrate may dissolve. The solubility of the hydrate-forming gas in the ambient water has a significant effect on the rate of dissolution: The higher the solubility of the guest molecule, the faster the hydrate dissolves (Rehder et al., 2004).

Most of the natural hydrate found at the seafloor is primarily methane hydrate. As a result the solubility of methane becomes of interest. The methane solubility in water increases with decreasing temperature and increasing pressure when hydrate is not present. In the presence of hydrate, hydrate-forming gas solubility decreases (Figure 2.5) with decreasing temperature (Servio and Englezos, 2001 and Subramanian and Sloan, 2002). The effect on methane solubility of increasing pressure at constant temperature when hydrate is present is unclear, but the effect is likely to be small (Egorove et al., 1999; Handa, 1990; and Servio and Englezos, 2001). If the temperature of a system containing hydrate, hydrate-forming gas, and water is increased, even within the hydrate stability zone, the hydrate will begin to dissolve due to the increased gas solubility (Subramanian and Sloan, 2002). If the equilibrium conditions of composition are not satisfied, hydrate can dissolve directly into the surrounding water. As hydrate is dissolved, the gas diffuses away from the hydrate surface and into the ambient water (Holman et al., 2004). The dissolution will occur even under pressure-temperature conditions at which hydrate is stable (Brewer et al., 2000). Another way to consider this process is by describing the surface as a dynamic equilibrium process. At equilibrium the rate of dissolution is approximately equal to the rate of formation. By diluting the HFG in the associated pore water, the rate of formation decreases, but not the rate of dissolution, and the crystal begins to have a net mass loss.

It is a possibility that dissolution of the hydrate may require less energy than dissociation. When hydrate dissolves, the gas molecules move from within the crystal lattice to the surrounding water resulting in the instability of the hydrate crystal structure (Max and Holman, 2003). The result is that the environment will not be cooled to as great an extent. This effect is related to the heat of dissolution of the HFG and can be as much as 25% of the total heat flow.

Recently, Rehder et al. (2004) conducted studies exploring dissolution rates of lab-synthesized methane and CO_2 hydrates in undersaturated, flowing seawater. The CO_2 hydrate dissolved significantly faster than the methane hydrate and, for the last 30 minutes, was positively buoyant, possibly due to a lattice structure containing only vacant cavities, which remain once the surrounding seawater has dissolved the gas. Periods of faster dissolution rates coincided with higher current velocities. The work states that the dissolution rates were limited by the diffusion rates into the bulk ocean, and the ratio of the dissolution rates should be equal to the ratio of the solubilities of the hydrate-forming gases. The ratio of the solubilities is 10.4 and the ratio of the dissolution rates was found to be 11. Egorove et al. (1999) believe that the natural hydrate deposits that exist on the seafloor require a saturated boundary layer in order to avoid dissolving.

Methane hydrate can dissolve into gas as well as water. Hydrate sediment samples that have been transported from the field to the laboratory or between laboratories following manufacture and pressurized by non-hydrate-forming gas are not stable even where they are kept under hydrate stability pressure-temperature conditions. The methane diffuses out of the hydrate over time into the surrounding gas, and the water molecules form liquid water where they do not partly reorder to ice structure if temperatures are below freezing. Although there has been no experimental confirmation, we suggest that methane will diffuse into the surrounding gas until it is in equilibrium with the hydrate. When similar natural gas hydrate samples have been similarly pressurized with a headspace of methane, the hydrate has been observed to remain intact (Winters 2005).

2.6.3. Dissociation and Dissolution: A Surface Phenomenon

Empirical experience over the years has been that hydrate that has been recovered from any pressure depth from seafloor or sub-seafloor sediments (as great as 3,300 m) and brought to the surface where it is substantially outside of its stability field, exhibits considerable physical integrity as it dissociates (Spence, et al., 2001). That is, hydrate appears to dissociate only at its margin where there is a strong solubility contrast and a diffusion gradient between the hydrate and the surrounding bulk water or gas environment. On the microscopic level, however, not much is known about hydrate under these circumstances.

Uchida et al. (2000) claim to have observed bulk breakdown of hydrate crystals through bulk dissociation that produced CO_2 bubbles throughout a crystal. In the final stages of dissociation, the crystals divided into small pieces, which is normal in the last stage of dissociation of hydrate; some of the pieces may be water ice. If their observation is correct, dissociation was not limited to an exposed hydrate surface. However, they do not observe early fracturing of the hydrate from internal gas pressure that could be expected to produce relatively larger fragments. It is also possible that the gas bubbles figured by Uchida et al. (2000) are essentially crystal surface phenomena, into which molecules of the HFG, which has diffused out of the dissociating hydrate, are coalescing. Also, internal striations almost certainly mark individual crystal domains within a polycrystalline assemblage. This effect would provide grain boundaries, or at least defect and dislocation trains along which HFG molecules could diffuse rapidly toward the margin of the hydrate crystals. Gas bubbles enlarging and causing fractures of the hydrate are not apparent. However, because hydrate is at least 20 times as strong as water ice (Durham et al., 2003), it may have the strength to contain considerable internally dispersed, pressurized gas.

There is much that is not well understood about the structure of hydrate, however, and there may be microscopic structures within what appear to be solid hydrate in which gas could coalesce from HFG molecules exiting their cages. Kuhs et al. (2000) have observed that hydrate they have fabricated from ice shows meso- and macroporous structures. Pore sizes were found to be between 100 to 400 nm with porosity of 25-40% for methane; other gases such as CO_2 had different pore sizes and porosities. The three-dimensional sponge-like microstructure was observed in single grains that were typically a few μm across as viewed with a field-emission scanning electron microscope. These observations may have relevance to hydrate that is naturally formed from ice, for instance in ice caps in polar regions, but there are unanswered questions regarding the similarities and differences between laboratory hydrate synthesized from ice and natural or laboratory hydrate grown from addition of reactants to a hydrate seed in either aqueous or gaseous media. If naturally occurring hydrate is found to have similar microporous structure, then the basis for a mechanism for the internal condensing of gas in pore spaces would exist, apart from intergrain accumulation.

Naturally occurring, apparently solid hydrate has been observed to be persistent well outside of its pressure-temperature stability field. Hydrate that is suddenly exposed to pressure and temperature conditions in which it is highly unstable, does not rapidly dissociate, expelling gas. Nor does the gas suddenly evolve, leaving behind an incompletely transformed ice structure that could be mistaken for hydrate. Although quantitative data on this point is sparse, gas exsolves from the hydrate over a period of time and apparently at a rate that is about proportional to the surface area of a mass of hydrate (Kim et al., 1987), which reduces as dissociation proceeds. Where a crystal mush is formed that has a very high surface area to mass relationship, however, dissociation is very rapid. Where solid hydrate has been formed that has a low surface area to mass relationship, the hydrate is much more persistent, with dissociation rates that are somewhat slower than growth rates (where the solid hydrate has been formed from water that is continually replenished with HFG).

One way to picture the surface dissociation phenomenon is to consider that the outer surface of the crystal holding the inner molecules in the appropriate position and orientation causes the hydrate crystal to maintain its internal structure. As each layer of molecules exits the crystal structure, the molecules lying underneath become able to diffuse away from the remaining crystal structure (Fig. 2.16). The left diagram depicts the dissociation of a crystal lattice. The lattice (left) begins to dissociate when the outermost water and gas molecules break away from the crystal structure and diffuse into the surroundings. The right most panel of Figure 2.16 represents the underlying water and gas molecules leaving the structure after they have been exposed to their surroundings.

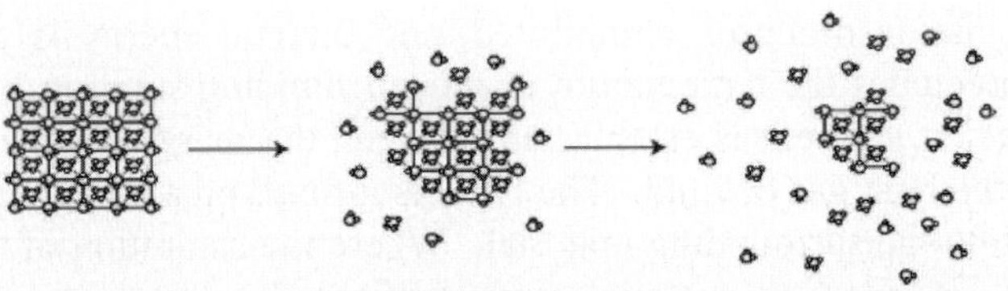

Figure 2.16. Left: An intact hydrate lattice. Middle: The hydrate lattice beginning to dissociate on the surface. Right: Continued dissociation occurring from the outer surface of the crystal lattice, inward.

Increasing the surface area of the hydrate will increase the rate of dissociation (Holman et al., 2004, Kim et al., 1987). Dissolution also appears to follow the same pattern and only occurs at the surface of the hydrate. The guest molecules on the outer surface of the hydrate crystal are able to diffuse into the surroundings. The empty cavities then become unstable, and the water molecule cage structure becomes unstable and breaks down. This peeling away of successive skins of surface hydrate clears the way for the guest molecules in the underlying layers to then repeat the dissociation process and diffuse out of what had been a 'deeper' cage structure in the hydrate crystal. Breakdown of hydrate either by dissociation or dissolution gives rise to concentration gradients where by the constituents of the hydrate are moving from an unfavorable, high-energy location to favorable, low-energy surroundings. Both of these processes occur along crystal surfaces. . Therefore, increasing the surface area of the hydrate will also increase the rate of dissolution.

Circone et al. (2004), have shown that when either the porosity is increased or the hydrate-hydrate grain contacts are disrupted by addition of a new phase, such as quartz, dissociation rates of methane hydrate increase as a function of surface area of the hydrate. This result conforms with empirical observation of recovered dredged and core samples, and suggests that highly porous hydrate deposits having low hydrate mass to hydrate surface area in communicating pore space will dissociate or dissolve faster than more dense or solid hydrate deposits. Exposed surface area of the hydrate is thus a key factor in controlling dissociation rate and gas production for any naturally existing hydrate deposit.

2.6.3.1. Hydrate Dissolution in a Nearly Saturated Environment

Hydrate that is exposed to conditions of slight undersaturation dissolves much slower than hydrate exposed to highly undersaturated solutions. In environments that are nearly saturated with HFG, the diffusion gradient that drives hydrate dissolution is small. This process causes the first step of the two-step dissolution process, described previously as gas diffusing from the hydrate structure, to proceed slowly, and thus, the entire dissolution process proceeds slowly.

2.6.4. "Self Preservation"

Where dissociation proceeds uninhibited, and thermal energy is supplied at a rate which precludes the temperature of the surrounding solution from reaching 0 °C, a diffusion gradient is established between the margin of the dissociating hydrate and the bulk gas or fluid. The HFG is infused on a molecular basis from the hydrate into the surrounding material. Where the saturation of the dissolved HFG in the surrounding material rises sufficiently, the gas molecules will coalesce into a discrete gas phase. Thus, the process of dissociation appears, at the molecular level, to be similar to the dissolution of hydrate. Even when hydrate is stable, if the HFG saturation of the surrounding material is low enough, a diffusion gradient will form that is strong enough to cause the HFG molecules to migrate away from the hydrate bulk. The water molecule latticework becomes unstable and transforms its phase from solid to liquid either just after or simultaneously with the migration of the HFG moving away.

Where heat is not provided at a sufficient rate to prevent the temperature of the surrounding solution from falling to 0 °C, individual small crystals of ice will form as a reaction to the thermal demand (dissociation-driven refrigeration).

The ice can coalesce into a semi-solid ice rind on the surface of the dissociating hydrate. The formation of ice on the surface of dissociating hydrate introduces a diminishment of the rate of dissociation that has been referred to as 'self preservation', a phenomenon that has been widely observed. Ershov and Yakushev (1992), Davidson et al. (1986) and Handa (1986a) have suggested that the water from dissociating hydrate may freeze and encase the hydrate, which will cause it to become temporarily stable.

The explanation for what is occurring on the molecular level is an area of debate (Davidson et al. 1986, Circone et al. 2004; Handa 1986a; Stern et al. 2001; 2002) as there is little well constrained observation or measurement. Handa (1986a) suggested that ice, forming on the outer surface of the dissociating hydrate establishes a pressurized cell because of the high mechanical strength of ice (Durham et al., 2003). In this model, the gas produced within the cell increases and the resulting increased pressure impedes dissociation. The mechanical strength of this ice layer, and its potential effectiveness to act as a physical pressure barrier, has been called into question by Stern et al. (2001), who imply that another mechanism must be responsible for the self-preservation effect. Ershov and Yakushev (1992) regard the ice rind as retarding dissociation by competition between the ice and hydrate crystal lattices such that, "near-contact ice layers must transform into a hydrate lattice with cavities partially filled by gas molecules diffused from hydrate". In other words, there are only so many available molecules of gas and an overabundance of water molecules isolated in a thin layer inside the ice rind. In this model, further dissociation producing further HFG molecules also produces more water, which freezes, further thickening the ice rind. They imply that gas molecules must migrate into the marginal dissociation zone across the ice barrier from the surrounding media for the stability of the hydrate to increase. In any case, the dissociating hydrate appears to create a positive feedback effect wherein further dissociation and thickening of the ice rind further diminishes the rate of gas diffusion.

Dissociation rates of pure methane hydrate held at atmospheric pressures have shown anomalous, non-linear dissociation kinetics (Stern et al., 2001). For samples held at temperatures between 242K and 271K, the dissociation rates were significantly depressed by as many as 4-5 orders of magnitude when compared to samples held at temperatures both above and below this range. The optimum temperature for this self-preservation effect was found to be 268K. Once the hydrate samples were warmed above the freezing point of water, the hydrate dissociated rapidly even though pressure was maintained at 1 atm. One CO_2 hydrate sample was dissociated at atmospheric pressure and 267.4K and showed the same self-preservation effect observed with methane hydrate, which indicates that the self-preservation effect is not unique to methane hydrate. A similar study (Shirota et al., 2002) also reported depressed dissociation rates in the vicinity of 268K at 1 Atm, as part of a study to take advantage of the self-preservation attribute for the transportation and storage of natural gas.

It has been suggested that an ice-rind formed as a function of the dissociation process acts to inhibit dissociation because it forms a diffusional seal. Formation of the ice rind introduces a physical barrier across what would otherwise be a methane diffusion gradient. This barrier effectively compartments the dissociation reaction so that the high diffusion saturation area inside the ice rind in the hydrate crystalline lattice becomes physically separated from the lower diffusion saturation area in the surrounding gas/liquid region outside the ice rind. The ice rind does not need to be solid or continuous to act as a diffusion barrier, whereas it would if it were required to form a pressure cell. The ice rind would likely be unstable and continuously melting and reforming in response to a balance of changing heat requirements of the dissociating hydrate while exposed to a region on its outside in which the hydrate would be unstable. The ice rind would hinder the HFG molecules from leaving the crystal structure as and its HFG saturation level would require diffusion through the solid ice layer, which is a slow process (2.7.3). If the ice were acting as a diffusion barrier, the pressure buildup within the ice rind would be minimal as a significant gas phase would not form within the ice rind, and thus expansion stress on the ice rind would not develop.

Diffusion of HFG molecules takes place as they move from an area of high concentration (within or at the margin of the hydrate) to an area of lower concentration. For example, in a 1 cm^2 area in a yz plane (Fig. 2.17), the net amount of material that diffuses through this plane can be expressed as the mass flux (J_x). If the concentration on either side of any plane is equal, the number of molecules moving in one direction through the plane should be equivalent to the number of molecules moving through the plane in the opposite direction. If there is a large difference in concentration on either side of the plane, the net movement of molecules will be from the high concentration side of the plane to the low concentration side. As diffusion continues, the net flux of molecules will be from the area of high concentration to the area of low concentration, and the solution will tend to become uniform.

Assuming the source of the molecules is located at the origin of the yz plane, the concentration and the change in concentration for a given distance will decrease as the distance from the origin (x) increases. This change is described by equation (2.1), which is referred to as Fick's first law of diffusion (Tinoco et al. 1995).

$$J_x = -D\frac{dc}{dx} \qquad \text{(Eq. 2.1)}$$

J is the mass flux in mol/cm^2, D is the experimentally determined proportionality constant in cm^2/s, c is concentration in mol/cm^3, and x is cm. As dc/dx decreases, the net flux will also decrease. Once the solution has become uniform with respect to dissolved components, the net mass flux will approach zero.

A layer of ice surrounding the hydrate crystal must impede this diffusion process. As the outer layer of the hydrate crystal begins to dissociate, or the gas molecules leave the hydrate structure, their escape will cause the hydrate lattice to collapse. Because the dissociation is an endothermic process, these water molecules are exposed to heat demands that cause the temperature to reach the ice point, 0 °C, at which further extraction of heat from the system will cause the water structure to pass from hydrate (cubic crystal form) to that of ice (hexagonal crystal form), without necessarily passing through a liquid phase. The driving force for crystallization of ice is that there is a thermal imbalance in the hydrate dissociation (54-57 kJ/mol hydrate) and freezing of water (in the range of 6-7 kJ/mol), which results in a dynamic for freezing once 0 °C is reached so long as dissociation continues, even considering potential heat losses to the environment.

If dissociation were to continue following the formation of an ice rind, additional guest molecules would need to diffuse away from the crystal. The newly formed ice layer, however, creates a barrier that prevents the gas molecules from passing through it to enter the bulk solution. This barrier would cause any molecules that were attempting to leave the structure to remain on their side of the ice rind barrier, and a methane diffusion gradient within the hydrate would be significantly weakened. If diffusion within and away from the hydrate crystal lattice were the rate-limiting step for hydrate dissociation, the rate of diffusion away from the hydrate would be determined by the diffusion through the solid ice layer, which would be relatively slow.

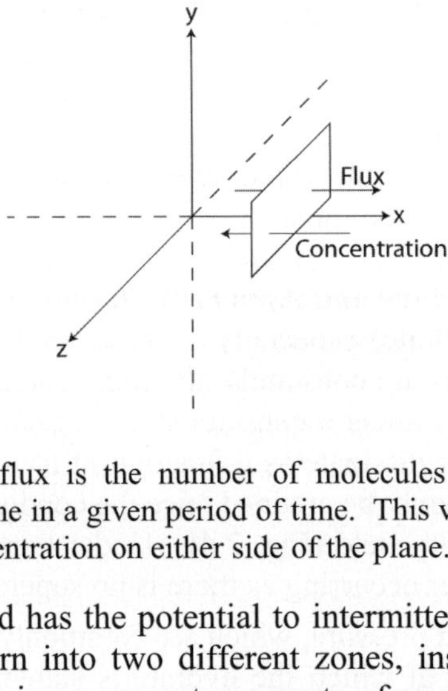

Figure 2.17. Mass flux is the number of molecules traveling through an arbitrarily defined plane in a given period of time. This value is proportional to the difference in concentration on either side of the plane.

Thus, an ice rind has the potential to intermittently separate the normal methane diffusion pattern into two different zones, inside and outside the ice rind. Diffusion through impermanent segments of an ice layer would be slow enough to be considered negligible. The observed self-preservation effect would be a result of the resistance to methane diffusion from the hydrate into the bulk surrounding by ice that was forming and melting locally, thus achieving a transient blockage.

A solid ice rind could result in the complete segregation of the hydrate crystal inside the ice layer from the bulk solution and could contain pressurized gas at least over a small area (Fig. 2.18). Although we have observed that 3 mm of CO_2 hydrate could be maintained at a side-to-side pressure difference of 220 psi, we doubt that this performance could be maintained over a large enough area to form a pressure cell because of the resulting lateral stress.

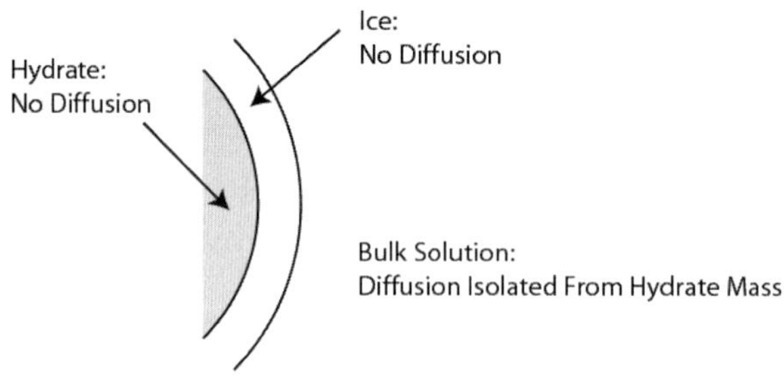

Figure 2.18. Simplified diagram of ice rind (shown as continuous) isolating hydrate from diffusion potential to the bulk solution.

The causes of the hydrate self-preservation effect must be explored further before a definitive and unequivocal explanation can be identified. However, when rapid dissociation is desired, for instance in the recovery of natural gas from oceanic or permafrost hydrate, managing heat and pressure of the dissociation should aim to avoid the temperature range within which the self-preservation effect has been observed. Furthermore, it is important to inhibit the establishment of ice rinds on dissociating hydrate as melting of ice can only be accomplished by application of heat.

2.6.5. The Phase Boundary and Apparent Stability of Hydrate

Naturally occurring hydrate, especially in open systems where gas flux and environmental conditions are constantly altering through time, may appear to be stable but are most likely under a constant state of change. Hydrate deposits are either tending to form or dissipate as a function of trying to reach a stable state. A truly stable state may only be attained once the conditions of the hydrate are at the surface of the stability field (Fig. 2.4). Under these conditions growth and dissociation are no longer occurring as there is no supersaturation.

Temperature and pressure, which are commonly used to define the phase boundary, or conditions at which the hydrate is stable, are not independent of each other in a constantly changing system. In addition, they give no information to the relative amount of each component of the system. In reality, there is constant feedback between the dissociating or forming hydrate and the surroundings. For example, if the temperature increases, the hydrate may begin to dissociate. As dissociation occurs, the concentration of the components, the salinity of the surrounding environment, and possibly the pressure (if the dissociation is occurring in an isolated environment) will change. The change in salinity and concentration of HFG, in an aqueous environment, or water vapor, in a gaseous environment, will have effects on the pressure-temperature phase boundary. A method used to describe this system should interlink all of these variables, and re-evaluate each variable as the system changes. In a method proposed by Xu (2004), the state of the system is described using pressure, enthalpy, and concentration.

According to Xu, it is common to incorrectly treat the existence of hydrate, gas, and liquid as a boundary as opposed to a region. By using enthalpy as one of the parameters to describe the system, information can be gained about the relative amounts of each individual component at given conditions (Fig. 2.19). Xu (2004) uses enthalpy of a hydrate system to explore the kinetics of hydrate formation and dissociation.

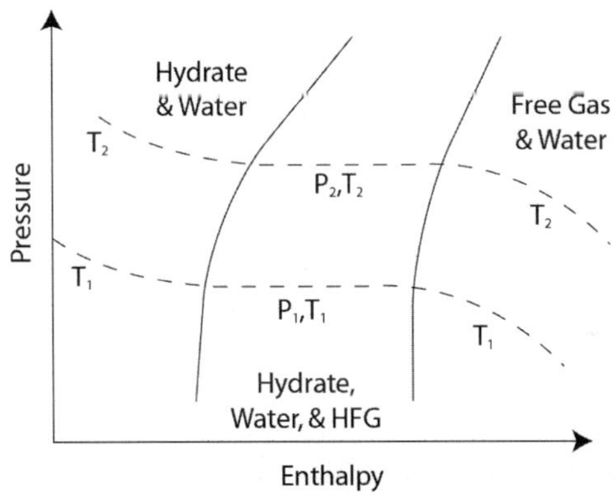

Figure 2.19. According to Xu (2004), when the hydrate begins to dissociate (or form) the system will remain at constant pressure until dissociation is complete. By using the Enthalpy vs. Pressure parameters, the volume fraction of each individual component can be determined.

As dissociation (or formation) occurs, there is a continual input (or consumption) of heat energy. By using enthalpy as one of the parameters to describe the system, the volume fraction of each individual component can be determined with respect to the amount of enthalpy gained (or lost), thus providing information on the reaction's progress. This treatment represents the phase transition as a region whose width is specified by the total enthalpy change needed in order to dissociate completely.

2.7. HYDRATE GROWTH MODELS

A positive gas flux is one of the main factors for the growth and persistence of a hydrate deposit. The gas flux refers, however, to the transport of HFG reactant in a form such that the reactants can contribute to the nucleation of new hydrate or the growth of existing hydrate, rather than actually bubbling gas into a GHSZ, in which case only hydrate shells on the gas bubbles may form (2.7.6).

Hydrate growth in the natural environment is mainly controlled by the rate at which dissolved HFG can be delivered into the vicinity of hydrate. Where HFG can be delivered rapidly, hydrate may grow rapidly. Where HFG is delivered slowly, hydrate can only grow slowly. The different mechanisms for transporting HFG have strongly different potential to form high-grade hydrate deposits. These transport mechanisms involve both gas and water environments. Although these models do not take into account all of the variability that may be found in rocks and sediments in which hydrate may form, they provide a framework that may be applied to identifying conditions under which economic deposits of hydrate may develop.

2.7.1. Circulation of HFG Enriched Groundwater

Where moving or percolating groundwater containing abundant HFG is brought into the presence of hydrate within its field of stability, the HFG reactants carried in the water will grow on hydrate if conditions are highly suitable for the formation of solid hydrate. In the natural environment, hydrate is typically produced from HFG dissolved in groundwater (Winters et al., 2004a) in contrast to laboratories, where hydrate is produced from gas bubble-phase HFG or by gas infusion of water ice. The rate at which the water can deliver the HFG is the controlling factor for maximum hydrate growth.

The most commonly observed large masses of solid hydrate have been observed on the seafloor near natural gas seeps in a number of locations, including the northern Gulf of Mexico (Fisher, 2000; Sassen, 2001) and on hydrate ridge on the U.S. Cascadia margin (Seuss, 2001), amongst other places. These are held to the sea floor by intergrowth with sea floor sediments, which normally keep them in place on the seafloor. Occasionally, either through natural processes or human interference, such as dredging or fishing, masses of solid hydrate may be dislodged and float to the sea surface. The large masses of natural polycrystalline hydrate are associated with profuse gas seeps and conditions of local natural gas saturation of the seawater. It is most likely that the large masses of solid hydrate have been produced by hydrate growing outward into water space surrounding the hydrate where substantial reactant (HFG) must be dissolved in the surrounding seawater.

Ginsburg and Soloviev (1998) have suggested that solid masses of hydrate, such as are found on the seafloor, are most likely to form when alteration of conditions within a field of hydrate stability causes a decrease of gas solubility in the pore water of a marine sediment in which hydrate naturally occurs. This is a special case where large volumes of HFG-bearing water effuse into the openly circulating cold seafloor water, and, combined with the growth dynamic, cause solid hydrate to crystallize near the source of the HFG because the mineralizing solution is chilled rapidly.

Formation of solid hydrate within seafloor sediments can best be accomplished where considerable warm HFG source solutions (where warm is used in the sense that the degree of saturation of HFG in the solution is sufficient to cause nucleation or allow growth when cooled) can migrate into the HSZ. This is the typical case for any HFG-rich solution near the base of the GHSZ where the temperatures are too high to allow hydrate to be stable (Chapter 3). Near the BGHSZ, the HFG-rich groundwater is essentially isobaric, and formation of hydrate depends more on cooling of the water. The solution is initially too warm to form hydrate (Fig. 2.5, A). The sediments cool the solution as it moves away from the warm conduit (Fig. 2.5, Arrow). Hydrate does not precipitate at the pressure and temperature phase boundary because there is a HFG deficit. The solution passes through the pressure, temperature, and concentration boundary at point B (Fig. 2.5). Some additional cooling will occur so that a sufficient driving force can be developed to overcome metastability, and form hydrate. Hydrate formed in this manner may be found considerable distances from the HFG-solution source region, and well away from where the phase boundary would likely be located if groundwater were stagnant. The framework of groundwater movement and simultaneous heating of sediments and cooling of the HFG-rich solutions becomes the controlling factor for further hydrate growth.

A deep saturation region with rising water containing dissolved HFG characterizes this model. The gas is saturated at high temperature and pressure. The temperature is sufficiently warm so that hydrate may not form at the theoretical base of the HFG reservoir. This solution moves upward into the stability zone, crossing the pressure/temperature BHSZ. However, because of the warm temperature where the saturation occurred, there is insufficient HFG for hydrate formation.

As the solution rises, the temperature and pressure begin to drop. The drop in temperature increases the solubility of HFG while the drop in pressure decreases solubility. Depending on the actual conditions in the sediment, these two factors may offset. A scenario for a thermal gradient of 34.3 K/km, 1.03 g/ml seawater density, a seafloor temperature of 3 °C, and hydrostatic pressurization only (Fig. 2.20) shows that the solubility of methane in water is nearly constant. The effect of lithostatic pressurization will be to increase the overall solubility of the methane with depth.

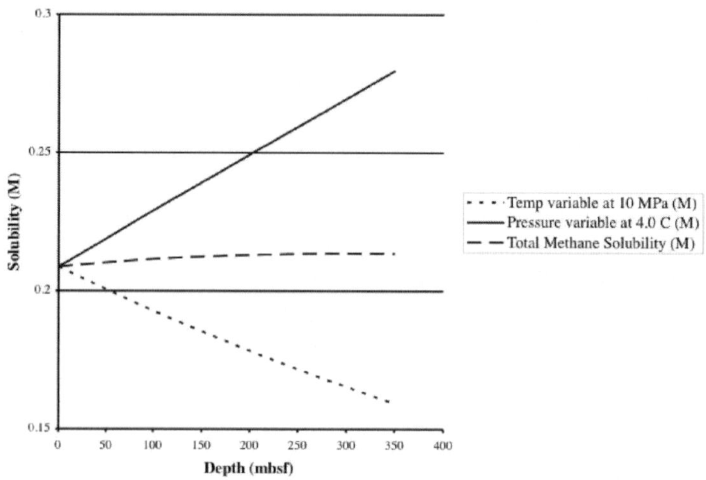

Figure 2.20. As the depth below sea floor increases, the temperature and pressure increase. The increase in temperature lowers the solubility of methane at constant pressure (dotted line). The increase pressure increases solubility at any given temperature (solid line). Calculations where both parameters can vary show that the opposing factors can cancel out (dashed line). Lithostatic pressure is very variable from site to site but would cause the top and central lines to curve upward. This is a generalization; the solubility of gas hydrate was not taken into consideration for this calculation. For the calculation in this table the following assumptions were made: hydrostatic pressure only, seawater is incompressible (see Peltzer and Brewer in Max, 2003 for why this may not be a good assumption), seawater density is 1.03 g/ml, the seafloor lies at 1000m depth, the isothermal gradient is 34.3 K/km (average of Sites 994, 995, and 997 from ODP leg 164), Henry's Law constant 1.3 x 10^{-3} M/atm, and Henry's Law thermal factor 1800 K.

This method of hydrate formation may form hydrate more quickly than where diffusion must be relied on to provide the transported reactant. When water can flow carrying the HFG with it, the HFG may be brought to the vicinity of the hydrate crystal's surface much more quickly than HFG can diffuse either through stagnant water or solid material.

2.7.2. Diffusion in Solution

This model concerns diffusion through stagnant water or, in a geological situation, of still groundwater in porous sediments. The rate of diffusion of the HFG through the groundwater is the controlling factor for hydrate growth.

As the hydrate crystal structure forms from a supersaturated HFG/water solution, HFG molecules are removed from the HFG/water solution at the face of the hydrate crystal. This creates a region of lower HFG concentration in relation to the bulk HFG/water solution. HFG molecules will tend to diffuse from the higher concentration bulk solution to the lower concentration hydrate interfacial region. As HFG is replenished at the face of the hydrate crystal, hydrate will continue to form, and HFG will continue to diffuse through the HFG/water solution as long as an HFG reservoir is available. A concentration gradient develops over a region that is commonly referred to as the concentration boundary layer. The thickness of the concentration boundary layer is defined by the distance from the crystal face to the region where the concentration of HFG equals that of the bulk HFG/water solution (Fig. 2.21) (Myerson, 2002).

In the concentration boundary layer, the thermodynamic driving force for the diffusion of HFG is the difference in chemical potential between regions of higher concentration and regions of lower concentration. Through diffusion of HFG, the system moves toward a state of lowest free energy. Fick's First Law of Diffusion (Tinoco et al., 1995) describes the rate of diffusion per area, or diffusive flux, as a function of the concentration gradient and the diffusion coefficient. For a given solvent, diffusion coefficients of solutes vary depending on how easily the solute can move within the solvent, which is determined by the molecular size and character of the solute.

In the case of stagnant groundwater in porous media, where the concentration boundary layer can be quite thick, the rate-limiting step for hydrate growth is likely to be the diffusion of the HFG through the groundwater rather than the growth of hydrate at the crystal/liquid interface. Mixing of the water solution around a crystal is known to reduce the thickness of the concentration boundary layer and increase the rate of crystal growth (Myerson 2002). Thus, hydrate growth is likely to be much slower within sediment pores of stagnant groundwater where there is virtually no water mixing than within sediment pores of flowing groundwater.

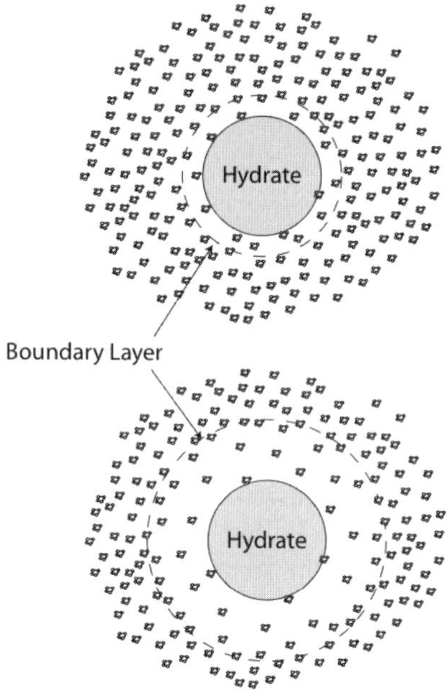

Figure 2.21. Concentration boundary layers of varying thicknesses.

Although water molecules as well as HFG molecules are removed from the HFG/water solution at the crystal/liquid interface during hydrate growth, there is not the same influx of water molecules as described for HFG. The hydrate crystal structure takes up more volume than its water molecules in liquid form, 1.25 times more in the case of methane hydrate. Thus, hydrate actually grows outward into the water around it, rather than drawing water molecules toward it. Slight mixing of the water may occur due to hydrate pushing on the surrounding water molecules in addition to slight mixing from the influx of HFG molecules.

Diffusion of HFG in stagnant water will provide slower growth than in moving water that is transporting the dissolved HFG into the immediate vicinity of hydrate for two reasons. First, percolating water will physically distribute the HFG in addition to the diffusion that will be taking place at the same time; the transport mechanisms will thus be additive and diffusion alone must be slower. Second, the diffusion 'halo' around hydrate, in which the growth diffusion zone will be broader where there is no physical transport by media, will introduce a double-diffusion situation which may introduce competition between the large-scale transport diffusion and the small scale growth diffusion, which could result in complex diffusion vectors.

2.7.3. Diffusion Through Hydrate and Other Solids

Where considerable porosity exists but permeability approaches zero, so that water-filled pores are separated by lithic material and hydrate, the only available mechanism for further growth of hydrate is through solid diffusion that will allow the formation of hydrate within the isolated pores. Diffusion of HFG through a solid hydrate mass is a slow process. The driving force for this phenomenon is reduction in concentration of the HFG trapped below the hydrate layer (Fig. 2.22). The mass transfer is accomplished by the migration of HFG molecules from one cage to the next, generally in the direction of lower concentration. A second diffusion process, water transport across the hydrate layer, can occur in two ways. First, when the system is vertical the primary driving force for water diffusion will be gravity reordering of inverted density fluids. In other words, the heavy fluid water resides above lower density HFG with the hydrate layer acting as a barrier. If the hydrate separates HFG rich/HFG poor areas, water may only diffuse through the hydrate layer by moving from cage to cage, where vacant cages exist, or by creating a diffusion gradient within the hydrate that could cause the methane molecules to exert a directional 'pressure' within the hydrate as a whole.

Molecular diffusion to lower concentration serves to lower the chemical potential and is a well-known driving force. It is the fundamental concept behind pressure-volume (PV) work that gas performs as it expands against pressure. For example, expansion of a gas is used as the energy source in internal combustion engines. In this case, however, the hydrate is serving as a limitation to the dilution of a high concentration gas reservoir (dissolved or free gas). This driving force allows for the forced migration of HFG across the hydrate plug to the far side. The HFG is becoming part of the hydrate layer, moving through the layer in a very slow diffusion process, and then emerging out of the top of the layer. However, if the solution at the top of the layer is already saturated, this newly added HFG molecule makes the solution supersaturated, and the hydrate layer becomes thicker. The recharge rate of the equilibrium solution is critical. If the recharge rate is rapid, a stable, high-density hydrate formation is not likely. However, if the upper side of the hydrate mass confronts a stagnant groundwater situation, a thick, relatively solid, slowly growing hydrate mass may develop. Where there is circulating water and dissolved HFG may be removed, diffusion of methane through the hydrate may only retard the dissolution of the upper hydrate contact.

Water that diffuses downward into the gas space will normally not result in hydrate growth because the hydrate base will commonly be at the BGHSZ. Where the gas is already fully saturated with water vapor at a higher pressure, as would be expected especially in an oceanic hydrate situation, there will be no significant diffusion of water into the gas space.

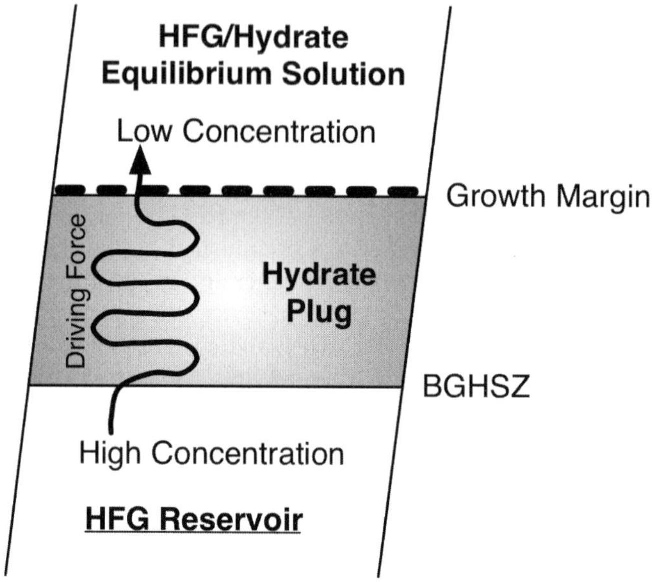

Figure 2.22. A single diffusion model for hydrate growth under stagnant conditions. The driving force at a growth margin is due to spontaneous diffusion of HFG through the hydrate layer toward lower concentration conditions. The opposite process, diffusion of water toward the HFG Reservoir with growth of hydrate on the HFG reservoir side of the plug, also occurs, but is not shown. Water diffusion processes will not lead to continuous hydrate production if the water that passes through the hydrate enters a zone where gas hydrate formation cannot occur (assumed to be below the BGHSZ).

Diffusion of either HFG or water reactant through solid hydrate will be the slowest of the transport processes and it is unlikely that large masses of hydrate will be formed where this is the only available transport mechanism.

2.7.4. Formation in Gaseous HFG Through Water Vapor Diffusion

There is one case in nature where delivery of reactant for the growth of hydrate takes place through a vapor phase.

Either HFG or water vapor, where both are in a gas phase, have better diffusion characteristics than gas dissolved in stagnant water. The collision frequency of molecules, and the likelihood that the molecules be in the correct orientation to attach to the hydrate lattice, have a significant control on the rate of formation. In the gaseous phase, the limiting reactant, water vapor, can diffuse much more quickly through the gaseous environment from its source to the location in which hydrate is most likely to grow.

The inter-molecular interactions in gas are very limited when compared to the interactions encountered between dissolved molecules and the molecules in a liquid phase. In the liquid environment, the limiting reagent, the HFG, moves essentially randomly, with respect to concentration gradients, constantly colliding with water molecules. These collisions impart the factor of dissolved HFG movement as they cause the HFG molecules to change direction and orientation. These molecular interactions limit the rate of diffusion and also limit the number of molecules in a given time period that are available to collide with the surface of hydrate that may be present. This increases the number of collisions in a gaseous atmosphere and therefore the frequency that the colliding water molecule is in the correct orientation.

The theory that hydrate can be grown in an HFG atmosphere and water vapor that diffuses through the gas from a source to a growth site has been supported by experiments (Max et al., 2004). The principal of the experiments was that in a vessel or space where both water and hydrate forming gas is present and where pressure is suitable for hydrate to form but temperature is too high, any surface that is cooled so that hydrate is stable upon it will allow hydrate to form. This situation is well known in the pipeline and marine hydrocarbon extraction industry where unwanted hydrate may form on the chilled walls of pipes carrying gas or petroleum rich in gas and where it constricts the pipe and restricts free flow. Experiments have been carried out both in gaseous and aqueous media in order to better understand the thermodynamics and kinetics of hydrate formation and dissociation under a variety of conditions. In a pressure vessel, kept at room temperature (about 20 °C), a solid metal surface was cooled to induce the formation of hydrate from HFG and water vapor. The apparatus (Fig. 2.23) was used in several experiments.

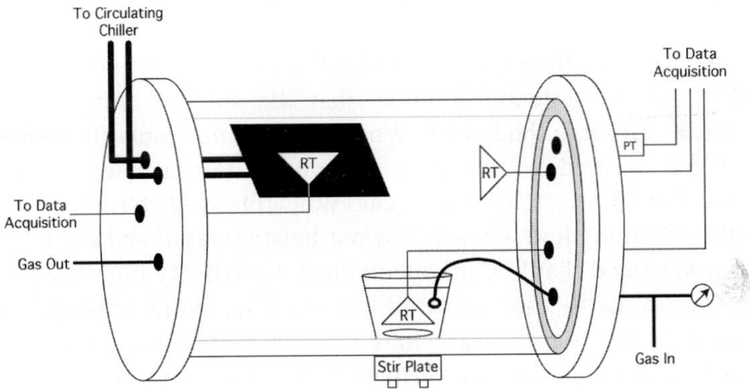

Figure 2.23. Diagram of the pressurized vessel used to grow hydrate in a gaseous atmosphere. RT indicates a resistance temperature detector, PT indicates a pressure transducer, and the black surface represents the chilled metal surface. Pressure never exceeded 3.35 MPa.

The vessel was pressurized by bubbling ethane through seawater in a small water containment vessel (Fig. 2.23). By bubbling the HFG through the agitated source water, the amount of water vapor in the HFG was maximized. The solid metal plate was chilled by an external circulating chiller to a temperature significantly lower than the bulk HFG (The low thermal conductivity of ethane makes this possible). The pressure was monitored along with the temperature of the plate and the gas surrounding the plate. The vessel was made of metal end caps and an acrylic tube of approximately six-inch internal diameter, which enable visual observation of the entire hydrate-forming region. The water vapor in the ethane reached a state of equilibrium with the source seawater. As the hydrate-forming surface was cooled, the water, which had been vaporized into the HFG, condensed on the metal surface. The condensed water was then within the hydrate stability zone, and hydrate formed on the metal surface within seven minutes of crossing the Pressure-Temperature phase boundary by supercooling. Two experiments (Fig. 2.24.) have demonstrated that solid hydrate mats up to 1 cm thick can be formed in an HFG atmosphere in a relatively short time (hours), so long as there is an abundant supply of water vapor. As water was removed from the ethane, it lowered the concentration of water vapor, which was then replaced by additional water evaporating from the source. In theory, if the experiment was continued, water should be constantly transferred from the source to the growing hydrate mat until either a maximum hydrate thickness would be reached, due to the limited thermal conductivity of hydrate, or the source water would be completely consumed, with the only free water existing as equilibrium water vapor. According to these data, hydrate can be formed on a selectively cooled surface, while the ambient conditions are not within the hydrate stability regions.

In addition, it appears that since the chilling of the hydrate mat was maintained to absorb the heat produced by crystallization, the surface of the hydrate grown into gas space has been observed to be only slightly irregular. The hydrate consistently appeared to distribute the rate at which heat could be absorbed across its growing front so that the solid hydrate layer remained approximately the same thickness. Where hydrate momentarily grew faster, the heat produced seemed to introduce a negative feedback effect, and growth slowed at that place until the heat was removed through the hydrate. Meanwhile, where hydrate growth had not been so rapid and the heat produced was less, it appeared that hydrate maintained a relatively high growth rate until the temperatures across the face equilibrated. This hydrate growth on a chilled surface supports the concept that where heat can be removed from a zone where hydrate is forming, heat transfer through the hydrate may moderate the hydrate layer thickness.

The type of water used for the source water also appears to have a significant effect on the rate of formation. Water vapor from natural seawater appears to have a much shorter induction time, seconds to minutes, than water vapor from normal fresh water sources, such as tap water, which appears to be hours to days. The reason for this is unclear but is probably related to one or more of the chemicals present that also evaporates or is carried in the water vapor. Although CO_2 hydrate apparently has a faster growth rate than ethane hydrate, this only shows that different HFG's form hydrate at different rates. However, the driving forces in each experiment may have been different. Further experimentation is underway to study this phenomenon.

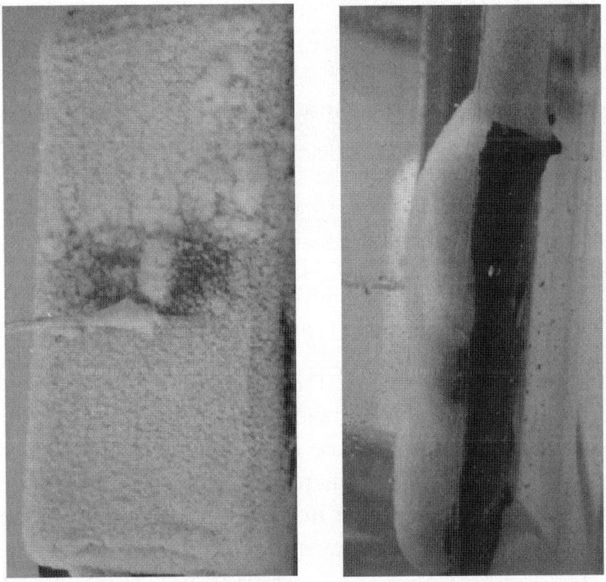

Approximately 1 cm

Figure 2.24. Photographs of hydrate grown on a cooled surface in a pressurized vessel containing water vapor and HFG. Left: Surface view, right, side view. Photographs courtesy of MDS.

When the hydrate was dissociated by simultaneous depressurization and warming, the HFG saturated water droplets remained on the plate. Upon cooling and re-pressurization, nucleation occurred in several locations on the chilled plate followed by the expansion of the hydrate layer. As the hydrate layer advanced and passed water droplets, the water droplets instantaneously changed from water to hydrate (Fig. 2.25). On the left is a photograph of the growing hydrate layer. The diagram on the right represents the silhouette of the hydrate layer. All of the water droplets below the hydrate layer had been converted to solid hydrate masses, while the water droplets above the edge of the hydrate layer remained as liquid water.

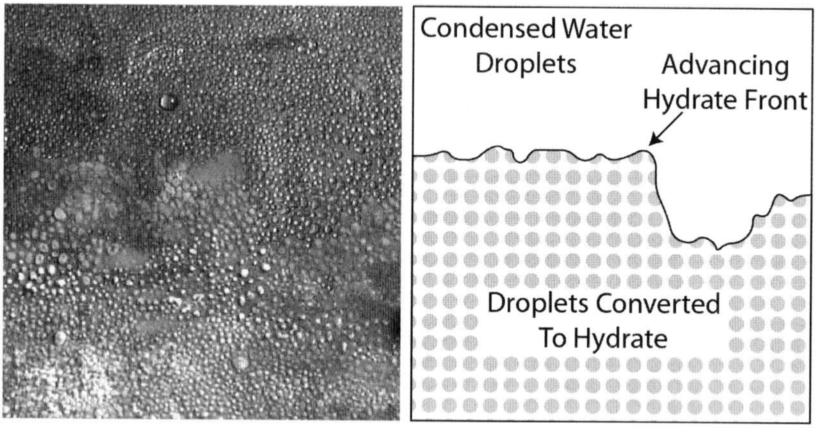

Figure 2.25. Left: A photograph of laboratory-grown ethane hydrate. Right: Representative diagram of the silhouette of the advancing hydrate layer.

In additional experiments, the same scenario was repeated with the exception of a constant flow of ethane, which was bubbled through the source water. Two different flow rates were used, 5 L/min and 10L/min. The growth rate was increased as the flow rate increased, which is interpreted as the result of increasing the supply of water vapor in the HFG atmosphere. These experiments support the concept of a boundary layer model for hydrate growth.

Effects of the ability of gas to absorb heat produced from hydrate formation, which can be significant over time, also act to control the location of hydrate growth. In these experiments, the HFG and its water vapor were both too warm to form hydrate unless chilled, and the hydrate only formed where a thermal exchange could be created between the hydrate growth front and the chilled land surface. These experiments provide an analog for the conversion of natural gas trapped in shallow geological reservoirs in permafrost terrain.

2.7.5. Variable Supersaturation

Once hydrate has formed in a particular region, its continued persistence can be expressed as a function of two-way gas flux. In a situation where a continuous supply of HFG falls below a certain level or ceases, especially in an open system where groundwater is in contact with at least one side of the hydrate deposit, the hydrate will dissolve as the HFG migrates away from the hydrate by either groundwater flow or diffusion. In a situation where the gas flux remains high enough to promote growth, a hydrate zone develops. In order to avoid decomposing or dissolving, this hydrate must have at least the equilibrium concentration of HFG on all sides and exist within the pressure-temperature conditions of hydrate stability. If the HFG concentration is high enough, hydrate growth will tend to continue until porosity fills, and the only transport mechanism that remains is solid diffusion. When hydrate fills and blocks porosity, further gas arriving in solution will tend to coalesce and pool, as is seen in the Blake Outer Ridge (Chapter 3). Only if the HFG concentration remains static, will the system be static, but even then, diffusion movement of HFG to a 'higher' or downstream location, is possible (Fig 2.26). Where faulting of the sediments-hydrate seal take place, for instance, subjacent gas can suddenly begin a focused migration through the otherwise hydrate-sealed lower part of the GHSZ.

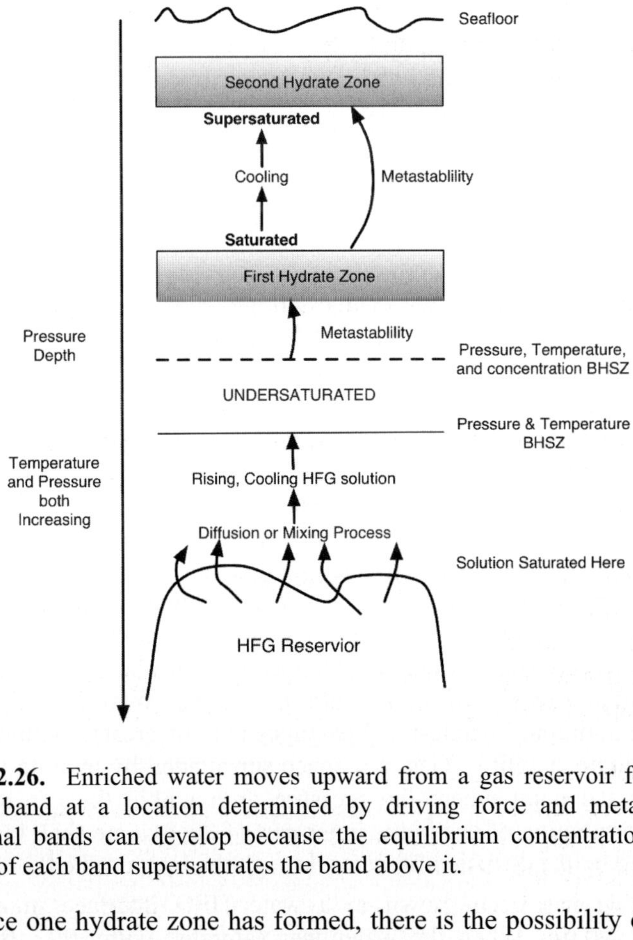

Figure 2.26. Enriched water moves upward from a gas reservoir forming a hydrate band at a location determined by driving force and metastability. Additional bands can develop because the equilibrium concentration of gas hydrate of each band supersaturates the band above it.

Once one hydrate zone has formed, there is the possibility of additional zones forming at shallower depths (Fig. 2.26) because of a natural gas flux 'forward' from the exiting hydrate. If the upper surface of the hydrate zone is exposed to water, the hydrate will release HFG into the water until an equilibrium state is reached. As the HFG/water solution flows away from the upper surface of the hydrate, the water will reach a state of supersaturation again. It is important to keep in mind that the concentration of HFG is not changing; this concentration is still below what Henry's Law would predict for HFG solubility. The state of supersaturation is re-reached because the equilibrium concentration of hydrate decreases with temperature and the temperature of the sediment decreases with shallower depth. If the temperature-pressure conditions remain conducive to hydrate formation, a new hydrate layer may form.

According to this growth model, several different hydrate zones may exist directly above one another. The conditions at which the top of the underlying hydrate layer exists will define the equilibrium HFG concentration. The equilibrium concentration at the top of the hydrate layer and the role of metastability, in turn, will define at what temperature-pressure conditions supersaturation will be reached again, inducing hydrate to form an additional deposit.

2.7.6. Direct Contact Between Gaseous HFG and Water

When HFG is introduced as a gas phase under conditions suitable for hydrate growth, the gas will either form a gas stream, where there is limited turbulence, or bubbles. The smaller bubbles may dissolve rapidly. Gas injected into a fissure in water saturated porous media will absorb water vapor as water in the surrounding porosity absorbs gas. This process continues until the water immediately surrounding each HFG mass becomes saturated with the HFG. Nucleation and rapid growth of hydrate takes place at the contact. A thin shell of hydrate will establish itself, and reactants may be extracted from both the liquid and aqueous phases. At the point where no reactant water or HFG is available on either side of the hydrate shell, the process of hydrate formation slows dramatically. The HFG will then be separated from the surrounding water by an essentially impermeable thin shell of hydrate. If no dissolved HFG were introduced to the water from another source, continued hydrate growth would depend on solid diffusion of either gas or water through the hydrate shell. This process is extremely slow. Considerable gas may thus be stranded in a zone of hydrate stability, unable to move into a physical matrix, such as porous sediment. In addition, hydrate-lined passages may be created within or through a zone of hydrate stability. Gas in a space surrounded by hydrate will only be consumed to the point where the pressure drop within the gas bubble falls, owing to incorporation of gas in the inner part of the hydrate shell (and only at a rate controlled by the diffusion of water through the hydrate shell).

Hydrate has been grown at a water-HFG interface in experiments conducted at MDS. When the water was saturated with HFG, but the bulk headspace was a non-HFG, no hydrate grew at the interface. When the interface was HFG, but the water was not pre-saturated with HFG, again, no hydrate grew at the interface until the adjacent water became saturated in HFG. Locally, hydrate formed in gas space where a diffusion path or a micropore watercourse was established. This indicates that both initial phases must have appropriate concentrations of reactants for hydrate to grow.

2.8. KINETIC CONSIDERATIONS

The kinetics of hydrate formation, like any other chemical reaction, are dependent on the rate of collision and proper orientation of the component molecules. In hydrate, the correct number of water molecules, in the correct orientation and position, must encapsulate an HFG molecule that must also reside in the correct position within a guest site, but probably without a requirement to be in a particular orientation. Additionally, all of the molecules must have the correct amount of residual kinetic energy or the collision complex will either fail to bond (not enough energy) or, if already bonded, come apart (too much energy). The essential energy component is the lower energy limit as a reaction cannot take place unless a sufficient amount of energy is invested (this is termed the *activation energy*).

The necessary parameters for the kinetic description of hydrate formation differ for the water molecules and HFG, but also change significantly if the supersaturated solution is a gas or liquid. The fundamental parameter will be the reaction rate, which, taken simply, is the product of the steric factor (Π) (Note: this symbol is usually written as *P*, which is easy to confuse with P [pressure]. We use Π in place of *P* to reduce confusion.), the rate of collision between two (or more) molecules that have energy equal to the activation energy (*f*), and the concentration of the reactants (M or mol/l). Crystallization of hydrate in a liquid medium is a function of interactions between encounter pairs: the HFG and water. Temperature plays a double role in that it increases the collision rate and increases the fraction of molecules that have enough energy to overcome the activation energy. Pressure has a similar effect on hydrate whereas it has virtually no effect on water ice, at least within the natural environment where permafrost and oceanic hydrates are found.

In the gas phase, the participatory molecules must approach each other to form the solid hydrate phase, but they may approach too rapidly. In addition to proximity the molecules must have proper orientation with respect to each other. The water molecules have very rigorous requirements for orientation; the HFG is only dependent on distance. This is due to the types of interactions. The water molecules are participating in covalent types of bonds that have a very high directional requirement. The molecules must be in very precise alignment in order to be stabilized by bonding. The HFG molecules need only be close as van der Waals bonding is non-directional. Π can be quite small, and if three or more bodies must come together in correct orientation as well as proximity at the same time in order to form an activated complex, Π can approach zero.

The orientation requirement for the water molecules is relaxed in solution because the motions are slow enough that reorientation can occur (Table 2.5.). As two molecules approach each other without the proper orientation and/or enough energy to form the activated complex, they may reach the proper orientation and energy to achieve the optimum bonding orientation.

Parameters	Water (gas phase)	HFG (gas phase)	Water (liquid phase)	HFG (liquid phase)
Orientation (θ, ϕ, ψ)	Critical (transient state)	Not Significant	Limited (can rearrange)	Not Significant
Position (x, y, z)	Important	Important	Limited (already close)	Important

Table 2.5. Classification of the relative importance of rate determining parameters.

Basic research into the significance of how these parameters affect hydrate formation under different environmental conditions is essential. Natural gas that is generated from hydrate for recovery (Chapter 7) can easily reform hydrate. There are certain techniques that can be used to reduce Π that will inhibit the formation of unwanted hydrate that would affect flow assurance of recovered gas (Max & Holman, 2003).

2.9. BEST CONDITIONS FOR HYDRATE CONCENTRATION

As Chen and Cathles (2003) have pointed out, one of the best natural examples of the maximum achievable growth rate of solid hydrate takes place near warm seafloor vents where dissolved HFG in venting groundwater cools rapidly in contact with cold seawater and provides a steady, high level dissolved reactant supply that promotes the growth of solid hydrate growing on or just beneath the seafloor. The best conditions of growth of large volumes of solid oceanic hydrate in porous sediments are similar. Once dissolved HFG is present in high enough saturation in groundwater, the main controlling factor is the ability of the groundwater to deliver the reactants (mineralizing solutions) so that solid hydrate can grow in such a manner to occupy a substantial portion of available porosity. Where HFG-enriched groundwater moves upward into a GHSZ, conditions favor hydrate formation (Ginsburg and Soloviev, 1997). Although the pressure in the GHSZ decreases, which introduces a weak tendency for hydrate to become less stable, the temperature falls, depending on the geothermal gradient (3.1, Fig. 3.1), and lower temperatures strongly favor hydrate formation.

In permafrost terrain, an ideal situation occurs where an existing shallow gas deposit is converted to hydrate during intensification of glacial episodes, especially where porosity and permeability are suitable for the rapid diffusion transfer of water vapor from a water-gas interface.

Identifying the most likely locations for economic deposits of gas hydrate in part will rests upon identifying locations where the ideal sedimentary and structural situation exists in geological strata. The hydrate growth models and geological analyses constitute a primary exploration technique.

APPENDIX A. BACKGROUND CHEMISTRY

A1. PHASE DIAGRAMS

Molecules of natural gases and water can be present in three forms; solid, liquid, and vapor, depending on the pressure and temperature conditions of their surrounding environment. In addition, they can react to form a crystalline structure called a clathrate hydrate. The main phase transformations of interest here are the location of the fields of stability for each of the different natural gas hydrates and compound hydrates formed from mixtures. For each field of stability, there is a boundary that marks the position of the transformation from the phase of mixed gas and liquids of water and natural gas to the solid crystalline form of hydrate. This phase boundary is usually shown on a phase diagram, which is a useful construct for understanding how one type of matter transforms into another. There are a number of phase diagrams in this chapter, and each shows a different specific aspect of physical behavior. For example, figure 2.7 is a the familiar pressure - temperature phase diagram (P-T diagram), while figure 2.4 is three-parameter phase diagram showing the 'hidden' variable of HFG concentration, which may be more appropriate for describing hydrate stability. Each diagram has different information, but each is based on the same concept. In general, a phase diagram shows the transition between one stable phase and another. The *first order* transition between water + HFG and hydrate can be shown as a relatively narrow line. There are other possible transitions, but they are not discussed here because these types of transitions do not seem to apply to hydrate.

The curving solid line (Fig. A1) is a diagrammatic PT phase boundary between hydrate and a mixture of the components that comprise the hydrate and assumes there is excess HFG reactant present. To the right or below this line gas hydrate is not stable. The water that makes up the host structure will be found as ice or liquid, not as hydrate. To the left or above this line hydrate is the stable phase, and water molecules can form the host structure if enough HFG gas is present. At the solid line, hydrate is in equilibrium with ice or water, and hydrate is only one of the stable phases. The chemical potential of all phases is exactly the same, and this is the set of conditions that must be met for hydrate to grow or dissociate. Technically, hydrate is only truly static when it is within the phase boundary, not on the phase boundary where it also could be growing or dissociating.

Chemists usually show P-T phase boundaries with both temperature and pressure originating at a common point, usually at the lower left of a diagram, and increasing in pressure upwards and in temperature to the right (Fig. A2). Kvenvolden, however, broke with this convention, and showed pressure increasing downward (Kvenvolden, 1988), so that the relation between pressure and depth below the Earth's surface were graphically evident (Fig. 2.3). This rendered the phase boundary of hydrate stability directly in relation to pressure-depth and allowed for more apparent relationships between hydrate and its environment. Most chemists show hydrate phase boundaries as an upward curving line with a single ordinate for both pressure and temperature while most geologists and geophysicists are more comfortable with Kvenvolden's manner of representation, which results in a downward curving phase boundary. Both representations are entirely correct.

For similar reasons of perception and the ease of relating hydrate to its environment, earth scientists and the rest of the scientific community tend to use the Celsius temperature scale whereas chemists use the Kelvin temperature scale, which allows for more straight-forward calculations.

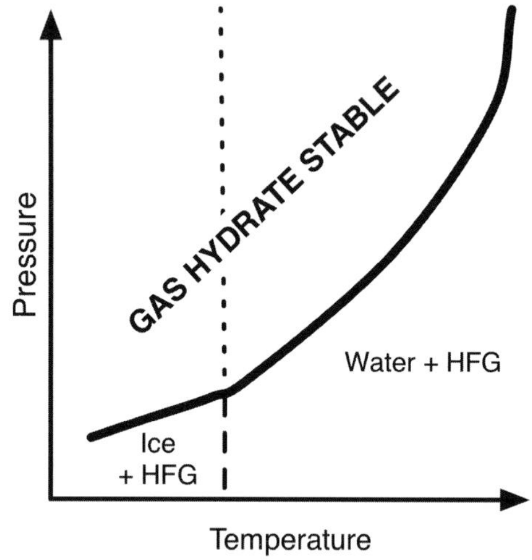

Figure A1. A phase diagram for gas hydrate. The thick, curving line represents the phase boundary for gas hydrate. Hydrate is unstable below the line. On the line, hydrate is in equilibrium with another phase, Water +HFG or Ice + HFG. Above the line, hydrate is the stable phase. The dashed line is the water/ice transition. The dotted line separates permafrost hydrate from oceanic hydrate and is not a phase descriptor.

The dashed line (Fig. A1) shows the approximate position and orientation (at precisely 0 °C and 273.15 K, also see Fig. 2.3.) of the pure water to ice phase transition. This line is important because it allows the physical stability relations of more than one solid material to the other to be located in temperature and pressure fields that are found naturally on earth. At higher temperatures than where the hydrate and the water-ice phase boundaries intersect, the solid line represents conditions where hydrate is in equilibrium with Water + HFG. At temperatures below this point hydrate is in equilibrium with Ice + HFG. Where water is present in excess of HFG in the region of hydrate stability, hydrate will coexist with water to the right of the water-ice phase boundary and with ice to the left.

A2. HENRY'S LAW

Henry's Law (Also see Glossary) is a useful tool for calculating the solubility of a gas in a solution although there are some limitations, and the calculated value can often deviate from an experimentally measured concentration. Henry's Law essentially states that, at a given temperature, the concentration of gas that will dissolve into a liquid increases linearly with pressure (Equation A 2.1):

$$M_P = k_H^o \times P_{applied} \qquad \text{(Eq. A 2.1)}$$

where M_P is the concentration of gas in solution (usually in mol/L), k^o_H is the Henry's Law constant at 298 K (M/atm is typical) and where $P_{applied}$ is the system pressure. Table A1 shows selected constants and thermal factors ($\Delta H_{solution}/R$).

HFG	Molecular Mass (g/mol)	Hydrate Structure[1]	k_H^o (M/atm)[2]	Thermal Factor (K)[2]	Solubility 0.1 MPa, 279 K (M)[3,4]	Solubility10 MPa, 279 K (M)[3]
Methane	16.04	sI	1.3×10^{-3}	1700	1.9×10^{-3}	1.9×10^{-1}
Ethane	30.07	sI	1.9×10^{-3}	2350	3.2×10^{-3}	3.2×10^{-1}
Ethene	28.05	sI	4.8×10^{-3}	1800	7.1×10^{-3}	7.1×10^{-1}
Ethyne	26.04	sI	4.1×10^{-2}	1800	6.1×10^{-2}	6.1
Propane	44.10	sII	1.4×10^{-3}	2700	2.6×10^{-3}	2.6×10^{-1}
Propene	42.08	sII	5.0×10^{-3}	3400	1.1×10^{-2}	1.1
iso-Butane	58.12	sII	1.1×10^{-3}	3100	2.2×10^{-3}	2.2×10^{-1}
H_2S	34.082	sI	1.0×10^{-1}	2200	1.6×10^{-1}	1.6×10^{1}
CO_2	44.010	sI	3.4×10^{-2}	2400	5.8×10^{-2}	5.8
SO2	64.065	sI	1.2	3000	2.4	2.4×10^{2}
Nitrogen	28.013	sII[(5)]	6.3×10^{-4}	1300	8.4×10^{-4}	8.4×10^{-2}
Oxygen	31.999	sII[(5)]	1.3×10^{-3}	1650	1.9×10^{-3}	1.9×10^{-1}

Table A 1. Henry's Law data for common natural gas constituents, [1]From Franks, 1973, [2]Averages of data collected in Sanders (1999), [3]At 279 K (6 °C). 4. 0.1 MPa is 0.99 atm, 5. Sloan (1998).

The solubility of gas increases as temperature increases. This occurs at a rate that is proportional to the heat of solution of the gas. The equation is usually reported as a perturbation of k^o_H (Equation A 2.2):

$$k_H = k_H^o \times e^{\frac{-\Delta H_{solution}}{R}\left(\frac{1}{T_{experiment}} - \frac{1}{T^o}\right)} \qquad \text{(Eq. A 2.2)}$$

Where k_h is the Henry's Law constant at a given temperature, $-\Delta H_{solution}$ is the enthalpy (or heat) of solution, R is the universal gas constant (8.314 J/mol/K), $T_{experiment}$ is the temperature of the experiment, T (K) is the temperature at which k_H^o is reported, and $-\Delta H_{solution}/R$ is the thermal factor. The new value for Henry's Law is used to calculate the concentration (Eq. A 2.3):

$$M_{P,T} = k_H P_{applied} \quad \text{(Eq. A 2.3)}$$

Where $M_{P,T}$ is the concentration at a given pressure and temperature.

A3. NUMBER OF WATER MOLECULES PER DISSOLVED HFG MOLECULE

The ratio of methane molecules to water molecules in ideal, pure methane hydrate is 1 to 5.75. However, the ratio in a methane/water solution is likely to be much lower. To calculate the solution ratio, as shown in Figure 2.2, the first step is to select or determine the concentration of hydrate forming gas. For example, the concentration determined by Henry's Law or the data from Servio and Englezos, 2001 and 2002 could be used (although that requires a conversion from mol fraction to molar to use the Henry's Law constants in Table A1). This concentration will be used to develop a ratio with the number of water molecules present (Eq. A 2.4).

$$\frac{\text{molecules methane}}{\text{molecules water}} = \frac{\text{concentration methane}}{\text{concentration water}} \quad \text{(Eq. A 2.4)}$$

The concentration of water is seemingly easy to calculate, and for pure water at atmospheric pressure and 298 K, it is 55.5 M. However, water is compressible. For this reason the density of the water must be determined or calculated for the desired pressure, and the concentration re-determined. The compressibility of a solution is usually represented by the letter *z* and is a factor multiplied by the standard state (water at 1 atm and 298 K) to determine the concentration at some other pressure. Brewer and Peltzer covered the significance of the compressibility factor in volume 1 of this series (Max, 2000, 2003).

The compressibility factor changes depending on salinity, sediment, and other dissolved species. For these reasons, the selection of an appropriate value of *z* is critical to any successful model.

A4. CHEMICAL POTENTIAL OF SALINE HYDRATE INHIBITION

The addition of salt or another material initially changes the location of the hydrate phase boundary in a similar fashion to salt lowering the melting point of ice. This is due to changes in the chemical potential induced by the increased ions in solution. The addition of any material to water dilutes and therefore reduces the chemical potential of the water because dilution of a concentrated material is spontaneous (Fig. A 2). This dilution is independent of the type of material and only is responsive to the amount (number of particles) of material added. The response is theoretically linear, and the freezing point constant for water measured to be 1.86 oC/m (or 1.86 K/m), where m is the molality (mol solute per kilogram of solution) of all non-water particles in the solution including ions. Deviations from linearity are attributed to chemical processes such as incomplete ionization of salts.

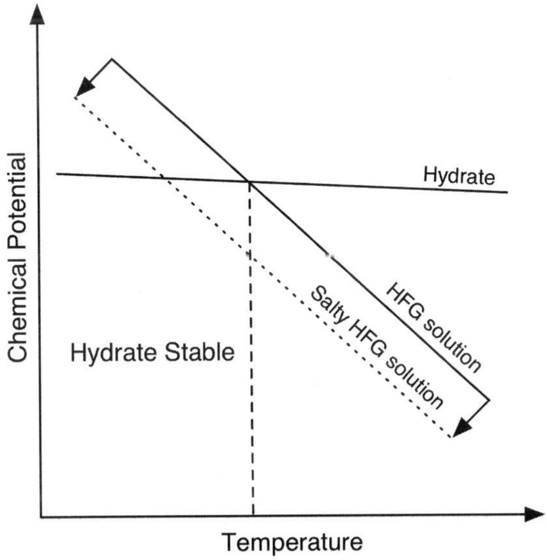

Figure A2. Chemical potential of hydrate decreases slowly with increasing temperature. Chemical potential of solutions decreases faster with increasing temperature. The phase with the lowest chemical potential is the stable phase. Dissolved salts linearly lower (to first approximation) the chemical potential of a solution, the effect is to lower the transition temperature (the temperature where the two lines cross) between the two phases.

At high concentrations of salt a different process, salting out, may occur. Salting out drives HFG out of solution by limiting the amount of water available to interact with the HFG. Hydrate formation from aqueous media can be completely prevented if the concentration of HFG possible in water is lowered to below the hydrate equilibrium concentration by the salting out effect.

A5. MOL OF GAS HYDRATE

There are two different methods to calculate the molecular weight of hydrate, both of which can be found in other sources. The first and most commonly used method, which is used here, relates one mol of hydrate to one mol of gas. The calculation (Eq. A 2.5) is:

$$MW_h = \sum_{j=1}^{n_c} x_i MW_i + \upsilon MW_w \quad \text{(Eq. A 2.5)}$$

where MW_h is the molecular weight of hydrate based on one mol of gas, MW_i is the molecular weight of type j guest molecule, n_c is the total number of different guest molecules, υ is the stoichiometric ratio of water molecules to guest molecules, and MW_w is the molecular weight of water. The fraction of the guest molecules (Eq. A 2.6) that are type i is defined as x_i such that:

$$x_i + x_{x+1} + x_{i+2} + ... + x_{n_c} = 1 \quad \text{(Eq. A 2.6)}$$

The second method used relates to the unit cell. While this method does not appear to be the method of choice in hydrate literature, it is commonly used in crystal chemistry. The calculation (Eq. A 2.7) for structure I hydrate is:

$$MW_{hu} = n_w MW_w + \sum_{j=1}^{n_c} MW_j \sum_{i=1}^{z} \omega_i n_i \quad \text{(Eq. A 2.7)}$$

Where MW_{hu} is the molecular weight of hydrate based on the unit cell, n_w is the number of water molecules within the unit cell, MW_w is the molecular weight of water, MW_j is the molecular weight of type j guest molecule, n_c is the total number of different guest molecules, ω_i is the fractional occupancy of type i cavity, n_i is the number of type i cavities within the unit cell, and Z is the total number of different cavity types found within the unit cell.

A6. DIFFUSION MECHANISM FOR HYDRATE BREAKDOWN

Hydrate breakdown appears to take place at the surface of hydrate, or on new surfaces that may develop penetrating in from the surface, which has the effect of increasing surface area. Both dissociation and dissolution involve HFG molecules in each water molecule cage individually leaving their stable positions and breaking out of the cage structure, an event associated with the breakdown of the water molecule structure. The actual mechanism of breakdown in dissociation and dissolution, however are probably different.

Dissociation of hydrate takes place when its crystal structure breaks down when either heat or lower pressure is applied. In either case, both the heat sink character and the immediate reaction of the surface dissociation region make it difficult to rapidly dissociate hydrate where the surrounding pressure-temperature conditions can be altered only slowly because of the feedback effects (2.6.4; 2.6.5; 2.7.4). In dissociation situations, hydrate breakdown takes place where both water and the HFG are sufficiently available in the surrounding media. Gas is often over-pressured, even with relation to the compression effect of the hydrate (Max and Dillon, 1998), and free water dominates the media. Therefore, diffusion gradients would flow from the surroundings into the hydrate and favor driving the HFG and water reactants into a hydrate structure. The tendency for dissociation overrides the diffusion gradient-driven growth dynamic (2.3.1). Formation of hydrate, however, can only take place where the chemical potential within the field of stability is suitable. The breakdown may take place either because the weak attractive forces of the HFGs and their cages are overcome and both components lose their geometrical as well as their physical energy coordination or because the cage structure distorts or is physically disrupted, which allows the HFG molecule to enter the surrounding media.

Dissolution takes place when hydrate is within its pressure-temperature stability field, but where the saturation of the HFG or water in the surrounding liquid or gas media is low. Methane hydrate and carbon dioxide hydrate dissolve in flowing seawater in proportion to the solubility of their respective guest molecules in seawater (Rehder et al., 2004). Dissolution can be expected to take place in groundwater, although at a slower rate, limited by the rate at which water movement or diffusion can remove the dissolved HFG from the vicinity of the dissolving hydrate.

Because the hydrate is usually developed in an aqueous groundwater, there is little impetus for the water cage to disaggregate. There is, however, a strong contrast in HFG saturation between the hydrate and the surrounding water. The driving force for hydrate breakdown through dissolution is likely the tendency for the HFG molecules to leave their cages owing to the establishment of a diffusion gradient between the hydrate and its surrounding media, especially in the surface and near surface layers. Dissolution can be described as a cascading effect wherein individual HFG-filled cages very near to or at the surface of a solid hydrate mass lose their coherence when existing in an environment that is too undersaturated in the HFG to allow the hydrate to persist. As each cage dissociates, it exposes further cages that lie deeper in the hydrate. The dissociation surface is a transitional zone where exchange of constituents between the hydrate and its surroundings takes place.

Although the relationship between the forces of the water cage and the guest molecule at the moment of dissociation remains controversial, it would appear that cage breakup involves a self activating physical movement of HFG molecules or a disestablishment of the van der Waals balance of forces that stabilize the hydrate cages. Diffusion processes (2.5.1; 2.6) would then appear to be the paramount mechanism for dissolution of hydrate.

Dissolution of hydrate through HFG diffusion would actually be a continuous or semi-continuous process. This is modeled here using three diagrammatic stages of how individual HFG-filled cage of water molecules respond (Fig. A3). Only one hydrate cage is shown but the entire hydrate accumulation is understood to be composed of closely packed HFG-water molecule cages. Where the domain of a hydrate cage is physically distant enough from the dissociation (Fig A3, 1a), the HFG position within the cage is physically located so that the lowest average energy conditions exist, and each cage's HFG-water molecule cage attractive field tends toward being balanced (Fig A3, 1b). Because there is a nearly three-dimensional gridwork of HFG molecules in their lattices, there is no diffusion stress with adjacent and nearby HFG.

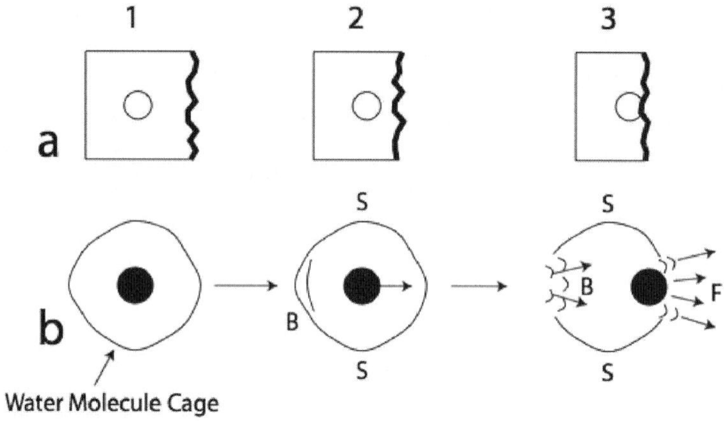

Figure A3. Schematic diagram of hydrate hydrogen bond strain and breakage owing to either dissociation or dissolution. a. Position of methane filled void or cell with respect to surface on which dissociation is taking place. b. Physical relation between water molecule cell wall or cage (water molecules not shown) and the relative position of an internal HFG molecule. 1. Cage surrounded by other methane filled cells. 2. Cage close to dissociation front. 3. Cage at dissociation front.

Where the dissociation zone approaches a cage (Fig A3, 2a), a perceptible disruption in the lowest energy state may develop owing to a tendency for the HFG to migrate toward the dissolution front. The incipient disruption increases with the nearness of the dissociation front and the diffusion gradient, as does the response of the HFG and the cage structure. The possibility exists that the HFG molecules in their cages develop an asymmetry with a vector toward the dissociation front (which on a molecular scale is highly irregular), to which they could migrate if they were in a gas or a liquid. The HFGs are physically restrained by the cell walls, but each cage becomes less stable as the dissociation front approaches. The increased dislocation of the HFGs from their lowest energy position puts stress on both the 'front wall' (F), which is the direction in which the HFG seeks to migrate, and a 'back wall' (B), which is the opposite direction (Fig A3, 2b). Sidewalls (S) are variably affected.

As the cage becomes the dissociation front, the potential energy of the encaged HFG increases to a maximum relative to dissolved HFG or free gas. As the HFG is escaping the structure, the water molecules are also experiencing similar potential energy maxima. The water molecules lose some of their cohesion and bonding/orientation and break away from the surface (Fig A3, 2a,

2b). The HFG then continues its diffusion mixing into the liquid or joining the gas phase, and naturally moves down the diffusion gradient, away from the hydrate that is supplying HFG to the undersaturated water. A high degree of coordination amongst water molecule may persist as they move from their fixed positions in the cage structure. These fragments may be metastable in the solution for some period of time.

A7. CONCENTRATION

Preferred descriptors of standard concentration units (Table A2) in this book are:

Unit	Description
Molarity (M)	Mol of a solute per liter of *solution*. Often reported as mM, or millimol (1/1000 of M). Dependent on volume.
Molality (m)	Mol of a solute per kg of *solvent*. This is an absolute concentration that is independent of volume.
Mol fraction (c)	Mol of a solute divided by the total of mols of all compounds in solution.
ppm (by mass, volume, or mol)	Mass of a solute divided by the total mass, volume, or mols of a solution. Similar to percent by as part of one million instead of one hundred. Volume function.
ppb (by mass volume, or mol)	Like ppm, except part of one billion. Volume function.
Weight percent (% Kg/Kg)	The mass of a solute divided by the total mass of a solution expressed as %.
Volume percent (% V/V)	The volume of a solute divided by the volume of the solution expressed as %. Also known as the volume fraction.

Table A2. Terms of concentration of materials as dissolved in a solute.

A8. CHEMICAL EQUATIONS

Chemical equations describe the chemical state, and physical and energetic transformations associated with a process. For example, the formation of methane clathrate hydrate can be easily written as:

$$CH_{4(l,g)} + xH_2O_{(s,l,g)} \xrightarrow{P,T,\chi} CH_4(H_2O)_{x(s)} \quad x \geq 5.75$$

A brief description of the chemical process would be one equivalent of methane in liquid or gas form will react with 5.75 or more equivalents of water in solid, liquid, or gas form at a set of appropriate pressures, temperatures, and concentrations to irreversibly form one equivalent of non-stoichiometric solid methane clathrate hydrate.

There are three special parts of every chemical equation (A3): reactants, the process (arrow), and product. Equations can also have different information such as free energy, enthalpy, or stoichiometric variables (as in the case above) associated with them. This extra information is generally written to the right of the equation. All chemical equations balance: all material and energy is accounted for on both sides of the equation. Some equations look like chemical equations but are not. They only show the essential or interesting portion of a reaction such as:

$$CH_4(H_2O)_{5.75(s)} \xrightarrow{\Delta} CH_{4(g)}$$ Not Balanced. • means added Heat.

Chemical equations have a formalism associated with them. To the left hand side of the arrow are the reactants (Fig. A4). The reactants are listed in their stoichiometric quantities. All of the atoms in the reactants must appear on the right hand side of the arrow as products. Next comes the process descriptor, the arrow. A single headed arrow indicates an irreversible process, such as forming hydrate deep within the stability field. Arrows pointing in opposite directions represent an equilibrium process (see Fig 2.19 for equilibrium reaction). Additional reaction requirements are usually written above the arrow, but may not be explicitly defined (such as *P* for pressure or D for applied heat).

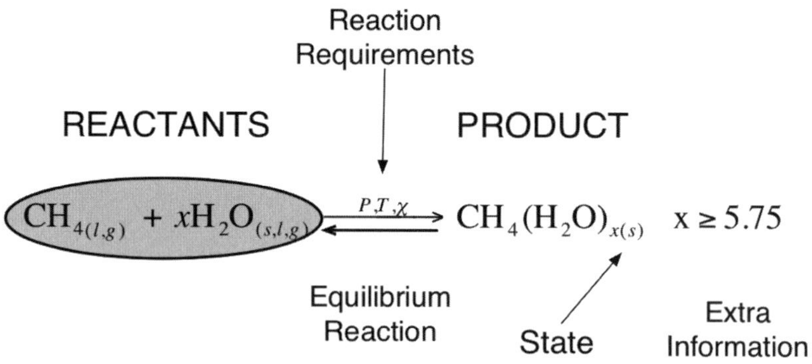

Figure A4. Schematic of a chemical equation.

Acknowledgement: Some of the theoretical and practical laboratory work relating to understanding nucleation and growth of natural gas hydrate took place at Marine Desalination Systems, L.L.C. while funded by the Defense Advanced Research Projects Agency (DARPA Contract NBCHC010003) and the Office of Naval Research (ONR Contract N00014-04-C-0237. Details of the work and experiments are contained in internal and formal reports.

Chapter 3

Oceanic Gas Hydrate Character, Distribution, and Potential for Concentration

3.1. THE CHARACTER OF OCEANIC GAS HYDRATE

Hydrates of many gases are stable in deep ocean conditions, but methane hydrate is by far the dominant type, making up > 99% of gas hydrate in the ocean floor. The methane appears to be almost entirely derived from microbial methanogenesis, predominantly through the process of carbon dioxide reduction. In some areas, minor amounts of gas hydrate are also created by other clathrate-forming gases such as hydrogen sulfide and carbon dioxide. Furthermore, thermogenic gas hydrates that include higher carbon-number hydrocarbons also exist. The thermogenic hydrocarbon gases are formed at great depths in the sediments, rise along faults and other pathways, and accumulate as gas hydrate deposits at or just below the seafloor (4.2).

3.2. WHERE GAS HYDRATE IS FOUND

3.2.1 Where is Gas Hydrate Stable?

Gas hydrate spontaneously forms wherever appropriate physical conditions exist - moderately low temperature and moderately high pressure (Fig. 2.1) - and the materials (water and HFG) are present in appropriate amounts. These conditions can be found in the deep sea, commonly at water depths greater than about 500 m or somewhat shallower depths (about 300 m) in the Arctic's colder water. Gas hydrate also occurs beneath permafrost on land in Polar conditions, but by far most natural gas hydrate is stored in ocean floor deposits. A simplified phase diagram for an ocean setting (Fig. 3.1), in which pressure has been converted to water depth in the ocean shows the pressure and temperature conditions that control hydrate stability in the marine environment. The heavy line is the phase boundary, separating regions in the temperature/pressure field where methane hydrate is stable to the left of the curve from regions where it is not, and thus, where gas + water would be stable. In figure 3.1, some typical conditions of pressure and temperature in the deep ocean were chosen to define the region where methane hydrate is stable. The dashed line shows how temperature conditions typically vary with depth in the deep ocean and underlying sediments. In this case, typical western North Atlantic Ocean thermal conditions were chosen, and a seafloor at 2 km water depth was selected. As indicated by the dashed temperature curve, near the ocean surface, temperatures are too warm and pressures too low for methane hydrate to be stable. As depth becomes

Michael D. Max et al. (eds.), Economic Geology of Natural Gas Hydrate, 105–130.

greater, temperature decreases and an inflection in the temperature curve is reached, known as the main thermocline, which separates warm surface water from deeper cold waters. At about 500 m depth, the temperature and phase boundary curves cross; from there downward, temperatures are cold enough and pressures high enough for methane hydrate to be stable in the ocean and thus, this represents the depth of the Top of Gas Hydrate Stability (TGHS). This intersection would occur at a shallower depth in colder, Polar waters.

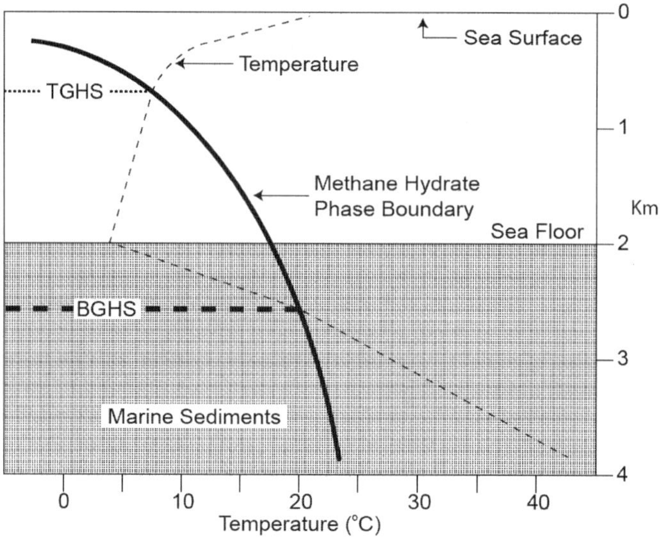

Figure 3.1. An example of physical conditions that control presence of gas hydrate in an oceanic setting is shown in this temperature versus depth plot. We assume a water depth of 2 km. The standard methane hydrate phase boundary (solid line) is plotted by converting depth to equivalent pressure. Gas hydrate will be stable to the left of this phase boundary curve (where temperatures are colder and pressures higher than the phase limit), whereas free gas and water will be stable to the right of the curve. A temperature/depth curve (dashed) typical of the western North Atlantic is plotted in the water column, which is connected to a typical geothermal gradient below the seafloor, indicating temperatures in the sediments. The intersection of the phase boundary and temperature curve in the water column defines the top of gas hydrate stability (TGHS) and the intersection below the seafloor, defines the depth of the base of gas hydrate stability (BGHS) at this idealized location. Courtesy of MDS.

Between the TGHS and the BGHS, if methane is sufficiently concentrated (near saturation), gas hydrate will form. Like ice, crystalline methane hydrate is less dense than water, so if hydrate forms in the water, it floats upward and breaks down (dissociates) at lower pressures and warmer temperatures, and thus it will not persist in oceanic waters. However, if the gas hydrate forms within sediments, it is bound in place. The lowest temperature occurs at the seafloor, and temperature increases downward through the sediments along the geothermal gradient toward the hot center of the Earth (Fig. 3.1). At the point where the curve of temperature conditions in the sediments (dashed line) crosses the phase boundary, conditions pass out of the region where gas hydrate can exist, and thus this depth represents the base of gas hydrate stability (BGHS). Thus, gas hydrate can persist in a zone (the gas hydrate stability zone - GHSZ) that extends from the seafloor down to a depth where the temperature has risen sufficiently to make hydrate unstable (even though the pressure increase that comes with greater depth will increase gas hydrate stability). The base of the GHSZ can extend down to a thousand meters below the seafloor at great water depths.

The precise location of the base of the gas hydrate stability zone under known pressure/temperature conditions varies somewhat depending on several factors, most important of which is gas chemistry. Gas hydrate in oceanic sediments is mostly (>99%) formed of biogenic (microbial) methane, but significant amounts of thermogenic gas hydrates that include higher carbon-number hydrocarbons exist in active petroleum areas (e.g. Gulf of Mexico, Caspian Sea). These locations, which frequently have active drilling and existing gas-handling infrastructure, must be considered important for issues of drilling safety and possibly gas resources. For example, in the Gulf of Mexico, at a pressure equivalent to 2.5 km water depth, the base of the gas hydrate stability zone will occur at about 21 °C for pure methane, but at 23 °C for a typical mixture of approximately 93% methane, 4% ethane, 1% propane, and some smaller amounts of higher molecular weight hydrocarbons. At the same pressure (2.5 km water depth) but for a possible mixture of about 62% methane, 9% ethane, 23% propane, plus some higher hydrocarbons, the phase limit will be at 28 °C. These differences will cause major shifts in depth to the base of the gas hydrate stability zone. Such mixtures of gases essentially make the formation of gas hydrate easier, and so can result in the formation of gas hydrate near the seafloor at shallower water depths (lower pressures) than for methane hydrate at equal temperatures. On the other hand, because by far most gas hydrate in the world is biogenic (biogenic gas is even very significant in the Gulf of Mexico), the analysis of biogenic methane hydrate accumulations is probably more important for both resource and drilling safety considerations. Below the base of the gas hydrate stability zone, gas plus water will be stable and methane hydrate will not be found, although hydrate formed from higher density hydrocarbon gases may occur.

The geothermal gradient tends to be quite uniform across broad regions where sediments and seafloor depth does not vary. For a given water depth, the sub-bottom depth to the base of the gas hydrate stability zone will be quite constant. However, as water depth increases (at constant seafloor temperature), the base of the gas hydrate stability zone will progressively extend farther below the seafloor (Fig. 3.2).

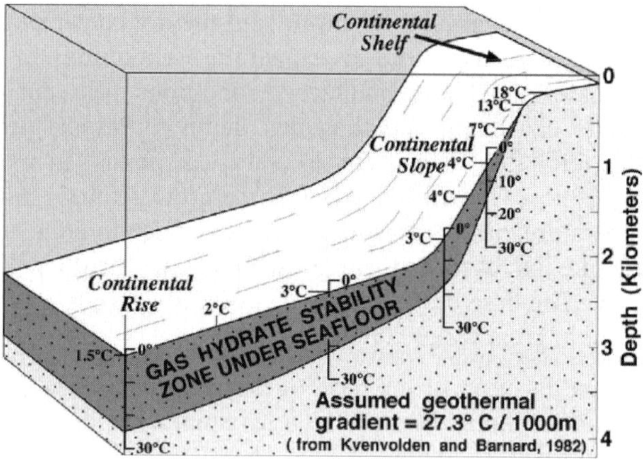

Figure 3.2. Conceptual drawing of gas hydrate stability zone on a passive continental margin.

3.2.2. Where Do We Find Gas Hydrate in Nature?

Methane hydrate occurs predominantly around the edges of the continents and in marginal marine basins, like the Mediterranean and Black Sea, and in some permafrost regions, as shown in the map of documented gas hydrate locations (Fig. 3.3; Kvenvolden and Lorenson, 2001; Kvenvolden and Rogers, 2005). This map was created in 2000, but still shows the basic pattern of worldwide gas hydrate distribution. The most notable find since the map's creation is the extensive discoveries off Oregon (Trehu, et al., 2004). Biogenic methane accumulates in these continental margin and basin sedimentary settings for two reasons: (1) the margins of the oceans and small ocean basins are where the flux of organic carbon to the seafloor is greatest because oceanic biological productivity is highest there, and (2) this is where sedimentation rates are highest. Rapid sediment accumulation serves to cover and seal organic material before it oxidizes, allowing microorganisms in the sediments to use it as food, forming the methane that becomes incorporated into gas hydrate. Thermogenic gas hydrate is also dominantly found in continental margin settings because thick sedimentary sections beneath a zone where hydrate would be stable, normally only are found on continental margins. Thick sections are required to produce high sub-bottom temperatures that are needed to generate thermogenic gas. In addition to sites in the marine environment, the map (Fig. 3.3) also shows some sites where gas hydrate has been reported associated with permafrost in the Arctic, the one location where it has been found in fresh water in Lake Baikal, Siberia and in intermediate salinity (between fresh and oceanic salinities) in the Caspian Sea. The tendency for gas hydrate to accumulate at continental margins means that much of it falls within the Exclusive Economic Zones of coastal nations, many of which are not presently energy producers, thus if gas hydrate proves to be a viable energy resource, then its presence in the EEZ of many nations will change global energy distribution and its economics.

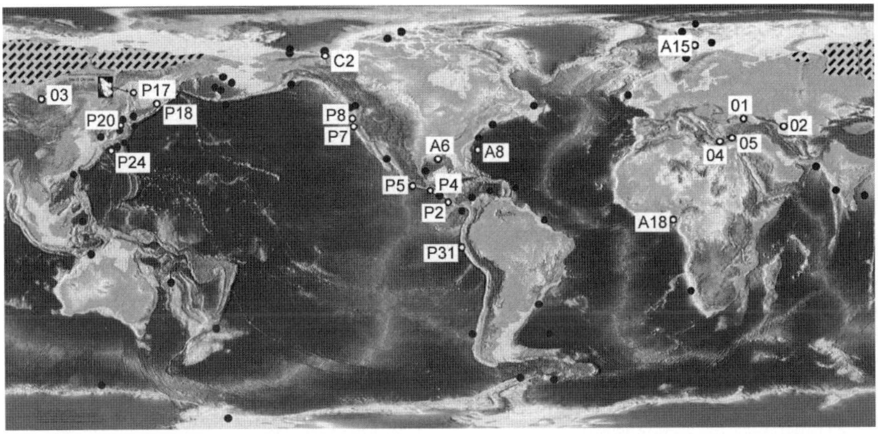

Figure 3.3. Worldwide known occurrences of gas hydrate by Kvenvolden and Rogers, (2005). Map originally compiled in 2000 for Kvenvolden and Lorenson (2001). Open circles mark sites of recovered gas hydrate samples. Solid circles represent sites of inferred gas hydrate presence on the basis of BSRs and well logs. Areas of possible gas hydrate occurrence in Russia are dashes. Characteristics of some locations (Table 3.1) are:

CODE	LOCATION	DESCRIPTION
P2	**Middle America Trench (Costa Rica)**	
	DSDP Leg 84, Site 565	Inclusions in mud & muddy sand
	ODP Leg 170, Site 1041	Disseminated and sheets
P4	**Middle America Trench (Guatemala)**	
	DSDP Leg 67, Site 497	Inclusion in sediment
	Site 498	Cement in coarse vitric sand
	DSDP Leg 84, Site 568	Inclusion in mudstone
	Site 570	Laminated ash; massive core
P5	**Middle America Trench (Mexico)**	
	DSDP Leg 66, Site 490	Laminated ash and mud
	Site 491	Inclusions in mud
	Site 492	Laminated ash
P7	**Eel River Basin (CA, USA)**	Layers, nodules in mud
P8	**Cascadia Basin (Oregon)**	
	DSDP Leg 146, Site 892	Aggregates, layers in silt
	Hydrate Ridge	Layers, massive in carbonate crust
P17	**Okhotsk Sea (Russia)**	
	Paramushir Island	Layers in ooze
P18	**Okhotsk Sea (Russia)**	
	Sahkalin Island	Layers in silt and clay
P20	**Japan Sea (Japan)**	
	ODP Leg 127, Site 796	Crystals in sand with clay
P24	**Nankai Trough (Japan)**	
	ODP Leg 131, Site 808	Fragment in wash core
P31	**Peru-Chile Trench (Peru)**	
	ODP Leg 112, Site 685	Fragments in mud
	Site 688	Grains in mud

A6	**Gulf of Mexico (Texas and Louisiana)**	
	DSDP Leg 96, Site 618	Nodules, crystals in mud
	Green Canyon	Nodules, layers in rubble
	Garden Banks	Nodules, layers in rubble
	Mississippi Canyon	Pieces in coarse sediment
	Bush Hill	Mounds at seafloor
A8	**Blake Ridge (Southeastern USA)**	
	DSDP Leg 76, Site 533	Fragment in mud
	ODP Leg 164, Site 994	Fragments in clay
	Site 996	Nodules, veins in mud
	Site 997	Massive core (~30 cm)
A15	Haakon-Mosby Mud Volcano (NO)	Inclusions and plates
A18	Niger Delta (Nigeria)	Nodules, dispersed in clay
O1	Black Sea (Russia)	Veinlets in silty clay
O2	Caspian Sea (Russia)	Laminated in clayey silt
O3	Lake Baikal (Russia)	Disseminated in sand, silt
Mediterranean Sea		
O4	Amsterdam Mud Volcano	Not described
O5	Kula Mud Volcano	Not described
C2	Mackenzie Delta (Canada)	Dispersed in sand, gravel

Table 3.1. Key to Figure 3.3 hydrate localities.

3.3. IDENTIFICATION OF GAS HYDRATE IN NATURE

Two basic issues that concern us about gas hydrate are: 1. where do concentrated accumulations exist and 2. how much is present. These questions have turned out to be very difficult to answer in a precise manner because gas hydrate persists beneath the ocean floor only in conditions of fairly low temperatures and high pressure. When gas hydrate is transported from the sea bottom to normal Earth-surface conditions, it begins to dissociate and will not survive for very long. Thus drilled or drop-cored samples cannot be depended upon to provide accurate estimates of the *in-situ* amount of gas hydrate present, its gas saturation, etc., as would be the case with most minerals. Even the heat and changes in chemistry (methane saturation, salinity, etc.) introduced by the drilling process, including the effect of circulating drilling fluids, affect the gas hydrate, independent of the changes brought about by moving a sample to the surface. Gas hydrate has been identified in nature generally from sampling and logging of drilled wells or by using remotely sensed indications from seismic reflection profiles.

3.3.1. Measuring Gas Hydrate in Wells and Cores

Drilled samples of gas hydrate often survive the trip to the surface from hydrate accumulations below the seafloor, despite the transfer out of the stability field, just because the dissociation of gas hydrate is fairly slow. Such samples have been preserved, at least temporarily, by placing them back into the pressure/temperature field of gas hydrate stability either by returning them to the pressure/temperature conditions where they formed, or, more commonly, by keeping them at surface pressure, but at the ultra-cold temperature of liquid nitrogen. Such preserved samples are valuable for many studies, but certainly they do not provide quantitatively accurate indications of the amount or distribution of gas hydrate, or its relation to sediment fabric that existed in the undisturbed seafloor sediments.

Several indirect approaches have been used to gain indication of the amount and/or distribution of gas hydrate in sediments. Some are as obvious as making temperature measurements along a core to indicate where gas hydrate existed, because the dissociation of gas hydrate, being an endothermic reaction, will leave cold spots.

The dissociation of gas hydrate leaves another indirect marker of its former existence, because, when gas hydrate forms, it extracts pure water to form the clathrate structure, excluding all salts as brine. Therefore, when hydrate dissociates in a core, the interstitial water becomes much fresher, and the amount of gas hydrate present before dissociation can be calculated. An example is shown in figure 3.4 from Ocean Drilling Program hole 997 off the South Carolina

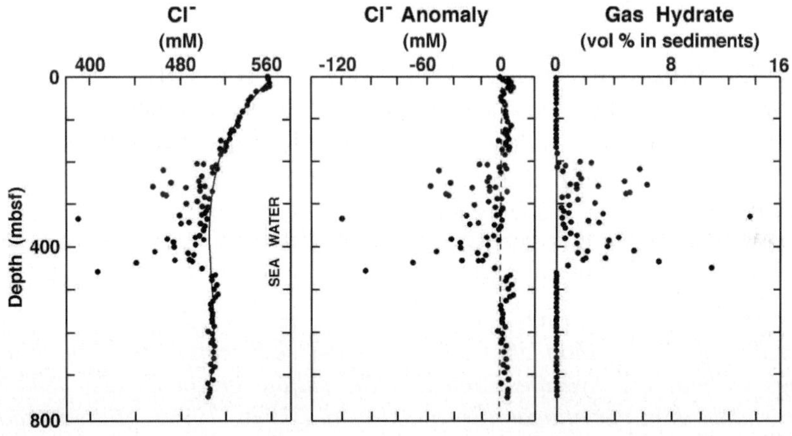

Figure 3.4. Chlorinity, chlorinity anomaly and calculated concentration of gas hydrate from a drill site on the South Carolina continental margin (Paull et al., 1996).

coast (Paull et al., 1996). Measured chloride content is shown in the left panel. The values near zero depth (depth at the seafloor) represent seawater chloride concentration (chlorinity). It is assumed that a smooth curve of chlorinity versus depth, following the main trend of data points (solid curve), represents the undisturbed (pre-drilling) chlorinity and that spikes of low chlorinity to the left of the curve represent the result of hydrate dissociation. The base of hydrate stability here is at 450 m below the seafloor (mbsf), and the top of significant gas hydrate concentrations is apparently at about 200 mbsf. The estimation of the base curve using this method is controversial. The second panel shows the base curve straightened out (dashed line) so that the plot is considered to represent the chlorinity anomaly that results from dissociation of previously existing hydrate. The third panel shows the calculated results in terms of proportion of sediment that had been occupied by gas hydrate.

Another approach that is important for identifying and quantifying gas hydrate in wells uses down-hole logging methods in which sensors are pulled through the hole, and measurements are made to assess sediment porosity and gas-hydrate saturation. Collett (2001, p. 193) notes:

"Since gas hydrates are characterized by unique chemical compositions and distinct electrical resistivity and acoustic physical properties, it is possible to obtain gas hydrate reservoir porosity and hydrocarbon saturation data by characterizing the electrical resistivity and acoustic physical properties of gas hydrates and assessing the chemical composition of the pore-filling constituents within a gas-hydrate-bearing reservoir. The well-logging devices that show the greatest promise of yielding gas hydrate reservoir data, including porosity, are the density and neutron porosity logs --, which primarily respond to the chemical composition of the pore-filling constituents. The electrical resistivity, acoustic transit-time, and neutron spectroscopy well logs -- can yield highly accurate gas-hydrate saturation information."

Piston core sampling of gas hydrate has produced a considerable body of data, especially regarding organic geochemistry, in the Gulf of Mexico, where much gas hydrate exists close to the seafloor (Milkov and Sassen, 2000; Sassen, et al., 2001). Coring in sediments can also provide data on sulfate gradients, which can be analyzed to calculate methane flux and thus predict gas hydrate development at depth. Walter Borowski (Eastern Kentucky University. pers. com., 2005) notes:

"In deep-sea, continental margin settings associated with gas hydrates, we observe that sulfate (SO_4^{2-}) in pore waters becomes depleted more rapidly (generally shallower than 50 meters into the sediments), as compared with settings without gas hydrates, and that the sulfate gradients also tend to be linear rather than curvilinear (Borowski et al., 1996, 1999). Sulfate gradients can be used to recognize localities prone to gas hydrate occurrence, because of linkage between the geochemical cycles of sulfur and carbon through the biogeochemical process of anaerobic methane oxidation (Martens and Berner, 1974; Barnes and Goldberg, 1976; Reeburgh, 1976; Reeburgh, 1982). Sulfate ions are present in seawater almost everywhere, and sulfate concentrations extend downward into the interstitial water of shallow sub-bottom sediments. At greater depths lies a zone where microbial methane production takes place and where high methane concentrations exist in equilibrium with gas hydrate deeper in the sediment column. At the contact between these biogeochemical depth zones, both sulfate and methane occur. At this boundary, known as the sulfate-methane interface (SMI), methane and sulfate are co-consumed:

$$CH_4 + SO_4^{2-} \text{ -----> } HCO_3^- + HS^- + H_2O, \quad \text{(Eq. 3.1)}$$

by a consortium consisting of methane-consuming Archaea and sulfate-reducing bacteria (Hoehler et al., 1994; Valentine and Reeburgh, 2000; Boetius, 2001; Orphan et al., 2001). Thus above the SMI, sulfate must increase upwards to oceanic concentration at the seafloor, and below it, methane must increase downward to concentrations near saturation, which will allow gas hydrate formation at depth. The consumption of methane by anaerobic methane oxidation is balanced by the diffusion of sulfate down to the SMI, creating linear sulfate profiles. The greater the rate of methane delivery, the greater the required rate of sulfate delivery and the steeper must be the sulfate gradient. Thus, high methane fluxes cause the SMI to move upward in the sediment column, allowing sulfate to balance methane consumption (Fig. 3.5) (Borowski et al., 1996). By measuring sulfate concentrations at various depths in cores, the depth of the SMI can be projected, and the methane flux inferred. Sulfate gradients have been shown to vary 16-fold over the Blake Ridge region (offshore southeastern United States) because of differing methane fluxes (Fig. 3.6) (Borowski et al., 1996, Borowski, 2004). Although shallow sulfate gradients may be a product of geochemical processes not associated with gas hydrate occurrences, sulfate gradients have the potential to be used to locate gas hydrate deposits that are not recognized by seismic surveys because of the lack of specific conditions that create bottom simulating reflections and other characteristic seismic signatures."

These seismic profiling approaches to identifying gas hydrate are probably the single most important exploration tool for first-order identification of gas hydrate (6.4.1).

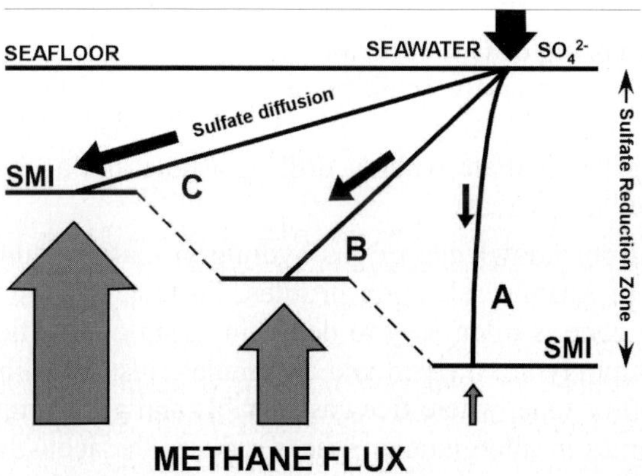

Figure 3.5. A. Schematic diagram shows how upward methane (CH_4) flux controls sulfate (SO_4^{2-}) profiles and the depth to the sulfate-methane interface (SMI). Arrow size is proportional to methane delivery. Typical sulfate profiles display convex-up curvature (A) reflecting oxidation of sedimentary organic matter through the biogeochemical process of sulfate reduction. Linear sulfate profiles (B and C) result when sulfate consumption is focused at the SMI by anaerobic methane oxidation (AMO). In order to balance the stoichiometry of the reaction, increasing amounts of methane cause the depth of the SMI to rise within the sediments. Thus, the rate of sulfate consumption and the steepness

of the sulfate gradients are dynamically controlled by flux of methane from below.

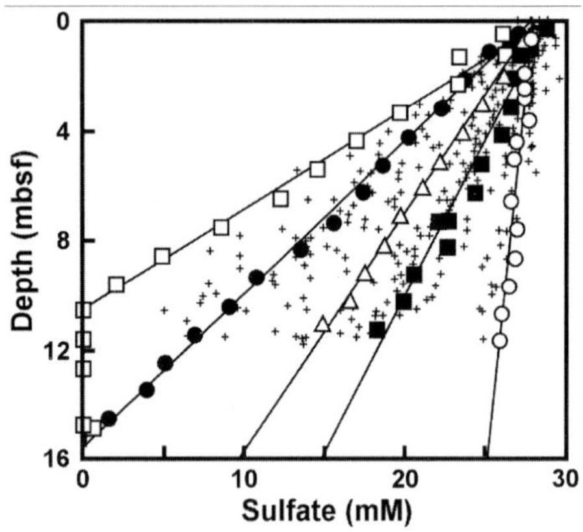

Figure 3.6. Pore water sulfate (SO_4^{2-}) concentrations versus depth in 27 piston cores taken in the Blake Ridge region (offshore southeastern United States). Concentration is in mM; measurement uncertainties are generally less than symbol size. Five selected cores are shown above with different symbols, whereas sulfate data from other cores are shown by crosses (+). Sulfate profiles are linear, and sulfate concentration gradients vary by a factor of 16. The data from other cores show that this range in sulfate gradients varies smoothly across the region. Sedimentation rate and organic matter delivery vary little across the region so that differences in sulfate depletion are most likely due to changes in methane delivery, affected by the presence of gas hydrates in the subsurface.

3.3.2. Remote Sensing of Gas Hydrate

Beyond the issue of precisely quantifying gas hydrate in the few spots on the seafloor where cores have been recovered, we would like to have a technique of remotely sensing gas hydrate without drilling a hole and a method that would allow us to survey large areas.

Much of our knowledge of gas hydrate in marine sediments has come from the study of seismic reflection profiles. Fortunately, the base of the gas hydrate stability zone is often easy to detect in seismic reflection profiles. Free gas bubbles commonly are trapped and accumulate just beneath the base of the gas hydrate stability zone, where free gas is stable and gas hydrate will not exist. Presence of bubbles in intergranular spaces reduces the acoustic velocity of the sediment markedly, even at low concentrations of gas bubbles. Conversely, in the gas hydrate stability zone the velocity is increased slightly by the presence of gas hydrate, which in the pure state has twice the velocity of typical deep-sea sediments and, furthermore, bubbles generally cannot be present because any free gas would convert to gas hydrate as long as water is present. The large velocity contrast that is produced by the contact of gassy deposits against non-gassy sediments generates a strong echo when an acoustic pulse impinges on it.

Thus, we can image the base of the gas hydrate stability zone in a seismic reflection profile. The base of gas hydrate stability, as revealed by this reflection, generally occurs at an approximately uniform sub-bottom depth throughout a restricted area because it is controlled by the temperature, as long as water depth, gas chemistry, and interstitial water salinity do not vary much, as is usually the case over a small area. In most places, thermal gradients across an area tend to be consistent, so isothermal surfaces have consistent depth below the sea bottom. Hence the reflection from the base of the gas hydrate stability zone roughly parallels the seafloor in seismic profiles and has become known as the "Bottom Simulating Reflection" (BSR) (Fig. 3.7). The BSR is easiest to see when stratal reflections are not parallel to the seafloor, and thus the BSR cuts diagonally through the reflections from strata. Note that the BSR may be a continuous reflection event in a seismic profile as shown at the center of the profile (Fig. 3.7), or it may be discontinuous, or even appear as a series of terminations of strongly reflective (gas charged) strata at a consistent depth.

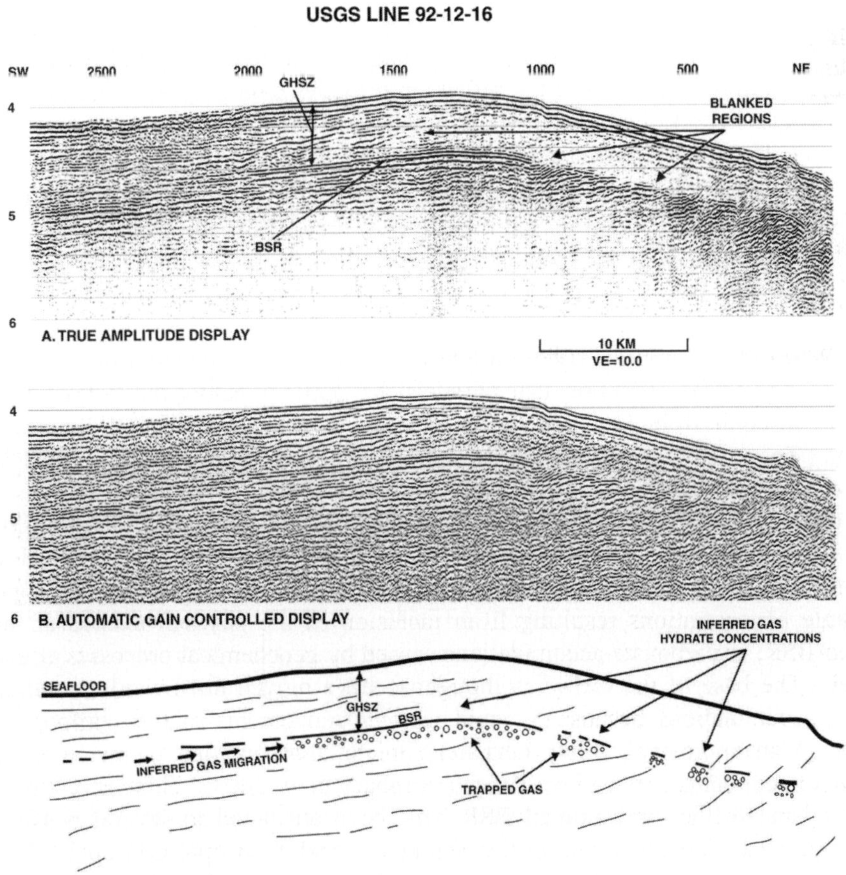

Figure 3.7. Seismic reflection profile across the Blake Ridge. The Ridge is a sedimentary drift that is migrating to the southwest. Also see figures 6.1 & 6.2.

Because the BSR occurs at a contact between higher acoustic impedance above and lower acoustic impedance below (a negative acoustic impedance contrast), it produces a phase reversal in a reflected acoustic wave compared to the reflection from the seafloor (6.4.1; Fig. 6.1).

The BSR is usually a good sign that gas exists trapped beneath the base of the gas hydrate stability zone, and strongly implies that gas hydrate is present, because free gas, which has a tendency to rise, exists just below and in contact with the zone where gas would be converted to gas hydrate. The coincidence in depth of the BSR to the theoretical, extrapolated pressure/temperature conditions that define the base of hydrate stability boundary and the sampling of hydrate above BSRs and gas below give confidence that this seismic indication of the base of gas hydrates indicates the presence of gas hydrate. However, gas hydrate can exist without generating well-defined extensive BSRs, especially where gas flow is focused into narrow zones (faults) as in much of the Gulf of Mexico, where BSRs are fairly rare, despite the large amounts of gas and gas hydrate. Furthermore, estimation of the volume of gas hydrate from BSR data is virtually impossible, because the reflection provides data only for conditions at the gas/gas hydrate boundary, where small variations in free gas content can strongly affect the reflection characteristics of the BSR.

The concept that BSRs result from a velocity contrast produced by gas, probably methane, trapped beneath the base of the GHSZ is probably correct in 99% of the places that BSRs are seen, and such BSRs might be termed "conventional" BSRs. However, there have been reports of multiple BSRs from many places in the world - the Blake Ridge, off Japan, off Norway, and the Black Sea, for example. By "multiple BSR" is meant a second (sometimes a second and third) reflection that parallels the seafloor below the bottom at a depth different from the methane BGHS and is independent of sedimentary strata. These nonconventional BSRs have been observed both above and below the conventional one and clearly must be produced by processes other than those for conventional BSRs. BSRs above the conventional BSR were identified on the Blake Ridge and off Japan, where they are considered to have formed at former levels of conventional BSRs. They may represent exceptionally high gas hydrate concentrations resulting from inclusion of the gas that existed at the paleo-BSR, or carbonate accumulations caused by geochemical processes at that level. The base of the GHSZ is thought to have moved downward, stranding these accumulations because of rapid tectonic movements of the seafloor off Japan (Matsumoto et al. 2004; Baba and Yamato 2004) and due to erosion of the seafloor sediments at the Blake Ridge (Hornbach et al. 2003). Nonconventional BSRs beneath the conventional BSR have been attributed to several possible mineralogical transitions within the sediments, opal A to opal CT, opal CT to quartz, or smectite to illite (dewatering) (Hein et al., 1978; Posewang and Mienert, 1999; Berndt et al., 2004). Suggestions also have been made that deeper nonconventional BSRs might be phenomena similar to the conventional BSR, but formed by petroleum gases other than methane, which would have phase boundaries that would be stable at higher temperatures, but it is unclear how the various gases would be fractionated. As of this time, the causes of multiple BSRs clearly are not settled.

A second significant seismic characteristic of gas hydrate-cementation is called "amplitude blanking" (Fig. 3.7). Blanking is the reduction of the amplitude (weakening) of seismic reflections, which appears to be caused by the presence of gas hydrate. Many observations of blanking in nature have been associated with gas hydrate accumulations. The explanation of blanking depends on the fact that the strata in the sediments have varying porosities. The variations in porosity constitute the physical changes that cause seismic reflections because variations in porosity create variations in velocity. Reflections result primarily from changes in acoustic velocity between strata and to a lesser extent to density variations (the product of velocity times density is called acoustic impedance, and contrast in acoustic impedance is proportional to change in reflection strength). The more porous layers are of lower velocity. However, gas hydrate has a tendency to form preferentially in the more porous layers, and as hydrate begins to form in those layers it will increase their velocity, reducing the impedance change between adjacent layers and reducing the strength of reflections (blanking, 6.4.1.1). Of course, if gas hydrate continues to form in the more porous layers, it can cause their velocity to increase to values much higher than adjacent layers (accentuation, 6.4.1.2), so that the gas-hydrate-rich layers could become **more** reflective than adjacent layers and create a reverse of the blanking affect (Lee and Dillon, 2001). This phenomenon (is only rarely observed, however, perhaps because the presence of gas hydrate commonly slows down the flow of more gas into the layer. Holbrook et al. (2002), in an excellent summary of seismic detection of hydrates, note that: "*Carefully calibrated amplitude blanking can be useful as an indicator of possible hydrate accumulations, but quantitative estimates of hydrate concentration are very difficult to obtain solely from reflectance.*" Use of blanking combined with other seismic indicators has been shown to be effective in identifying the presence of gas hydrate in seismic profiles (Hornbach et al., 2003).

Several other methods using seismic data have been employed to attempt to characterize gas hydrate deposits in the sediments, including vertical seismic profiling, in which a seismic source is fired near a well and the receiving hydrophone is moved through the well-bore, and so-called "walkaway" seismic profiling in which shot-receiver distance is increased during firing (Holbrook, et al. 1996; Korenaga et al., 1997; Holbrook, 2001; Pecher et al., 2003). Amplitude versus offset analyses and full waveform inversion have been attempted (e.g. Hyndman et al. 2001; Hato et al., 2004). Direct imaging of gas hydrate has been attempted by "seismic impedance inversion analysis" (Inamori and Hato, 2004) using well log data with seismic data. Use of deep-towed seismic reflection systems can provide greater resolution in gas hydrate studies (Gettrust, et al., 1988; Rowe, et al., 1993, Wood, et al., 2001).

In areas where gas flow through faults is strong, accumulations of gas hydrate in mounds at seeps on the seafloor are common. In the Gulf of Mexico, standard 3-D oil-exploration seismic data have been used to image such mounds by mapping variations in acoustic reflection at the seafloor. Areas of strong reflectivity have been shown to correlate well with seep-related features such as authigenic carbonates, thick accumulations of chemosynthetic organisms (tube worms, clams, mussels, etc), and hydrates. Thus, mapping of this parameter apparently can be used as a survey tool (Roberts, et al., 2000; Jesse L. Hunt, Jr. U.S. Minerals Management Service, personal communication, 2005).

The methods mentioned so far have generally been associated with P-wave data, but the presence of a gas hydrate cement within the sediment is likely to have even more effect on S-wave velocity than on P-wave velocity, so it is likely that multi-component ocean-bottom cables and the measurement of S-wave velocities may become an extremely important approach to delineating gas hydrate accumulations. Finally, an approach to remote detection that departs from seismic methods may be the use of electrical resistivity measurements in seafloor sediments because gas hydrate displays relatively high electrical resistivity.

3.4. CONCENTRATION OF GAS HYDRATE IN NATURE

Worldwide, the mass of methane in hydrate is commonly stated to amount to twice as much (on a carbon basis) as all the fossil fuels on Earth, or a volume of about 21×10^{15} m^3 (Kvenvolden, 1998, 2000). From a resource perspective, this is an interesting, but in a sense useless number (it sets an upper limit), because it is nearly certain that much of the natural methane hydrate is dispersed in the sediments, as are much of the oil, coal, and conventional gas. As with all natural resources, including petroleum, we will extract methane from gas hydrate at naturally formed concentrations, so the issues of how and where gas hydrate is concentrated are of primary significance.

3.4.1. Two Modes of Gas Hydrate Concentration

Gas hydrate in nature seems to be present in two different styles of occurrence. One style is represented by gas hydrate distributed through the sediments as pore fillings (especially in sands) or small nodules or veins (more common in muds). This is the sort of broadly distributed gas hydrate that is best identified in seismic profiles. A second style of gas hydrate occurrence is highly concentrated in narrowly defined deposit, often manifest as surface mounds. These deposits have been successfully studied using deep-diving research submersibles. Considerable amounts of gas hydrate can become concentrated in either style of occurrence. The two styles of occurrence lead to the concept that there are two basic means of concentrating gas hydrate, which might be described as two models, the diffuse gas-flow model and the focused gas-flow model, which can be roughly correlated with the result of the different hydrate-forming gas flows, high- and low grade hydrate deposits (4.6). These are end-member concepts, however, and probably no area can be completely characterized as one type or the other. The diffuse gas-flow model typically is dominant in many passive continental margin settings in which sedimentation is relatively slow across a broad continental shelf/slope/rise with little tectonic activity (Figs. 3.2; 4.4).

3.4.1.1. Diffuse Gas-flow Model

The great preponderance of biogenic methane in gas hydrate probably is generated at fairly shallow subbottom depths by bacteria dispersed in the sediment. Methanogenic bacteria require anoxic conditions, thus they exist in sediments only below levels where sulfate reduction takes place. The level of most active methanogenesis may be just meters or tens of meters below the seafloor, although some methanogenic bacteria exist deeper in the sediment. In contrast to the shallow depth of methane generation, seismic results and drill hole data show that gas hydrate concentration generally increases downward through the GHSZ to its greatest concentration near the base of the zone, several hundred meters below the seafloor. For example, note the gas hydrate distribution described by chlorinity (Fig. 3.4) and the increase in blanking downward toward the BSR (Fig. 3.7).

We also observe that the normal situation is to have free gas trapped beneath the base of the GHSZ. Low velocities caused by the trapped gas are the source of the velocity contrast that creates the BSR, as noted above. Evidence for the trapping of gas beneath the gas hydrate zone, in addition to the widespread presence of the BSR, includes directly measured low velocities and the logging and direct sampling of gas in drill holes. Formation of gas hydrate requires the presence of gas near saturation in the pore water, so the presence of gas bubbles (supersaturation) is favorable for gas hydrate development.

The presence of gas trapped at the base of the GHSZ requires a means of supplying gas to that level in the sediments. Furthermore, the high concentrations of gas hydrate just above the base of the GHSZ suggests that the gas trapped beneath is acting as a source for the formation of gas hydrate in this adjacent region and that there are mechanisms that serve to transfer gas from the sub-GHSZ reservoir upward into the gas hydrate zone to form hydrate there.

Normal geological processes, both at passive and active continental margins can account for the supply of gas to cause saturation of pore water beneath the base of the GHSZ. Passive continental margin settings often have ongoing sediment deposition. Consider a location at perhaps several tens of meters below the seafloor, within sediments where bacteria are actively generating methane. The seafloor above the chosen location is apparently moving upward relative to this fixed site in the sediments owing to sedimentary disposition. As the seafloor builds up, the geothermal gradient tends to remain constant, so the isothermal surfaces must rise with the accreting seafloor. A sediment grain or bit of gas hydrate in the shallow sediments effectively sees the base of the gas hydrate stability zone (GHSZ) migrate upward toward it and eventually pass it as the seafloor builds up. Eventually that bit of gas hydrate ends up sufficiently far below the seafloor that it is just beneath the base of the GHSZ and thus outside the range of gas hydrate stability. The gas hydrate then dissociates and releases its methane, which tends to rise through the sediments and accumulate at the base of the gas hydrate stability zone, where the presence of gas hydrate in the sediments above may provide a permeability barrier (a seal). Ultimately the gas may diffuse upward into the gas hydrate stability zone, where it forms more gas hydrate. As indicated diagrammatically in Figure 3.8, seafloor at time 1 is associated with a base of GHSZ for that time. At a later time 2, when the seafloor has moved upward by sediment deposition, the base of GHSZ also must have moved upward as the isotherms migrate up. The result will be the dissociation of the gas hydrate that had existed between the base of the GHSZ at time 1 and its base at time 2. A great deal of gas will be released from the gas hydrate and will tend to migrate upward through the sediments to the new base of GHSZ, where its concentration can generate new gas hydrate. We have considered this as a stepwise process, but, of course, it is essentially continuous, potentially resulting in a continual supply of gas to the base region of the GHSZ, enriching this region with gas hydrate and accounting for the commonly observed increase in gas hydrate concentration downward to the base of the GHSZ.

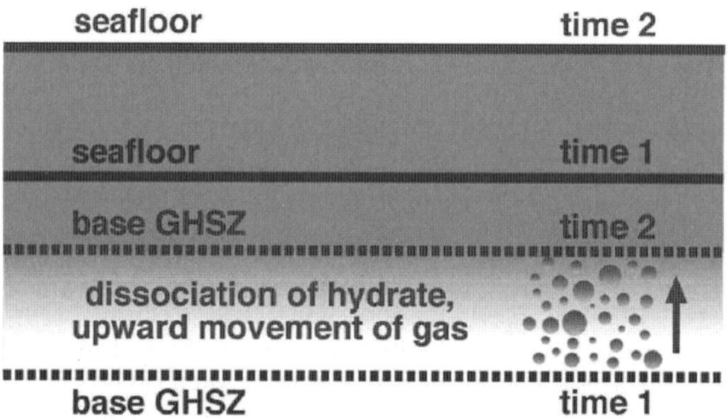

Figure 3.8. Conceptual diagram summarizing the effect of seafloor accretion on hydrate within the sediment, which produces release of gas at the base of the GHSZ.

Another possible mechanism has been proposed to provide free gas to the region directly below the GHSZ. This depends on the change in solubility of methane with depth (Paull et al. 1994). Methane has a solubility minimum at a depth just below the base of the gas hydrate zone (Fig. 3.9), so upward migration

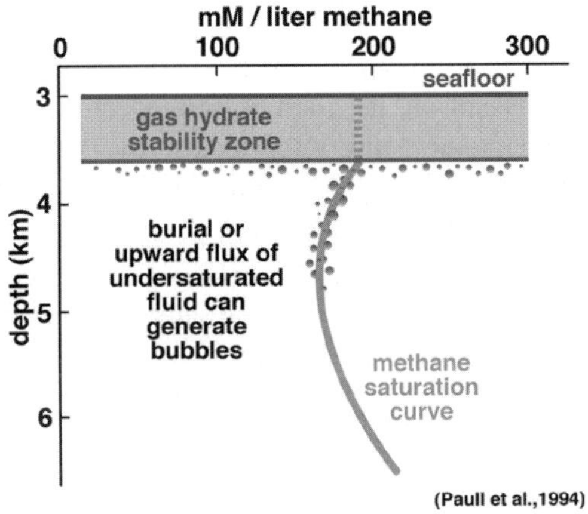

Figure 3.9. Possible release of free methane from upward-flowing waters as a consequence of the shape of the methane saturation curve.

of the zone during sedimentation or upward flow of interstitial fluids carrying gas will tend to release gas bubbles at the depth of this solubility minimum and provide the saturation concentration of gas needed to form hydrate.

Most of the easily digested organic matter in sediments is converted by methanogenic bacteria at shallow depths, leaving only refractory material, which initially might suggest that all methane is generated at fairly shallow subbottom depths. However, studies have shown that bacteria are present and active to subbottom depths of more than 800 m and that the source of their food is the breakdown of refractory material in the "oil window" where temperatures are in the range of about 100-150°C (Wellsbury and Parkes, 2000; Parkes et al., 2000). The concept is summarized in Figure 3.10. Thus, the source of some gas supply

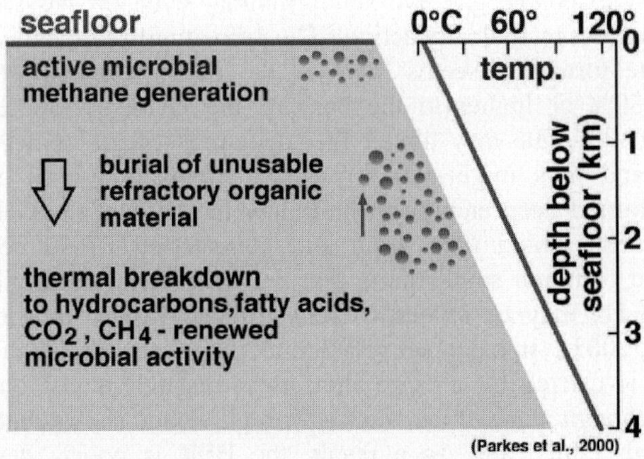

Figure 3.10. Conceptual diagram indicating that, although much methane is generated by bacteria at shallow subbottom depths, the thermal breakdown of refractory material at depth can also provide nutrients for bacteria that are active at greater depths. Adapted from Parkes et al. (2000).

to the base of the GHSZ may be both from the zone itself, and from below the zone and dependent on deep methanogenesis. In support of this concept, field studies from the Lima Basin off Peru strongly support the concept that deeply-sourced biogenic methane is being supplied to the base of the GHSZ and forming gas hydrate (Pecher et al. 2000; Pecher al. 2001).

Thus far we have made two assumptions, as implied in Fig. 3.8. One is that the base of the GHSZ is flat and chemical conditions are uniform so that gas hydrate dissociation or crystallization will be laterally uniform. The second is that the movement of gas through sediments is by diffusion. If the base of the GHSZ is not flat and conditions are not uniform, then conditions for lateral migration of gas may exist, which will result in lateral concentration of gas hydrate. We will consider structures and processes that can generate lateral gas hydrate concentrations below, but first we should question the validity of a purely diffuse gas-flow model.

3.4.1.2. Focused Gas-flow Model

Diffusion is a very slow molecular process, and flow of fluids carrying dissolved gas through sediments (advection) is also slow (2.5.1; 2.7). We have cited seismic and drilling evidence for the presence of free gas as bubbles in the sediment pores. Presence of bubbles can actually plug pores to hinder flow. Much gas may be transported through very porous strata (e.g. flow through sandy layers as identified in the Nankai Trough off Japan; Baba and Yamada, 2004). However, a great deal of gas-carrying fluid moves as channelized flow through fractures in the sediments (4.7). Faults are to be expected in areas having active tectonism and also in places where sediments are disturbed by salt or shale flow, slumps, and slides. However, faults appear to be common in many areas of extensive gas hydrate formation, even on passive continental margins that are not tectonically active. The porosity of sediments in some areas of gas hydrate formation seems high (in the Blake Ridge, porosity remains generally at 50% or higher to the base of the GHSZ). Speculatively, the presence of gas hydrate may hinder the movement of grains needed to allow compaction, and thus might partially deter compaction until a site in the sediments is buried deep enough to go below the base of the GHSZ. At that point dissociation would take place and compaction would occur, perhaps abruptly. The common small faults that are seen in locations like the Blake Ridge (Fig. 3.11) may be related to such effects (Rowe and Gettrust, 1993; Gorman et al., 2002). In many actively subducting plate margins, like the Lesser Antilles, gas is carried by a large fluid flow through faults, but this is not universal. Although work in the Nankai Trough wedge shows that much gas is carried through faults, and, as a result, the BSR is poorly developed near seafloor seeps (Morita et al. 2000), in other areas of the same sandy, clastic-dominated wedge, the BSR is not affected by faults, and gas flow is inferred to be through matrix permeability (Moore et al., 2000; Baba and Yamato, 2004).

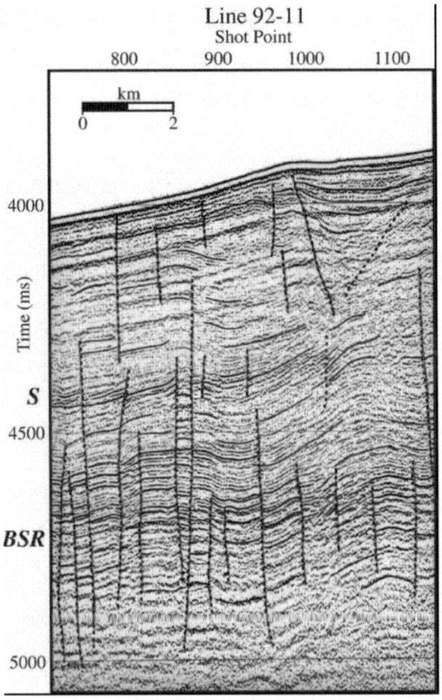

Figure 3.11. Seismic reflection profile off the southeastern United States with dashed interpretation of the pervasive minor faulting that may be related to gas hydrate processes. The base of the gas hydrate stability zone (GHSZ) is marked by an indistinct bottom simulating reflection (BSR). The BSR represents the top of free gas trapped beneath gas hydrate-bearing sediments. In this case, it is irregular and discontinuous because of the disruption of gas-bearing strata. Because the base of the GHSZ is a physico-chemical boundary that is controlled by temperature and pressure, it actually is not faulted, and the fault offsets are actually larger than the irregularities in the BSR. Also see 4.7.

3.4.2. Lateral Variations that Create Trapping of Gas and Gas Hydrate Concentration

In addition to vertically upward migration of gas driven by concentration differences, lateral concentration must also occur in order to create deposits of gas hydrate that might be economically valuable.

3.4.2.1. Structural Trapping

The simplest mechanism for causing a lateral motion of gas is a buoyancy drive. For example, let us imagine a system like Figure 3.8, but rather than modeling a flat seafloor, consider what would happen at a ridge or dome created by sediment accretion. A closure forms. Again, as the seafloor rises by accumulation of porous sediment from its level at time 1 to its level at time 2, the isotherms must rise and cause the base of the GHSZ to move upward and the hydrate in the deeper zone must break down.

But owing to the slope of the seafloor on the flanks of the ridge and resultant slope of the base of the GHSZ, which acts as a seal, the gas will tend to migrate laterally along the base of the GHSZ toward the crest of the hill. The result will be the concentration of gas at the shallowest part of the BSR and ultimately its slow upward migration to form a gas hydrate concentration above the gas accumulation (Fig. 3.12).

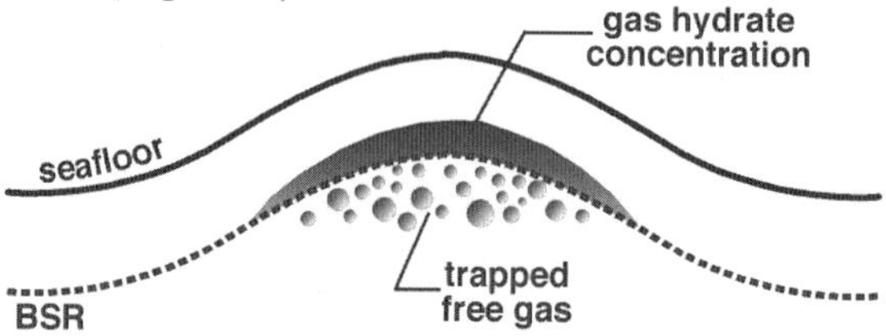

Figure 3.12. Idealized diagram of gas hydrate concentration (shaded area) at a seafloor flexure. As the base of the GHSZ, marked by the BSR, tends to follow an isotherm and the thermal gradients generally remain fairly constant, the base of the GHSZ will tend to parallel the seafloor. Thus a flexure will cause a culmination at the base of the GHSZ and create a gas trap and hydrate concentration.

Many other circumstances that could create gas concentrations at the base of the GHSZ and result in gas hydrate concentrations above those sites can be imagined. For example, where dipping strata are composed of alternating permeable and impermeable layers, these layers may be intersected by the base of gas hydrate stability so as to create a seal for the permeable layer. The strata will become gas-charged (Fig. 3.13). A gas hydrate concentration will be expected above the trapped gas (Fig. 3.14). Such a situation appears to have generated huge gas hydrate concentrations in the Nankai Trough area off Japan.

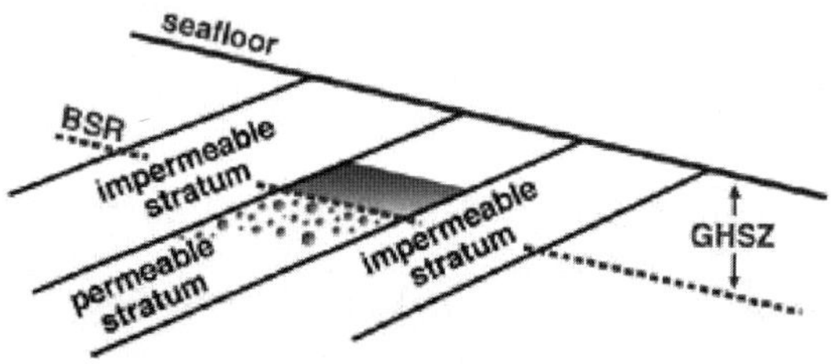

Figure 3.13. Idealized diagram showing gas hydrate concentration (shaded area) in a situation of interlayered permeable and impermeable strata. Strata dipping into the seafloor are sealed at their up dip end by gas hydrate.

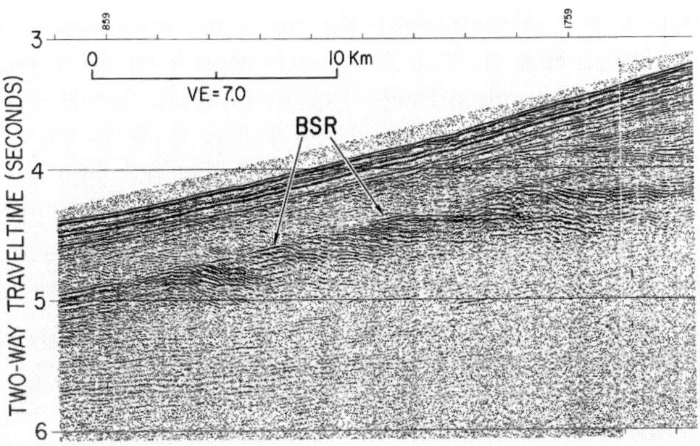

Figure 3.14. Seismic profile on North Carolina continental rise showing a situation comparable to Figure 3.13.

3.4.2.2. Physical Variations that Cause Gas Hydrate Concentrations

So far we have been considering that migration of gas depends on diffusion, buoyancy, etc., and is occurring in laterally uniform conditions of temperature and pressure. Temperature was assumed to increase uniformly downward along a consistent geothermal gradient, and pressure was assumed to increase uniformly downward and remain the same over time at a given depth. We have considered the effect of sedimentation, but other geological processes can also complicate all of these assumptions. It is imperative to keep in mind the principles discussed in Chapter 2, the phase relationships (Fig. 3.1), and relationship these to geological processes that will have an impact on the stability of gas hydrate. These geological processes include any that change the temperature and pressure near the BGHSZ, and the gas, water and hydrate compositions. The possibilities for influence on gas hydrate are endlessly varied. We consider just a few examples here.

3.4.2.2.1. Fault-controlled Gas Flow. When sediments are fractured by faults there is a tendency for considerable flow of fluids through the fractures (Fig. 3.15). When the fluids come from greater depths they are likely to be much warmer than the shallow sediments of the GHSZ.

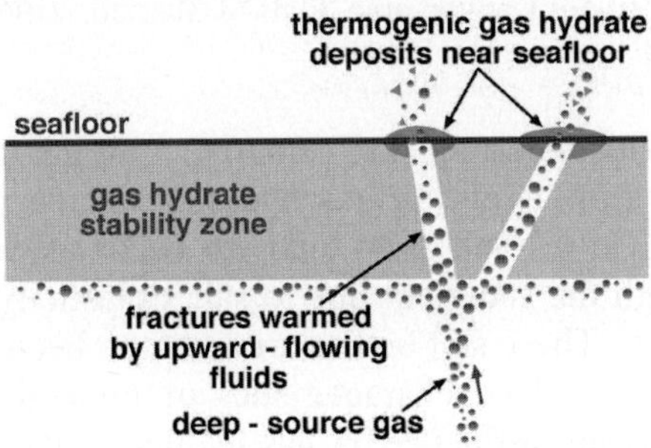

Figure 3.15. Effects on gas migration and gas hydrate formation by fluid flow through fractures.

When these warmer fluids circulate through the faults, they will warm the adjacent sediments so that the GHSZ is disrupted, and the region of gas plus water stability can extend upward even as far as the seafloor due to the warming. The fluids that migrate through the fault can transport gas to the seafloor without converting to gas hydrate. When the gas reaches the seafloor, it is rapidly chilled by the mass of colder ocean water, and gas hydrate may form in the ocean water, on the seafloor, or within sediments just beneath the seafloor. Methane hydrate and most other hydrocarbon hydrates are less dense that the seawater, so hydrate that forms on the seafloor may float away. In the Gulf of Mexico, sedimentary rocks are highly disrupted by salt flows at depth that results in extensive faulting in the upper strata (Fig. 3.16). This process has resulted in escape of large amounts of the gas that might have collected at the base of the GHSZ and formed near-bottom gas hydrate accumulations (Fig. 3.15). Although there are some indications of stronger reflections just beneath the GHSZ (Fig. 3.16; 6.3), which might be explained by gas-charged strata sealed by gas hydrate-bearing sediments above, this situation and particularly the presence of well-defined BSRs is rare in the Gulf of Mexico. The presence of relatively weak BSR is possibly due to the pervasive escape of shallow gas in chimneys and vents (4.7). Typically, gas hydrate accumulations observed in the Gulf are expressed as seafloor mounds (Roberts, 2001; Sassen, et al., 2001).

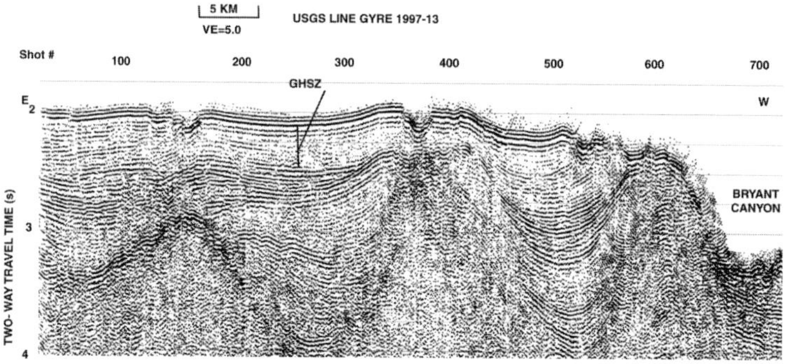

Figure 3.16. Seismic reflection profile in an area of sediment disruption due to salt mobilization, Bryant Canyon area, Gulf of Mexico. USGS profile collected by David Twichell.

3.4.2.2.2. ***Influence of salt diapirs.*** In the previous section, we considered the affects of faults on gas hydrate, and, in particular, considered those related to salt flow. Salt flow tends to occur as diapirs, which are salt plugs that rise through the sediment due to salt's plasticity and relatively lower density than sediment. These salt bodies are likely to become detached from the deep salt at their base. Two characteristics of mineral salt cause these salt diapirs to have significant influence on gas hydrate in the shallow sediments in addition to the fracturing they create in the sediments. First, salt has a thermal conductivity that is higher than sediments. This tends to create a warm area above a salt diapir by focusing heat flow. The warming will restrain formation of gas hydrate. Furthermore, the salt will dissolve in interstitial water and release sodium, chloride and other ions. These chemicals in solution create an inhibitor (anti-freeze) effect on gas hydrate and suppress its formation. The effect of the focusing of heat and inhibitors above a salt diapir (Fig. 3.17) shows that the base of the GHSZ will be domed or pushed up above the diapir. This creates a site where gas is likely to be trapped, and thus where the trapped gas is likely to be fed into a hydrate concentration above (4.8). For example, in Figure 3.18, strong reflections to the left of the salt diapir are interpreted to signify trapped gas, and very weak reflections above the trapped gas are considered to represent blanking of normal reflection strength, which implies presence of concentrated gas hydrate in the strata (Taylor et al., 2000). In addition to the interpreted trapping of gas and concentration of gas hydrate at this diapir off the

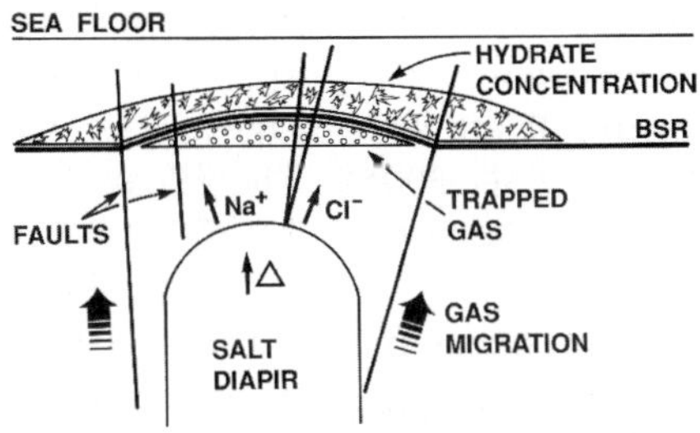

Figure 3.17. Diagram showing a gas trap formed by doming of the base of the GHSZ above a salt diapir. The greater thermal conductivity of salt compared to sediment, which creates a warm spot, and the inhibitor (antifreeze) effect of the ions dissolved from the salt both serve to raise the base of the GHSZ over the diapir to form the gas trap.

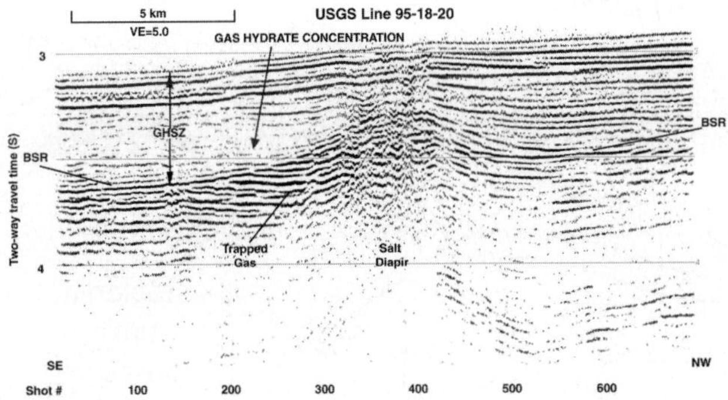

Figure 3.18. Seismic reflection profile across a salt diapir off the southeastern United States. The BSR rises above the diapir for thermal/chemical reasons and creates a gas trap at the base of the GHSZ as diagramed in Figure 3.17. Main gas trapping is indicated by the strong reflections beneath the BSR to the southeast (left) of the diapir, and greatest gas hydrate development is interpreted to exist above that gas concentration on the basis of the weaker reflections (blanking).

North Carolina coast, it is apparent that the diapirism has resulted in faults that penetrate to the seafloor above its central area (Fig. 3.18). Observations have disclosed gas hydrate, gas escape, and methane seep communities on the seafloor at this location (Paull et al., 1995, 1996). Using reasonable values of heat flow perturbation and salinity changes above the diapir, Taylor et al. (2000) concluded that the shallowing of the BSR (base of GHSZ) above the diapir could have resulted either solely from the temperature effect of the heat flow alteration or solely from the inhibitor effect of the salinity increase, or, most probably, some combination of the effects.

3.4.2.2.3. Tectonic uplift. Vertical, tectonically-caused movement of the seafloor certainly can change pressure at depth within the sediments. The effects of tectonic uplift are diagrammed in Figure 3.19, which indicates that the thrusting of a subducting plate creates a wedging effect, causing uplift of thrust anticlines. This uplift causes reduction of pressure within the uplifted sediments, resulting in dissociation of gas hydrate with release of gas at the base of the GHSZ, a shallowing of the base of the GHSZ, and a rising of the BSR toward the seafloor (Bangs et al., 2005). The released gas will tend to migrate upward forming gas hydrate concentrations at shallower depths. An example of such a situation is shown in Figure 3.20, where a BSR is seen on the thrusted northern boundary of the Caribbean plate north of Haiti. The thrust margins of the world, of course, cannot be looked at as simple escalators that just lift sediments to shallower depths, because tectonic interactions also create higher pore pressures and complex fault patterns that will affect fluid flow and gas hydrate phase stability (e.g. Dahlen, et al. 1984; Davis et al. 1983; White and Louden, 1982). Rapid fluid flow and expulsion through the tectonic accretionary wedge will result in faster transport of gases (e.g., Davis et al. 1990: Baba and Yamada, 2004.

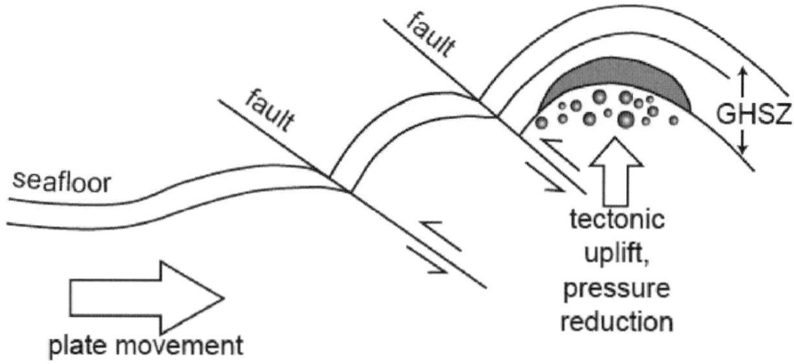

Figure 3.19. Diagram showing that plate motion at collision zones will continuously wedge newly arriving material under earlier-arrived sediments, resulting in uplift as well as folding of earlier sediments. The uplift results in reduction of pressure, which causes dissociation of gas hydrate, release and trapping of gas at anticlines, and concentration of gas hydrate above the trapped gas (shaded area).

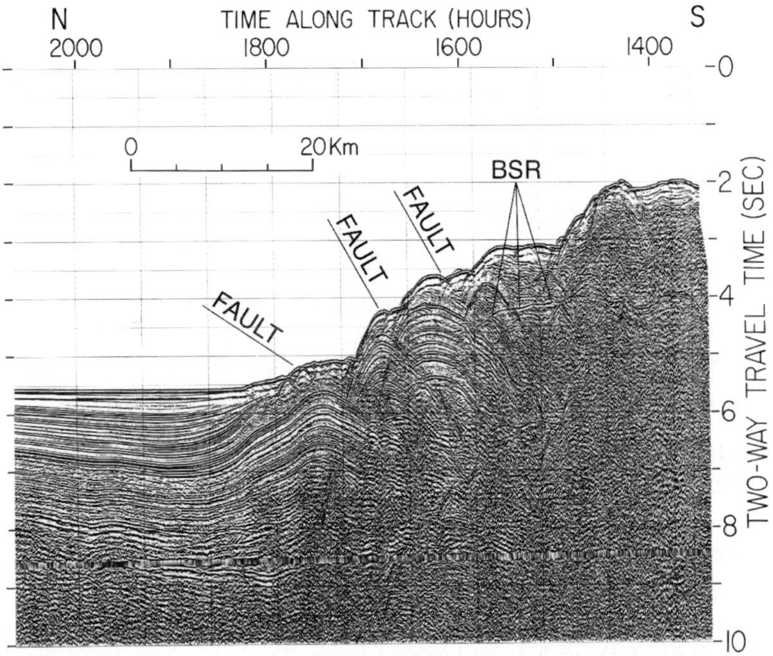

Figure 3.20. Seismic reflection profile across the tectonic accretionary wedge north of Haiti. The situation is like that diagramed in Figure 3.19.

3.4.2.2.4. Tectonic Subsidence. The reverse situation to the tectonic uplift model was identified by Pecher et al. (2001a, b) in the Lima Basin off Peru. There, the seafloor is subsiding, which will increase pressure at the seafloor and result in gas hydrate becoming stable to higher temperatures. Thus, this process will cause the base of the GHSZ and the BSR to move downward in the sediments and result in the inclusion of free gas trapped beneath the GHSZ into the hydrate phase at the base of the zone. In much of the area this process has resulted in the absence of a seismic BSR due to the loss of free gas. Despite this presumed absorption of gas at the base of the GHSZ, in some places BSRs are present and gas is believed to be passing through the GHSZ and escaping at the seafloor (Fig. 3.21). This interpretation supports the concept that the generation of gas can take place at considerable depths below the base of the GHSZ (Fig. 3.10).

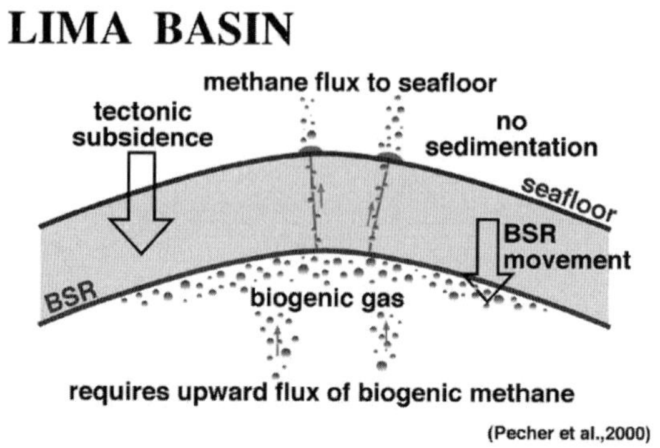

Figure 3.21. Diagram suggesting processes where the seafloor is subsiding, but deep source biogenic gas still creates a methane flux at the ocean bottom.

CONCLUSION

We have reviewed the character of gas hydrate in marine sediments, considered briefly how gas hydrate is studied, and examined some processes that cause concentration of gas hydrates in nature, with a few examples. In order to locate concentrations of gas hydrate to use as exploitable resources, we must be prepared to evaluate how the gas hydrate-controlling processes interact with the infinitely variable geological situations that we see in nature. Our examples are not expected to cover all conceivable possibilities because there are innumerable geological situations that will lead to concentration of gas hydrate. The hope is to prepare ourselves to anticipate the behavior of gas hydrate in nature under a variety of situations. Louis Weeks, a renowned oil finder, once was asked how to find oil - his answer was "think like an oil droplet". We need to think like gas hydrate.

Chapter 4

Natural Gas Hydrate: A Diagenetic Economic Mineral Resource

4.1. INTRODUCTION

Natural gas hydrate, which is synonymous with hydrocarbon hydrate composed mainly of methane, is often referred to as an ice-like crystalline material because it looks and superficially behaves like ice when it 'melts'. But it is different in two main ways. Besides containing hydrocarbon molecules that are compressed within the solid crystal lattice to about 164 times that in gas (at STP), varying the pressure on hydrate may dramatically alter the temperature at which it becomes unstable, whereas varying the pressure does not significantly change the temperature at which ice will melt (Fig. 2.1). Depending on pressure, the temperatures below which different gas hydrates are stable can significantly exceed the melting point of ice. This attribute of hydrate stability allows hydrate to coexist within water-saturated sediments wherever the pressure - temperature field is suitable for the particular hydrate forming material, and where gas occurs in great enough abundance to allow crystallization to proceed.

Hydrate forms in sediments as a solid crystalline material that may occupy virtually all pore space, and as solid hydrate in veins and pods. Oceanic hydrate most commonly forms in sediment that may have been only recently deposited (on a geological time scale) and in which many of the diagenetic and lithification processes have not yet begun to develop. Compaction caused by the weight of overlying, subsequently deposited sediment, may only slightly have reduced porosity and restricted permeability. The sediment host for oceanic hydrate is not deeply buried (Chapter 3) and is relatively weak mechanically and liable to be involved in large-scale collapse (Dillon et al., 2001) or redistribution as mass flows (Paull et al., 2003; Dillon and Max, 2003). Formation of hydrate in these sediments may increase the strength of the sediment, and dissociation of the hydrate may weaken the sediment (Dvorkin et al., 2003). The geological setting is thus very different from that of both conventional and other unconventional gas deposits, which normally occur in geologically strong reservoirs.

When hydrate forms it displaces pore water and commonly forms preferentially in beds where both porosity and permeability are higher. Where hydrate forms in secondary porosity, it is usually in the form of veins or pod-like, irregularly shaped bodies, as well as disseminated in small blebs. Hydrate has been identified in a wide variety of geological environments and host

Michael D. Max et al. (eds.), Economic Geology of Natural Gas Hydrate, 131–190.

sediment types in permafrost and oceanic environments (Chapters 3, 5). Where natural gas flux (a measure of the gas and dissolved gas transported to a location) at the seafloor from subjacent sediments is high enough, essentially solid hydrate forms on seafloor sediment, held down from rising under its own buoyancy by bonding with the seafloor. Once the hydrate-forming gas is bound in the crystalline hydrate, physical movement of either the sequestered water or gas is usually prevented, whether the hydrate is freely floating in pore space, participating with the sediment particulates in forming the sediment framework, or forming on grain surfaces and cementing them while filling porosity.

In the presence of a positive gas flux, the diffusion driving force for crystallization causes hydrate to form (Chapter 2). So long as dissolved hydrate forming gas exists in surrounding pore water at levels that either promote growth or at which a rough equilibrium with the hydrate is maintained, the hydrate will remain stable. Where levels of dissolved gas in surrounding fluids becomes too small and the relative saturation of the groundwater in contact with the hydrate decreases to a point below which the diffusion gradient reverses, the hydrate will dissolve. Dissociation of hydrate, of course, will take place whenever pressure-temperature conditions are changed so that the hydrate is no longer stable.

The same elements that are responsible for the formation of conventional gas and mineral deposits apply to gas hydrate. There has to be a source of the materials. These materials have to be transported in great enough quantities to allow for an economic concentration to be formed in any trap. Traps may be purely geological, as in the case of most conventional petroleum and gas deposits where porous rocks are enclosed by non-porous rocks in such a manner that the hydrocarbons migrating into them cannot migrate elsewhere. Or, the traps may be mainly the result of a change in groundwater chemistry, which caused the minerals to precipitate or crystallize. This is mainly the situation of most low-temperature stratabound mineral deposits in which metals such as Cu, Pb, and Zn migrate in groundwater and precipitate ore minerals when the local water chemistry and pressure-temperature conditions dictate. Hydrate deposits may have attributes of both conventional and unconventional gas and metaliferous and non-metallic mineral deposits.

In one major respect, however, hydrate differs considerably from conventional mineral deposits. Whereas conventional deposits tend to be relatively insoluble and do not significantly change form without structural or metamorphic assistance even long after the mineralizing solutions have stopped flowing, the continued existence of individual hydrate deposits depends on the maintenance of a positive gas flux. Hydrate deposits will dissipate into surrounding pore water that is undersaturated in the HFG. An exception is those permafrost hydrates which have formed in existing conventional shallow gas deposits and will again revert to conventional gas deposits if the hydrate dissociates in interglacial periods.

The formation of economic deposits of natural gas hydrate will be largely controlled by (1) an abundant supply of hydrate forming gas, (2) migration of the gas into suitable host sediments, and (3) the nature of the sediment with respect to continued supply of gas in the most suitable form for obtaining a high percentage of porosity fill with hydrate, and (4) the maintenance of conditions for the preservation of a hydrate deposit.

4.2. THE SOURCE OF HYDRATE: GENERATION OF HYDROCARBON GASES

Methane constitutes over 99% of the hydrate forming gas in most natural gas hydrate (Kvenvolden, 1995) and can be generated on Earth in three main ways. Juvenile methane may be derived from the mantle by the process of fractionation of the original materials that condensed to form the solid Earth. It may form through bacterial reduction of buried organic matter during which methane is produced as a metabolic byproduct, or through thermal maturation (essentially cooking) of buried organic matter. Juvenile hydrocarbon gases that are known on Earth and elsewhere in the solar system, and which may reflect similar gases that condensed with the early earth, for instance (Hoyle, 1955), are not regarded as significant on Earth now. Many of these early-formed hydrocarbon condensates still exist in the outer planets such as Jupiter and Neptune. Of special note are the oceans of methane or ethane first imaged from the surface of Titan, a moon of Saturn that is regarded as an analog for an early formed earth and imaged for the first time on 14 January 2005 by the Huygens space vehicle (ESAS, 2005; Kerr, 2005). Vast quantities of methane or ethane might also have been formed on the surface of an early Earth or trapped within the Earth's mantle and crust, but have escaped from Earth. Tectonic out-gassing of the mantle and crust and biosphere action and oxidation over at least the last three and a half billion years (Bleeker, 2002), would have caused the loss of these early condensates, if they originally were present.

Methane production through magmatic or fractionation processes, and other naturally produced gases that often occur with the hydrocarbon gases of interest (i.e., He, H, CO_2, SO_x, Xe, etc.) are not discussed here, although they may contribute peripherally to the local mode of formation or composition of hydrate (see Selley (1998). Even though the concept of azoic conditions pertaining to deep rocks and sediments caused many scientists to look elsewhere than to biogenic hydrocarbon generation for the source of gas and petroleum (DeLong, 2004), buried organic matter appears to provide the best source of methane and other hydrocarbon gases in a manner such that they can be easily captured, particularly in oceanic hydrate.

The source of methane at a particular location may not be entirely certain but the overwhelming evidence from analysis of methane carbon isotopes (Coffin et al., 2003: Kvenvolden, 2003; Sassen et al., 2000, 2004; Wellsbury and Parks, 2003) is that the greater part of the methane in oceanic gas hydrate appears to be mainly produced by biogenic microbial alteration of buried organic matter (Fig. 4.1) rather than by thermogenic gas generation that requires higher temperatures (Fig. 4.2). Biogenic methane accumulates in continental margin sediments because the continental margins are where the flux of organic carbon to the seafloor is greatest (Buffet and Archer, 2004). These sediments contain the preponderance of organic material of dominantly plant matter that is washed to the sea from the continents. In addition, oceanic biological productivity is high along the shelf edge, especially where upwelling of nutrient-rich deep ocean waters is greatest. Characteristically, the continental margins are also where sedimentation rates are highest in the oceans, resulting in more rapid burial of a larger part of the organic matter than in lower sedimentation rate, deep water areas farther from the continents. Rapid accumulation of sediment covers and seals the organic material within the sediments before it can be oxidized within the sea or on the seafloor, which allows microorganisms in the sediments to use it as food and form methane as a byproduct. At temperatures up to about 80 °C, biological degradation by microorganisms consumes organic matter including hydrocarbons and produces methane as a by-product (Head et al., 2004). Biogenic methane has $\delta^{13}C$ from about -55 to -85 per mil whereas thermogenic methane has $\delta^{13}C$ from about -35 to -60. A deep, methanogenic biosphere is being widely recognized (D'Hondt et al., 2004; Lanoil et al., 2001; Mikucki et al., 2003; Newbury et al., 2004), and it is probable that the anaerobic, methanogenic biosphere is ubiquitous in marine sediments.

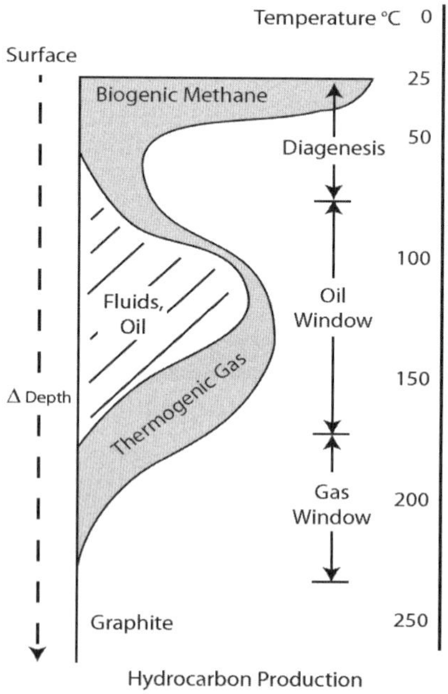

Figure 4.1. Diagram of gas and petroleum generation as a function of pressure-depth and temperature. After Max and Lowrie (1993).

Katoh, et al. (2000) proposed a one-dimensional model for the generation of biogenic methane and the formation of hydrate. The model includes a methane generation ratio, rate of hydrate formation, migration of reactants, dissolved pore water and gaseous methane, and increasing seal character that results in slowing gaseous methane migration. A two or three-dimensional model is more appropriate, however, as considerable gas is transported from source beds to the GHSZ and related traps that concentrate the gas. In addition, the gas is recycled in formation and dissociation episodes.

Although most methane to date on which carbon isotopic analysis has been done comes from biogenic sources, the presence of ethane and other higher density hydrocarbon gases and biodegraded crude in some areas, such as the northern Gulf of Mexico (Milkov et al., 2004; Roberts, 2001; Sassen et al. 1999; 2001a; 2001b), indicate thermogenic sources for the gas production. This gas

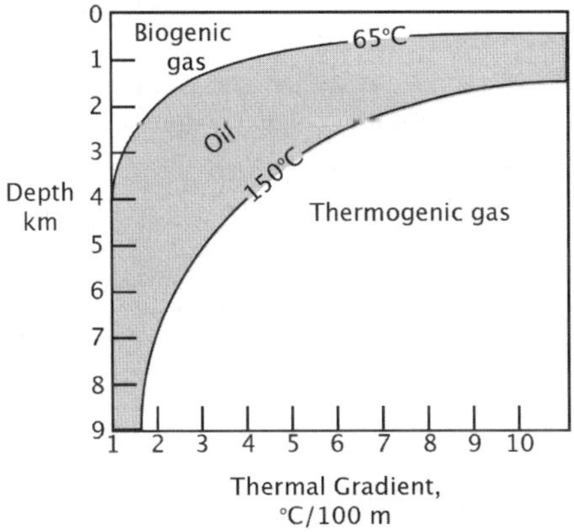

Figure 4.2. Diagram of thermal gradient vs. depth showing temperature regions for biogenic and thermogenic gas. After Puesy (1973).

would have been produced at higher temperatures either deeper in the seafloor than biogenic methane or at relatively similar depths where higher geothermal gradients occur (Fig 4.2).

Deep biosphere that produces biogenic natural gas will extend to different depths depending on the tectonic framework and geothermal gradients. In collision margins, thick sediment prisms and a depressed geothermal gradient that allows biogenic gas to be produced often from considerable depths, are cut by abundant thrusts and faults that provide conduits for gas migration from hotter zones underlying the backarc of the mobile belt (Curry et al., 2004; Hyndman et al., 2005). Sediment porosity, however, can still offer pathways for gas migration (Piñero et al., 2005). Such margins as are found along the west coast of North America and elsewhere along the Pacific Rim such as the Cascadia margin (Kvenvolden, 1988; Trehu, et al., 1999, 1995, 2003, 2004), along the eastern Indian (Curray, 1989; Max, 2003a; McCaffrey, 1996) and the

Indo-Australian and SW Asia (Sunda, Burma subplates, etc.) collisional plate margin (Lay et al., 2005), along the NW side of the Palmer Peninsula of Antarctica (Lodolo and Camerlenghi, 2003), and other places where subduction zones underlie thick accretionary prisms of marine sediments (Fig. 4.3). As well as biogenic gas, thermogenic gas generated deep below the prism from subducted sediments, or produced from near heat sources associated with igneous rocks or the roots of volcanic arcs in these accretionary prisms, can migrate as focused gas along these conduits into the GHSZ at the top of the prism. In active continental margins, gas of thermogenic origin is common (Chapman et al., 2004). In the northern Gulf of Mexico, oil is associated with hydrate deposits (Roberts, 2001) and thermogenic gas is common in hydrate (Sassen et al., 2000, 2001a, 2001b, 2001c, 2003).

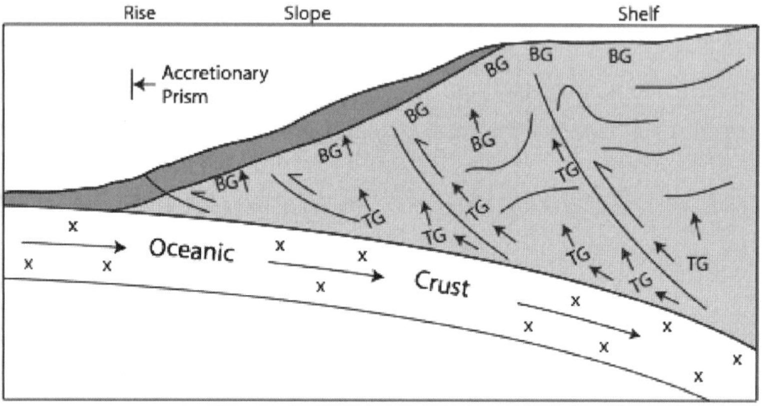

Figure 4.3. Schematic of sedimentary wedge overlying an active or collisional continental margin. Continent to right, ocean to left. Geothermal gradients in the lower sedimentary wedge vary laterally depending on the relative age of the buried oceanic crust and its igneous history. Isotherms (not shown) may overturn in the lower sedimentary prism and temperatures may decrease toward the subducted oceanic crust slab. Thermogenic gas (TG) is not necessarily generating in the lower sediments, but is traveling up faults from deeper sources.

Where passive margin sedimentation has occurred atop an essentially tectonically quiescent contact between oceanic and continental crust, such as is seen generally along United States Atlantic Ocean margins and the Atlantic margins of Africa and South America (Fig 4.4), a generally unfocused gas source for the GHSZ exists because there usually are no major structural breaks extending through the entire sediment wedge. Thus, there are few easy pathways that will allow deep generated, thermogenic gas to migrate into the GHSZ. Often, however, salt deposits within the sedimentary succession may form diapiric structures through considerable thickness of sediment and even, in the case of the northern Gulf of Mexico, breach the seafloor. These structures provide near-vertical passageways for fluid and gas transport from deep in the sediment pile to the near-surface GHSZ. Where sediments overlie cold continental or oceanic crust, the thermal gradients will tend to be lower than where the underlying crust is hot, although local groundwater may significantly modify the geothermal profile if it is actively flowing through the sediments rather than being static. Also, some sedimentary rocks, such as evaporates, effectively can duct and concentrate heat (Dillon and Max, 2003, and provide conduits for thermogenic gas to rise). Geothermal gradients, however, on both tectonically active (collisional) and passive (tensional) continental margins are suitable for allowing considerable thicknesses of sediments below the base of gas hydrate stability to provide a deep biosphere habitat. Below that, deeper in the sediments at virtually any geothermal gradient, conditions are suitable for the production of oil and thermogenic gas (Fig. 4.2, 4.3). As the sediment warms from below, temperatures at depth may rise to a point where gas and oil production cease, although dry gas (pure methane) may be produced from organic matter by very high temperatures.

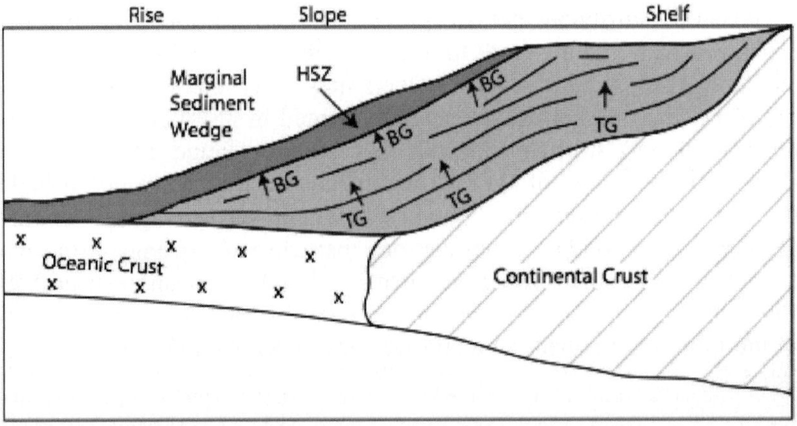

Figure 4.4. Schematic of sedimentary wedge overlying a passive continental margin. Geothermal gradients in the lower sedimentary wedge vary laterally depending on the relative age of the buried oceanic crust and the igneous history during the events that established the oceanic/continental crust boundary. Isotherms in sediments tend to be roughly parallel to the seafloor.

Whereas older passive margins, such as the Blake Ridge, have contained some beds of organic-rich sediments capable of generating methane since Jurassic times, the gas hydrate system of the Nankai Shelf has evolved only since early Tertiary times. Gas from gas hydrate bearing zones of the Nankai is mainly microbial methane, although it overlies a subduction zone in which the generation of thermogenic gas can be expected.

It does not matter, however, what the source of the methane and other hydrocarbon gases may be. Nor does the isotopic composition of any of the hydrocarbon gases appear to affect the formation of hydrate. The rate at which the methane arrives within the HSZ and the manner in which it arrives, for instance dissolved in groundwater or in gaseous form (Chapter 2), are the major controlling factors for hydrate formation and concentration. Formation of the hydrate usually takes place in the natural environment slowly enough so that heat produced during formation is dissipated. If hydrate formation takes place faster than its heat of formation can be dissipated, the reaction would be impeded at some temperature for any depth.

4.3. THE ROCK AND SEDIMENT HOST

The nature of the sediment in which the hydrate forms and the general statigraphic and structural framework of the sediment host are fundamental to the likelihood of formation of economic hydrate concentrations. Porosity and permeability are the most important attributes. The chemistry of the sediment and its lithic constituents do not appear to have a noticeable affect on the formation of hydrate, except where they directly influence the salinity of groundwater. A combination of the original sediment porosity and only the very early diagenetic attributes are of primary interest in oceanic hydrate host sediments in which hydrate tends to form near the top of sediments piles, even in accretionary prisms where tectonic effects may reach the seafloor. Permafrost hydrate reservoirs, in contrast, are usually formed in reservoirs having histories of lithification and structural events that are not found in passive margin sediments. The character of the host sediments of a hydrate reservoir is defined by porosity and permeability, both of which have original or primary characteristics or secondary modifications that can vary strongly from place to place, even in originally similar sediments. Hydrate deposits can probably be characterized as stratigraphic traps because the hydrate is much more profuse within the more porous beds in a stratigraphic succession (Winters et al., 2003; Chapter 5).

4.3.1. Porosity

Porosity is the first of two essential attributes of any conventional gas or hydrate reservoir. The pore space of sediment is that portion of a subsurface volume that is not occupied by solid material. This open space is usually expressed as a percentage of the total volume (may also be referred to as the void ratio). Porosity can either be primary in origin, in which case it is inherent to the type of sediment or rock and its depositional history, or of secondary origin, which involves the action of other process such as cementation, dissolution, or structural activity. The pore spaces usually contain water but natural gas and petroleum may displace the water and occupy discrete portions of the porosity under certain conditions.

Primary porosity of a sediment host is mainly a function of the type of sediment deposited and its degree of compaction. Sands, which even when compacted may retain porosities as high as 30 to 40% (Nur et al., 1998), can withstand considerable compressive stress because the mechanically strong sediment grains hold porosity open, and the porosity can be modeled. Porosity will not significantly alter within the pressures common to natural gas hydrate deposits (Winters et al., 2004). Gravels may have an even higher porosity. Silts and clay-rich sediments, however, may initially have extremely high porosities but can be compacted into a relatively low porosity material. Shale compaction curves show that porosities of 60% near the surface can be reduced to 30% at 1 km depth and 20% at 2 km depth (Magra, 1968). Mixtures of sands, silts, and clays will have a wide range of porosities locally, depending on their depositional and burial history. The nature of the packing of well sorted, moderately rounded and spherical sediment grains is also important and may vary between cubic (high porosity, 48%) and rhombohedral (low porosity, 26%). There are a wide variety of different schemes for classifying sediments, their textures, and their porosity for geological, engineering, and other perspectives. Selley (1998) provides a good selection of references on porosity pertaining to hydrocarbon reservoirs, which will not be discussed further here.

There are three main types of pore systems. Fully communicating, complex matrices of interconnected pores (Fig 4.5, a) constitute the most common porosity in clastic sandy and more course sediments. These strata are the most likely potential hosts for economic deposits of hydrate as well as for other types of diagenetic mineralization whose components are transported by groundwater (Robb, 2004). In pore matrices, fluid or gas that enters a pore from one source has the potential to migrate out of that pore in at least one or many other courses. Pore matrices offer the greatest opportunity for fluids and gases to percolate through sediments under the influence of gravity of fluid pressurized from some other causes. Cul-de-sac pores (Fig 4.5, b) are linked by a common conduit, but each is essentially a dead-end. Migration of fluid and gas through these pore systems may also occur, depending on the nature of the conduits, but each pore only has a single entrance or exit that does not allow fluid and gas to flow out except where a pressure difference may be induced between the pore and the conduit. These pore systems may be possible hosts for hydrate formation, which may fill the pore slowly, because the gas produced from hydrate dissociation will almost inevitably be over pressurized (Max and Dillon, 1998) and will tend to force itself out of the pores. Closed pores (Fig. 4.5, c) are isolated and not interconnected. Some rocks such as volcanic pumice may be almost entirely composed of isolated pore space formed by volcanic rock forming rapidly around many gas bubbles. Some carbonate rocks can have very high porosity that is essentially not interconnected. Closed pore sediments and rocks are not likely hosts for economic hydrate deposits.

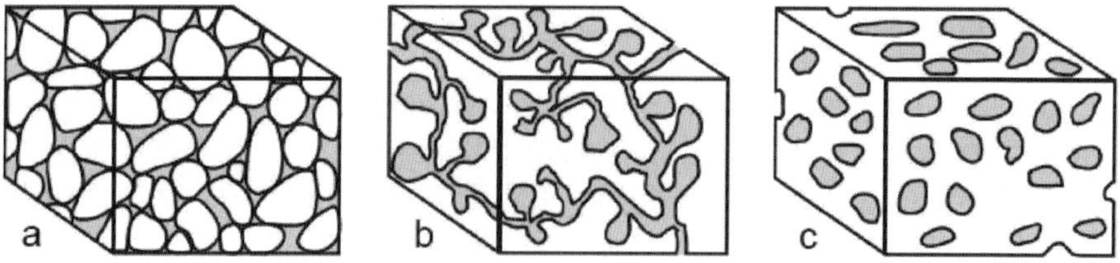

Figure 4.5. Diagram of basic types of pores shown by shaded areas. a. Matrix pores. b. Cul-de-sac pores, c. closed pores. Narrow passages between pores are pore throats.

From the point of view of porosity being available to store gas and liquids, total porosity is less important than effective porosity. Total porosity can include isolated vugs, which may be either fluid or gas filled and between which there are no open pathways. Gas in this isolated porosity are not recoverable without artificial fracturing, and the porosity is thus of no importance to reservoir potential of the rock or strata.

Effective porosity is that porosity which is intercommunicating and into or through which groundwater and gas can move. Effective porosity can be regarded as the total porosity less the isolated porosity. In effective porosity, capillary-bound water, which adheres to the sediment grain surfaces and will not traverse pore throats below some particular size, may somewhat diminish the volume available for recoverable fluid and gas storage. In this case effective porosity is synonymous with free gas or fluid. Effective porosity, which is primarily an economic term, can be measured in volume/volume, percentage or porosity units (p.u) in some special applications.

Reservoir porosity is directly related to sediment grain size and appears to be directly related to hydrate pore filling. At the Mallik 5L-38 well, for instance, pebble and sand beds displayed 80% or greater hydrate pore fill (Uchida et al., 2003). The Nankai hydrate is developed within a 100 m thick interval above the BSR at 270 mbsf (Matsumoto et al., 2003). Standard Archie equations have indicated that gas hydrate saturations at Nankai were in the range 40 to 80% within preferred, more coarse-grained porous stratigraphic horizons.

4.3.2. Permeability

Permeability is a measure of the ability of fluids and gas to migrate through a rock or sediment. It is the most important attribute of any gas reservoir. No matter how large the porosity of a rock or sediment and no matter how full of gas the pores may be, if they are not naturally connected in a manner that allows fluids and gases to migrate easily, or if a suitable permeability cannot be induced, the energy resource cannot be extracted. The unit of permeability is the Darcy (1856), which is defined as the permeability that will allow a fluid of 1 centipoise (cP) viscosity to flow at a 1 cm/s velocity for a pressure drop of 1 atm/cm. Darcy's original work was modified by Muskat and Botset (1931) and Muskat (1937), who formulated Darcy's law as:

$$Q = \frac{K(P_1 - P_2)A}{\mu L}$$

where

Q = rate of flow
K = permeability
(P1 – P2) = pressure drop across the sample
A = cross-sectional area of the sample
L = Length of the sample
μ = viscosity of the fluid

Permeability in geological hosts of hydrocarbon reservoirs is usually substantially less than one Darcy (d). Thus, the millidarcy (md) is commonly used. Permeabilities in hydrocarbon reservoirs are in the range of 10 to 500 md, although where faults have formed in a tensional stress framework, permeabilities may be considerably higher. For reservoirs, permeability can be regarded as: fair 1.0-10 md, good 10-100 md, and very good 100-1000 md. Most commercial hydrocarbon reservoirs have permeabilities in the range of 100 to 500 md Lower permeabilities are suitable for gas production than for petroleum.

In general, porosity decreases with burial. Beach sands, for example, have a porosity of about 40%, whereas that of compacted sandstones of petroleum reservoirs is around 5 to 25%. Porosity and permeability may also be decreased by secondary precipitation of minerals as cement or grain overgrowths. On the other hand, solution of material by migrating groundwaters can increase porosity and permeability, such as the preferential solution of fossil fragments and widening of fractures. About 60% of the world's petroleum reserves are held in sandstones, and 40% in limestones.

The character of the original sediment determines primary permeability, although it may be diminished by the diagenetic processes of compaction and cementation. In general, permeability decreases with decreasing grain size, especially where compaction has yielded mature sediment. Sandy sediments dominated by tectosilicate minerals, such as quartz and feldspar will tend to press against each other and provide a strong mechanical framework within which open pore spaces are held open. In contrast, shales and clays that are dominated by phyllosilicate minerals can compact closely together and form many random zones of essentially zero permeability that restrict the passage of fluids, even though substantial porosity may exist. A sand of 10% porosity, for instance, will likely have higher permeability than a shale having considerably higher porosity. Most of the aquifers that would form the best hosts for hydrate formation are unconsolidated porous media such as sand and gravel. Some very porous materials are not permeable. Clay, for instance, has high porosity, but very small pore throats do not permit free movement of water.

As the sediment particle size decreases, the size of the pores and the pore throats, which are the narrower passages between larger pores, decrease considerably. This decreases the ease with which a fluid may pass through the matrix as the capillary pressure rises. Capillary pressure can be calculated from interfacial tension and radius of pore throats and is a function of porosity and grain size distribution. The fluid film on the wetted surface of the sediment grains is also not amenable to movement, and at some particular size of pore throat for a fluid of a particular viscosity, the rate of flow decreases significantly.

Some estimation of the rate of fluid flow and transport of dissolved or microbubble HFG reactant has been made by Bünz et al (2003) for the vicinity of the southwestern Norwegian continental shelf that was disturbed by the Storegga slide (mass flow system) at about 8.2 ky. A large volume of sediment was removed by the mass flow in an area where a BSR is commonly seen. Since the disturbance, and the deepening of the seafloor of the slide system, the BGHS and its position as revealed by the BSR has been reestablished about 80 m lower (with respect to the BSR depth on either side of the disturbed region) than the position of the BSR before the slide event. Thus, in that time, penetration of the cold temperatures of the seafloor water has reestablished the thermodynamic equilibrium that defines the depth of the BSR. In order for the hydrate seal to again form at the BGHS through the formation of hydrate, and for gas to build up by a process of fluid flow, the local groundwater fluids have had to have been on the order of several cm/yr, which was regarded as high when compared to theoretical modeling for the relatively fine grained sediments involved. Coarser grained sediments will have a proportionally higher percolation or flow rate (Broadbent and Hammersley, 1957).

4.3.3. Secondary Porosity and Permeability

Non-sedimentation processes imposed upon sediments and rocks may result in increased permeability. The formation of secondary porosity may either diminish or enlarge primary porosity. Secondary permeability may increase overall permeability by orders of magnitude. For instance, where joints or tensional fractures or shear faults develop, the result is interconnected open fissures that provide much larger conduits for water and gas migration than exist in any normal sediment. Capillary pressure decreases to almost zero and fluid can flow freely, as if it were in a stream or river, rather than in intergrain porosity.

There are two main mechanisms that cause enhancement of permeability and porosity: solution activity and tectonic/structural activity. Solution activity enhancement almost always causes porosity enhancement. Dissolving some element of the rock or strata and carrying it away in the groundwater can only be done where there is significant permeability. Chemical effects involving significant solution are not likely to be part of an oceanic hydrate sedimentary host, however, because hydrate sediment hosts are mainly clastic sediments that are not amenable to short-term dissolution. Early solution affects are usually only associated with carbonates, in which some of the solution cavities could be regarded as original porosity. In addition, the nature of groundwater flow in recent marine sediment in a compaction environment is also not very suitable for inducing significant solution effects. Solution activity, however, may have been part of the geological history of host strata of a permafrost hydrate deposit, which will have had a more complex geological history prior to accumulation of gas that is subsequently converted to hydrate. Flowing groundwater systems in caves, which are commonly found in karst terrains, are the ultimate expression of solution porosity/permeability.

The most common type of secondary porosity, however, are joints, fractures, or faults that are usually related to structural or tectonic activity. Some joint and fracture systems may be caused by dehydration or by dilation caused by processes of diagenetic recyrstallization. The nature of secondary porosity and permeability depends on the mechanical attributes of the sediment or rock. Extensional and compressional forces and the orientation of a triaxial strain ellipsoid are the main factors controlling the orientation of fractures, which can also be used to describe compaction effects. Joints or faults may commonly open in one orientation as they close in another. Fault porosity may be transitory in that extensional zones may open and conduct large fluid and gas flows only for relatively brief periods. Fractures can be artificially induced (fraccing) to increase porosity where flow enhancement of reservoirs is carried out as a recovery technique for petroleum and gas deposits (6.6.4). Better means of inducing artificial porosity are continually sought. Baker et al. (2005), for instance, describe a natural laboratory for the study of fluid and gas movements through fractured sediments on a large scale.

Structural mechanisms that induce secondary permeability usually affects whole bodies of sediment and rock, but may be more pronounced within certain strata in a succession. A particular example of structurally controlled jointing occurs where joints are opened within the convex side of brittle beds in a folded succession while porosity is commonly diminished on the convex side of the bed that is in compression. Depending on the stresses applied, jointing may provide almost open watercourses in a preferred direction, or a network of joints and open microfractures through which groundwater or very fine gas bubbles can percolate rapidly. Opening of strata-bound joints may take place in relatively recent sediments undergoing folding because the process is based on the elastic limits of the materials and strain rates, not on the degree of lithification. Near vertical faults may also penetrate the entire GHSZ (Figs 3.11, 6.6), reaching down into the gas-rich layer beneath (Fig. 4.6).

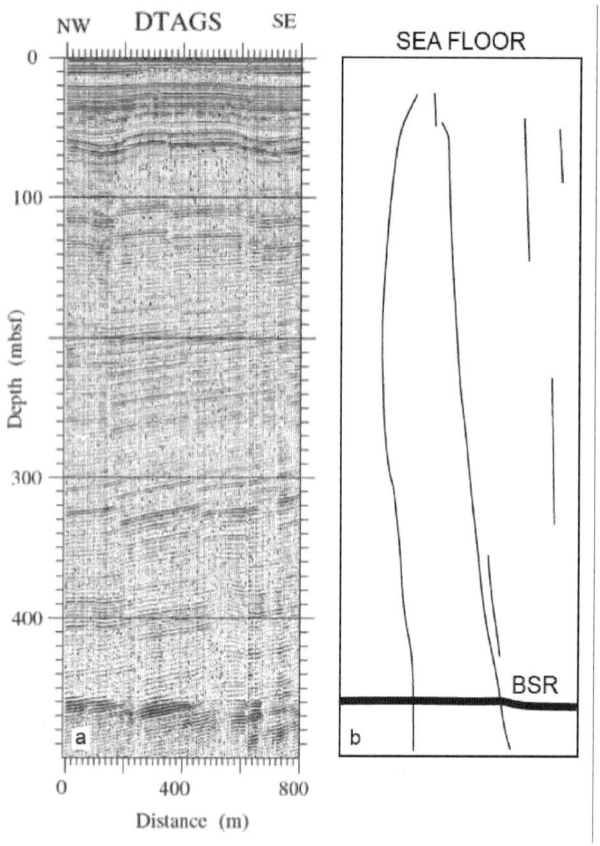

Figure 4.6. High-resolution seismic section produced using deep-towed acoustics/geophysical system (Wood & Gettrust, 2001; Wood et al., 2003). Provided by W. Wood, Naval Research Laboratory. Left, seismic record. Right, interpretive section showing faults. Seafloor at 0, mbsf (meters below seafloor). BSR shown approximately.

Faults may also provide a transient high permeability path for the movement of fluids from beneath the GHSZ to seafloor (Dillon et al., 1996; Dillon et al., 1997). In addition, these faults appears to be confined to the GHSZ, especially where considerable dispersed hydrate (Chapter 7, Class 4 hydrate) is developed, mainly in the lower third of thick stacks of fine-grained sediments (Dillon et al., 1997) in the GHSZ, which have similar mechanical attributes. Structure that has resulted in the development of tensional secondary fracturing may, in part, control the localization of hydrate. The orientation and morphology of hydrate filled fractures, for instance, appears to be controlled by structural position on Hydrate Ridge on the Cascadia Margin (Weinberger & Brown, 2004).

4.4. HYDRATE GROWTH REGIMES

Normally, the term 'groundwater' is used for water in rock and sediment pores beneath land areas and 'pore water' is commonly used for water-filled porosity in marine sediments. This difference in terminology, however, does not mean that there is some significant difference in the way water moves through sediment or the role it plays in carrying mineralizing solutions beneath land and sea areas. Modeling groundwater movement in the marine sediments allows direct application of the body of descriptive and predictive groundwater knowledge to describe subsurface water movements in the vicinity of GHSZs. Because understanding the ways in which mineralizing solutions move through marine sediments is one of the keys to describing the processes by which oceanic hydrate deposits may be formed, it is appropriate to use the term groundwater in relation to the hydrate system in this book.

Hydrate naturally develops and grows in both aqueous and gaseous media. In groundwater, the common growth media for oceanic hydrate, the most likely conditions to produce high hydrate filling of pore space occurs where dissolved HFG is carried into the GHSZ by circulating groundwater (2.7.1). In permafrost hydrate, the growth media is usually the hydrocarbon gas, through which water is brought to the site of the developing hydrate by diffusion processes.

4.4.1. Hydrate Mineralization: The Role of Water in Porous Strata

When the sediment is not permeable and HFG reactants cannot be transported to sites where hydrate can accumulate, it can only form from natural gas produced in-situ by the degradation of organic material buried with the sediments. Even where sediment would consist largely of biodegradable carbon compounds, the amount of gas produced would not form a significant volume of gas or hydrate and would not constitute a likely economic hydrate resource.

Growth of solid hydrate requires a surface of crystalline hydrate to be in contact with pore water (connate or groundwater, Robb, 2004) in which dissolved HFG is transported to the region where hydrate is growing. The source of the mineralizing groundwater appears to be from below the GHSZ (McGee, 2004). In order to fill most of the water-filled pore space, hydrate must grow outward, into the water space, and the rate at which HFG is delivered must be suitable for continued growth with a minimum of secondary nucleation or dendritic growth. Ideally, growth will be epitaxial, or layeritic, with few open pores or inclusions, resulting in solid hydrate. The difference in chemical potential between the aqueous solution and the hydrate crystal, where the presence of solid hydrate influences the apparent solubility of the HFG in the surrounding water, is the fundamental driving mechanism for hydrate growth (2.3.1). The interfacial water thus has a high potential to provide the HFG to allow growth of hydrate directly from solution. The process of accelerated hydrate growth from gas-saturated water will continue so long as there is an abundant supply of gas-saturated water. The concentration gradient that forms in HFG enriched water in the presence of hydrate is a powerful driving force.

Understanding the groundwater system in a deep gas province is the key to predicting where likely hydrate concentrations may exist. This involves knowledge of both the strata and its orientation, and the hydraulic system that is operating locally. Unconfined aquifers are those that can receive fluids from below and are bounded by the seafloor, where exchange can take place. Some aquifers, however, lie beneath layers of impermeable materials. These are referred to as confined aquifers, or sometimes artesian aquifers when a breach is made and they can flow under pressure. Artesian aquifers commonly pressurize permafrost hydrate deposits.

Porous media and fractured aquifers constitute two main types of aquifers. Where groundwater or very fine gas bubbles can percolate through sediments, an excellent means for distributing dissolved HFG reactants exists. Mathematical approximations for the movement of water through a reservoir/aquifer having almost an infinity of paths of movement in both steady state and non-steady state situations have been proposed by many authors. For instance, Stauffer & Aharony (1994) allow some predictions to be made about groundwater flow. Hydrogeologic modeling of Missippi Valley type Pb-Zn ore deposits are directly analogus to the mineralization of hydrate deposits as they are governed by the same physical chemical laws of solubility and crystal growth. In these ore deposits, ancient groundwater migration controlled ore crystallization. Garven et al. (1999) present two and three-dimensional models for flow-fields in porous marine sediments that control brine (mineralizing solution) movements and show that both density-driven and topography-driven fluid flow were important for ore genesis. They also note that some of the diagenetic mineralization in areas of high groundwater flow rates where the systems were open to the seafloor, fed ore deposition by fluid mixing, in some cases very near the seafloor.

The best condition for substantial, rapid growth of solid hydrate is provided within a hydraulic system composed of pore water that has gas dissolved in it to near saturation. Where this groundwater can migrate from below the BGHS into the GHSZ, hydrate will nucleate, and more importantly, grow most rapidly. When HFG occurs as bubbles, the movement of these bubbles into the GHSZ is problematic. Where bubbles pass into the GHSZ, hydrate shells will form on the bubbles, movement will cease and no HFG will pass into the groundwater. Where bubbles are very small, they are more likely to adhere by surface tension to sediment grains than move freely through porosity. High concentrations of hydrate in the McKenzie Delta area of Canada and offshore SE Japan appear to be characteristic of hydrate deposition in well bedded sandy or coarse sediments having good porosity and high original permeability (Chapters 2 & 5). This potential for groundwater movement, fine HFG bubble migration, or diffusion of HFG in the porous beds appears to be directly related to the localization of the hydrate.

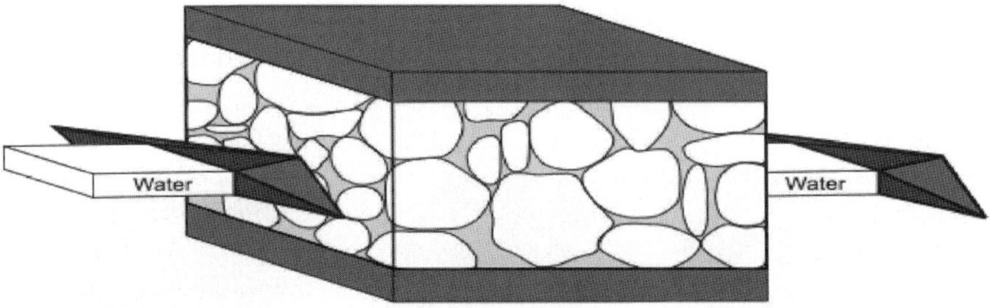

Figure 4.7. Flow confined to porous beds. Dark shading, impermeable strata, light shading, permeable strata in clastic sediment.

Although gas hydrate can occur in a wide range of sediment types, the porosity and permeability of the host material influence the type and quantity of hydrate formed and, therefore, dictate whether the deposits in a particular area may be economically recoverable in the future. Katsube et al. (2003) report, along with virtually all studies, that increased effective porosity is associated with an increased gas hydrate content in these sediments. However, they find that the increased effective porosity is not directly in proportion to increased potential storage porosity (which would normally be considered to be the same as effective porosity). A certain percentage of the porosity must therefore not be capable of sustaining mineralization. They also show that increased gas permeability is associated with increased gas hydrate content, suggesting that gas permeability is an important factor in both gas-hydrate accumulation and will likely also be so in production. This observation also supports the current model for accumulation of gas in permafrost hydrate deposits in that gas appears to have migrated into the vicinity of the present permafrost hydrate and provided the basis for the hydrate deposit. Increased gas permeability is mainly a result of increased connecting pore-sizes due to the presence of coarser textured sediments.

So long as HFG reactants can be supplied to the hydrate, an entire pore volume has the potential to become hydrate (2.5). The best way of maintaining a good supply of HFG is by a process where water circulation can be maintained. As hydrate grows in pore water space, which has the effect of reducing porosity and possibly permeability, maintaining a good supply of HFG in circulating pore water becomes increasingly difficult. Where water circulation is substantially reduced, but strong, local gas fluxes are present, diffusion of HFG reactants through the pore water can also occur, but much slower hydrate growth can be anticipated than where there is circulating transport of HFG.

Hydrate can occur in two primary manners in porous sediment. It can nucleate and grow in pore water in a freely floating mode (Fig. 4.8, a) or it can nucleate on sediment grain surfaces and grow outward into pore water or gas space (Fig. 4.8, b). In either case, groundwater is replaced with crystalline solid material that behaves structurally, acoustically or seismically, and from a porosity and permeability viewpoint, as and additional component of the original lithic sediment.

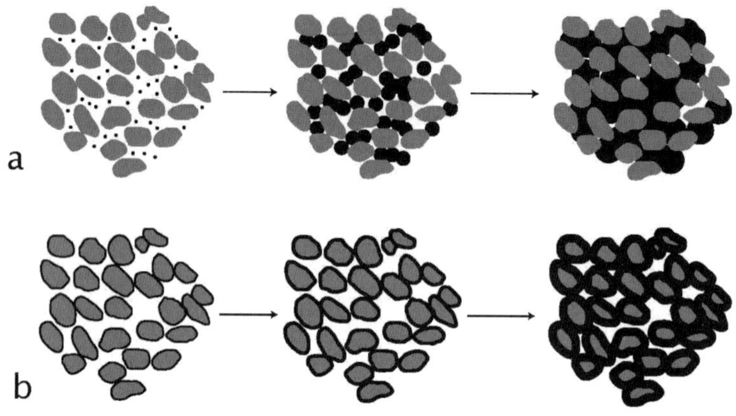

Figure 4.8. Evolution of hydrate in clastic sediment. a. Hydrate nucleates and grows in free pore water space. b. Hydrate grows as an encrustation on sediment grains.

Nucleation of floating hydrate (Fig. 4.8, a) progresses through intermediate stages until only relatively minute water passages exist through pore throats and in pore space that has been cut off from supply of dissolved HFG reactant. Once homogeneous nucleation has taken place, HFG saturation levels will fall and relatively few nuclei or crystal domains that survive an early period of competitive growth will act as growth sites (2.4). Hydrate will enclathrate dissolved HFG to grow outward into the water space until the supply of HFG is cut off through growth of hydrate and restriction of porosity. Under the right conditions (excellent porosity and permeability, large pore throats, varying pressure and temperature conditions, and good dissolved HFG supply) nearly the entire effective porosity may be filled. Where water circulation is substantially reduced but strong local gas fluxes are present, diffusion of hydrate-forming reactants through the groundwater matrix can occur, and substantial, but slower hydrate growth can be anticipated.

Where hydrate forms as a film on sediment grains (Fig 4.8, b) it is likely that permeability would be substantially reduced after only minimal hydrate growth. Hydrate growing primarily along grain boundaries would seal narrow grain-to-grain constrictions, which could result in the sediment-hydrate 'rock' becoming impermeable, while considerable isolated groundwater remaines trapped in larger diameter pores (Winters et al., 2004a). In this instance, further growth of hydrate within the trapped groundwater will depend on solid diffusion of hydrate-forming reactants through existing hydrate and into the trapped blebs of groundwater, which would result in much slower hydrate growth. Winters et al. (2004b) report that acoustic velocity analysis of the sandy and more coarse sediments encountered in both the Mallick 2L-38 gas hydrate drilling program and in sandy sediments used in laboratory experiments, indicated that hydrate did not commonly cement significant numbers of sediment grains. This observation suggests that most hydrate nucleation in natural reservoirs achieve homogeneous nucleation in water rather than as a cement on sediment grains. However, visual and tactile observations of methane hydrate formed in sediment under certain laboratory conditions did yield significant cementation of sediment grains. Considerably more observation of natural and laboratory pore filling will be required before the precise nature of hydrate nucleation and pore filling in clastic sediments will be determined. Nonetheless, the water suspended growth model rather than the cementation model offers the greatest likelihood of essentially complete pore filling.

When hydrate is formed, salt from the reaction is rejected into the interstitial pore water, which was seawater buried along with the sediment. This has the result of increasing salinity of the remaining water. Because dissociated oceanic hydrate shows salinities lower than seawater (Jager and Sloan, 2001; Maekawa, 2001; Mitts et al., 2004; Torres et al, 2004), the more saline residuum produced during hydrate formation must have equilibrated with normal seawater prior to the recent dissociation event. Thus, it is likely that the majority of the hydrate forming process has taken place in a dynamic fluid situation where groundwater allowed mixing of the more saline water in the immediate vicinity of the growing hydrate.

Specific surface area (SSA) measurements are important in a number of industries (unrelated to marine sediments) such as cement (Jennings & Thomas, 2004), coal fly ash, wood particles, and fine coal particles used for combustion. Where petroleum fills pore space, the pore volume to wetted surface (SSA) ratio can be important as petroleum will remain sorbed on particle surfaces and is lost to recovery. Where the SSA is relatively small, with respect to solid particle mass and pore volume, petroleum recovery is potentially higher. Where gas fills the pore volume in a sediment reservoir, water flush has the potential to remove most the gas from the pore meshwork, except for stranded pockets.

The SSA of sediment in which hydrate is likely to form is important only when the sediment particle surfaces are used as a host for nucleation, which is followed by growth, that essentially coats the particles rather than forming in intergrain areas. There are not yet many measurements of SSA that bear on hydrate formation. Gilbert et al. (2004), however, have examined cores of hydrate-rich sediment from the Mississippi Canyon region of the Gulf of Mexico and have found that the SSA of the sediment was not directly related to hydrate formation.

Hydrate may form particles in pore space that become large enough to press against more than one sediment grain. Because the hydrate is a solid crystalline material, and indistinguishable from sediment grains in a mechanical sense, the hydrate essentially become part of an extended sediment framework (Waite et al., 2004). Because hydrate is quite strong (Durham et al., 2003a; 2003b), its presence will act to prevent compaction of the sediment. Hydrate cementation can also act to prevent the original sediment from being compacted. Sediments having significant framework hydrate, which may allow for very large percentages of the bulk sample to be composed of hydrate, however, have the potential to allow sediment collapse during dissociation.

4.4.2. Permafrost Hydrate: Water Vapor diffusion in an HFG Atmosphere

Whereas oceanic hydrate is rarely associated with water ice, except in rare cases in Arctic regions or in glaciers (Kuhs et al., 2000), permafrost hydrate is found in a compound water ice - hydrate cryosphere (Fig. 4.9). Water ice is stable from the surface downward to near the zero degree isograd (depending on salinity) whereas hydrate is stable from some depth at which total pressure (hydrostatic plus lithostatic) and temperature allows hydrate to form down to some depth determined by temperature, pressure and salinity. This same model may pertain to hydrate formed extratrerrestrially, particularly on Mars (Max and Clifford, 2000).

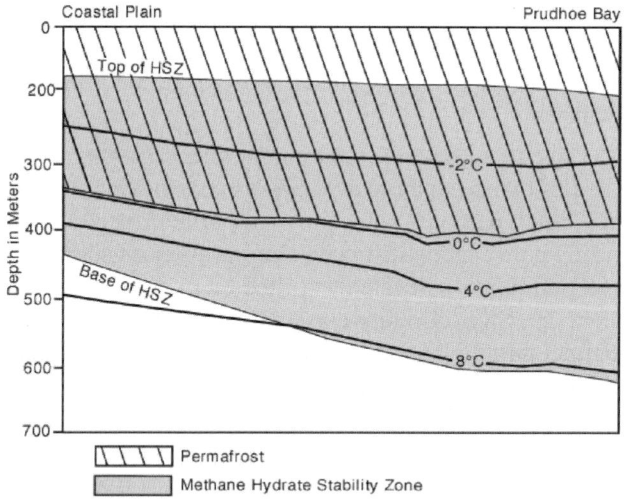

Figure 4.9. Diagram of compound water ice – hydrate cryosphere, all of which can be described as 'frozen ground'.

In the case where a natural gas deposit would fill most of the pore space of a geological trap prior to establishment of conditions of hydrate stability, hydrate may form very expeditiously from water vapor dissolved in gas, if sufficient water vapor can be provided. This growth model would apply for the development of very high pore filling, which appear to be common for most of the shallow high-grade permafrost hydrate deposits examined to date (Chapters 3, 5). In these, hydrate saturation from 20% to 40% of total rock mass and exceeding 80% of pore space is common (Collett et al., 2003). Prior to the formation of these hydrate-rich deposits, a gas zone would have existed over water in the reservoir. Both the water and the gas would have been present and not communicating freely through permeability with groundwater systems outside of the geological gas trap. For instance, in the Mallik 2L-38 well in the Mackenzie Delta, the gas hydrate system, including hydrate, gas, and groundwater, is closed. The residual waters are characterized by high concentrations of chloride up to 3200 mM (which is within the range of normal sea water). This high chlorinity inhibits the formation of gas hydrate and promotes its dissociation (Sloan, 1998). Brines produced through hydrate formation did not equilibrate readily with other groundwater systems. In contrast, in the Nankai oceanic hydrate deposit, strata are substantially open to the ambient groundwater circulation system.

It is likely that many, if not most, permafrost hydrate accumulations formed mainly from pre-existing, conventional natural gas deposits, especially in North American Polar regions (Chapter 5). If this were not the case, the gas would have had to migrate into the present zone where the hydrate is located following the development of the water ice permafrost. This scenario would require the movement of an unusually large volume of gas (in most cases) migrating in a very short time. Further, where water-ice has already frozen, the gas in hydrate that was structurally or hydraulically higher in the trap would have had to form hydrate by a process of gas diffusion, which would probably have been much too slow to allow any significant hydrate to form.

Although one of the "givens" concerning Arctic hydrate accumulations, such as Mallik, is that the structures originally contained free gas, which converted to hydrate during the Pleistocene, but there may be exceptions to this manner of gas concentration. Logging results at Mallik showed that the hydrate saturation of the pore space is >80% at the BHSZ, but decreases higher in the section and becomes patchy in the upper zones. If that part of the stratigraphy containing the richest hydrate were related to gas that had been already trapped, the richest zone might have been expected to reside in the highest part of the trap.

Two conditions, however, could explain the apparently anomalous distribution of hydrate grade (degree of richness of the hydrate mineralization). First, the present hydrate deposit is polycyclic in its formation and represents the concentration of hydrate over a number of cycles of hydrate formation and dissociation. The original concentrations formed in the earliest phase of hydrate formation were not as rich as the richest zones seen now. The grade would have depended on the ambient pressure of the gas that has been converted to hyrate. If the pressure were such that less than 164 volumes of gas were available for any particular volume of hydrate in pores space, then hydrate formation would sequester all the free gas and remaining pore space would either fill with water or with gas at such a pressure that it was in equilibrium with the hydrate. Enriching the hydrate deposit in its lower part would be an almost inescapable function of the action of more than one cycle for hydrate formation and dissociation (4.4.3). Alternatively, the hydrate 'trap' may not be closed to gas diffusion in its highest part and the lower values could represent dissolution (2.6.2) of hydrate and diffusion up-structure of the gas (2.7.5).

Permafrost hydrate deposits mainly appear to have formed in existing gas deposits in shallow geological traps; the gas appears to have been generated largely thermogenically and to have migrated into the traps where the gas constituted a conventional gas deposit before it was converted to hydrate (Dieckmann et al., 2003). The gas hydrate in the Mallik 5L-38 is composed mainly of thermogenic methane (Lorenson et al., 2003). The gas has formed elsewhere and migrated into the trap in the same manner as any conventional gas deposit.

Hydrate may form in permafrost areas where the base of the local BGHSZ reaches downward to an existing gas deposit during the intensification of a glacial period. Formation of hydrate in the already concentrated gas would likely promote rapid growth of solid hydrate. Formation of hydrate in saline groundwater requires either colder temperatures or higher pressure conditions than in either fresh water or where water vapor is the delivered reactant, because higher salinity inhibits hydrate growth (Sloan, 1998). In addition, the transport of reactants and the ease with which they will join the growing crystal lattice are optimum.

These methane deposits commonly consist of free gas in between overlying hydrate and underlying water with hydrate in direct contact with subjacent water. Prior to formation of the hydrate, before the onset of glacial conditions, water and gas in the geological trap comprised a conventional gas deposit. The hydrate has formed in the upper part of the conventional gas deposit. At present, the base of gas hydrate, as exemplified by the Mallik deposit (Taylor et al., 2003), is an active front of upward moving hydrate dissociation. This dissociation reflects the downward propagating warming of the present interglacial period. During the intensification of a glacial episode, in contrast, the BGHS will be an active front of downward propagating hydrate growth. Ginsburg and Soloviev (1998) suggest that in sediments having high pore filling, hydrate will crystallize over a considerable volume, rather than at a migrating front. This may be especially true where the hydrate zone becomes colder upward, which would allow progressively more water vapor to be drawn from the gas that remains to be converted in progressivly colder areas of the reservoir. Where chilling is rapid, the formation of a hydrate front that rapidly extracts water vapor from the gas and becomes impermeable is more likely.

Gas in a geological trap may be either wet or dry depending on its source, transport, and local groundwater conditions. If the gas is contains little or no water vapor, all of the gas concentration can exist within the field of hydrate stability without the formation of hydrate. Where gas contains substantial water vapor, but where there is no water in contact with the gas outside of the zone where hydrate will form, hydrate will form until the water has been extracted to some low level of saturation and then hydrate formation in the now relatively dried gas will cease. However, where water is in contact with the gas, which is the normal case in a geological gas trap where subjacent water provides the pressurization, an abundant supply of water vapor is continuously available. Where crystallization of hydrate is made possible by migration of water vapor, the reaction or phase change will take place close to the fresh water phase boundary (Fig 2.15), regardless of the salinity of the water from which the water vapor has evaporated. However, increasing the salinity of the water lowers the vapor pressure of water vapor (Raoult's Law), which may slow down mass transfer and lower the hydrate growth rate. In the extreme of very high salinity, the vapor pressure may be lowered to such an extent that an insufficient amount of water is present to form or maintain hydrate.

The results of the HFG atmosphere experiments (2.7.4) and the empirical experience of others can be used as a direct analog for the manner in which hydrate formation may take place in an existing shallow geological gas deposit which is chilled gradually from the top. Natural cooling of sediment grains by conduction during intensification of a glacial episode is the reason why permafrost extends into the ground as a response to greater cold at the surface. Interstitial water is frozen, producing the permanently frozen ground. Thus, in a porous sediment gas reservoir, where grains in contact provide the cooling mechanism that promotes the formation of hydrate, crystallization will tend to form from the grain boundaries out into the gas space.

Where porosity is occupied by thermogenic gas, its temperature may remain elevated, because it is part of a larger reservoir of warmer gas in the geological trap. Because the gas will tend to remain warmer than the sediment grains, pore throats may remain open longer as warm gas flowing through the throats (to replace the gas consumed by volume change in hydrate formation), may inhibit hydrate from forming at the points of greatest constriction, where gas velocity is greatest. Hydrate deposition will likely take place first in wider pore space and continue until growth of hydrate by water vapor diffusing through the gas ceases when the remaining tiny pores fill or when water vapor is cut off for some other reason.

In contrast to the formation of oceanic hydrate, where hydrate forms in the water saturated marine sediments wherever suitable gas flux and concentration reaches the HSZ, hydrate in an existing gas pocket in permafrost terrain can only form where water dissolved in the HFG migrates to the region where hydrate is stable. In this case, the hydrate forming reactant that must be transported is water vapor, which is delivered through the molecular diffusion up from the bottom of the reservoir to the top of the gas space. This process is diffusion limited, but when compared to a hydrate growth scenario existing in an aqueous environment, the process is both relatively quick and should tend to produce essentially solid hydrate which has the potential to fill the porosity.

As the conditions of hydrate stability deepen with intensification of the glacial cold at the surface, the BHSZ moves downward into the gas space. The cooling rate of the sediment and the diffusion rate of the water vapor are the main controls of hydrate growth. If sufficient water is present to recharge the vapor space, nearly all the HFG vapor can be converted to gas hydrate. As hydrate forms and extracts gas from the pocket, pressure will tend to drop unless water pressure from below allows gas pressure to recover or to be maintained, depending on ease of groundwater circulation.

Where there is gas in pore space in a geological trap (Fig. 4.91, a) and conditions of hydrate stability have not yet reached down below the reservoir trap's roof, hydrate will not form in the reservoir. As the base of the GHSZ deepens in response to colder conditions on the surface, conditions of hydrate stability penetrate into the reservoir (Fig. 4.10, b). At this point, hydrate will begin to form in the uppermost part of the reservoir porosity (Fig 4.10, c) where temperatures fall into the region of hydrate stability. The two way diffusion of hydrocarbon gas and water becomes biased by diffusion of water vapor into the gas as water vapor in the upper part of the reservoir is converted to hydrate. As time progresses and the cold penetrates deeper into the ground, the BHSZ continues to descend into the reservoir (Fig 4.10, d).

As the hydrate forms, water vapor will be removed continually from the hydrate-forming gas and a diffusion gradient of water vapor will be established in the gas pocket. In order to maintain the equilibrium concentration of water vapor within the hydrate-forming gas, water will evaporate on a continuous basis and enrich the superjacent gas in an effort to maintain an equilibrium concentration for the local pressure-temperature conditions (Fig. 4.10, large arrowheads).

The process of diffusion of water vapor to the hydrate 'roof' area may be interrupted by the condensation of water on sediment grains at levels intermediate between the hydrate and water interfaces as increasingly cold ground is encountered upwards. Water-ice will not form, however, because the BGHSZ is found below the base of water-ice permafrost (Fig, 2.27). Wetted grain surfaces, which could normally be anticipated, and any droplets of water condensing on the chilling sediment grains will provide an excellent host for hydrate nucleation in that the water can be expected to become saturated quickly with HFG, allowing nucleation to take place when temperature falls.

Where the pressure-depth of the gas deposit is less than about 1.64 km (a shallow gas reservoir), which is about the depth equivalent of the compression factor of hydrate (164, Kvenvolden, 1988), there will be a positive flow of gas molecules into the hydrate, and pressure will decrease in the gas reservoir. Where the pressure-depth is greater than 1.64 km, the formation of hydrate will result in increased pressures within the gas reservoir because the formation of hydrate compresses the surrounding gas. Because the BGHS is rarely, if ever, found below 1.64 km depth in permafrost terrain on earth, this effect of crystallization increasing the pressure and causing a positive feedback effect that would further promote the growth of hydrate is unlikely.

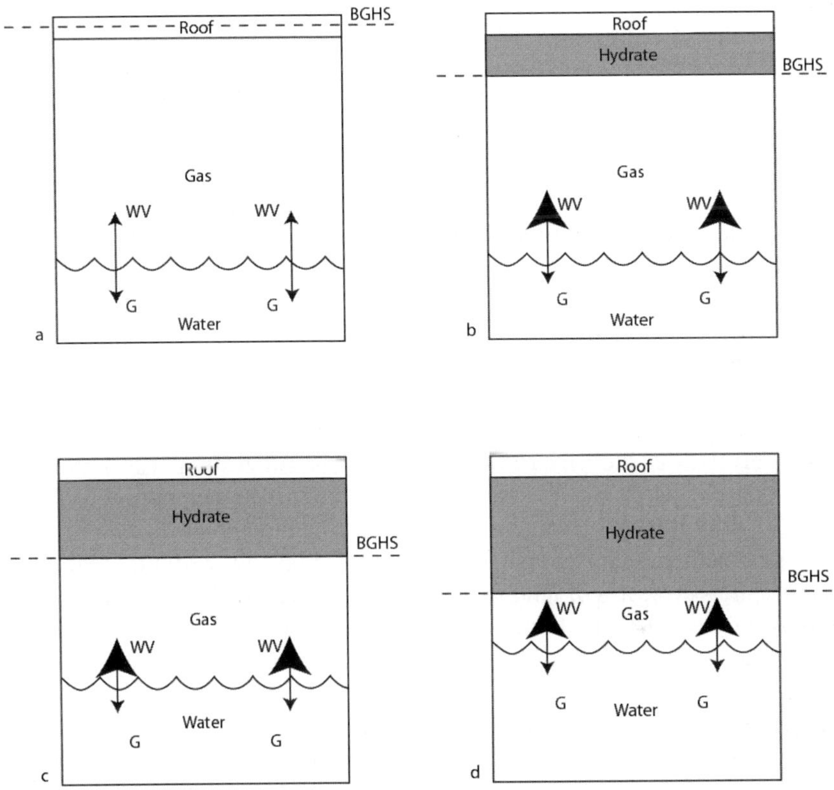

Figure 4.10. Diagram illustrating hydrate growth from water vapor dissolved in HFG. Reservoir sediments and rock not shown. As the BGHS descends from above the roof of the gas deposit (a), hydrate begins to form (b). As hydrate continues to form, water vapor is removed from the HFG. The consumption of HFG and water vapor causes the pressure to drop and allows the water to rise (c). As the level of the BGHS continues to deepen and the water level continues to rise (d), the water level may eventually reach the bottom of the hydrate mass.

On colder planets and moons in the solar system, however, where methane or other hydrocarbon gases and water are present (Max & Clifford, 2000; 2001), this positive feedback compression effect may be important. At the correct pressure and temperature at the surface of a distant planet or moon, the formation and dissociation of hydrate could be a primary climate modifier.

As hydrate formation continues in the gas reservoir and gas is extracted into the solid hydrate, pressure in a shallow gas reservoir will tend to decrease. Where groundwater in the geological trap does not move upward to maintain nearly the original pressure in the gas, pressure in the reservoir will drop and hydrate formation may cease even though temperature continues to fall. Where the water (containing very fine gas bubbles or not) continues to exert pressure and can percolate into porosity that was occupied by the gas (Fig. 4.10, c), and where the heat produced by the hydrate formation can be dissipated, hydrate growth will continue. Eventually, if the BGHS extends through the gas zone and into the water, all the gas may be transformed to hydrate, except for HFG dissolved in the contact water, which must be maintained at levels of saturation sufficient to preclude hydrate from dissolving into the water.

Because diffusion through gas in pore throats is likely to be more efficient than through pores saturated with water, it is possible to envisage that virtually the entire region may become solid hydrate. Virtually all of the gas in the reservoir may be converted to hydrate.

4.4.3. Implications for Concentration of Hydrate near the Base of the GHSZ

Max and Lowrie (1996) described a gas hydrate conservation cycle, they regarded as providing a long-term process for concentrating natural gas in and immediately below the GHSZ over time. This suggestion was based on the process of free gas production through dissociation of even diffuse hydrate at a BGHSZ. The free gas would concentrate in sediment porosity immediately below sediment whose porosity remained sealed by hydrate. Once gas is generated, however, it will tend to concentrate below the BGHS if it cannot find a conduit to the seafloor. Bangs et al. (2004), for instance, describe an upward shift in the BGHSZ in Hydrate Ridge (Cascadia, Oregon) with a consequent formation of a gas-rich horizon following postglacial warming of bottom water. Although the rate of gas production and delivery to the HSZ (4.2) might be slow in some places, the process of forming and dissociating hydrate over and over again in response to changing environmental conditions of seawater warming and cooling and sea level rising and falling, would tend to concentrate natural gas in and immediately below the BGHS. The formation of hydrate from dissolved HFG in groundwater would provide the primary gas concentration mechanism of the cycle and dissociation would be a secondary mechanism. The concentrated gas was envisaged to form concentrated hydrate as it rose, by unexplained means, into the HSZ.

Once free gas forms and is trapped beneath the GHSZ, a significant proportion of it apparently does not rise advectively or by diffusion into the GHSZ to contribute significantly to the formation of further solid hydrate. Most of the gas trapped below the BGHS either remains in place or escapes to the surface without significantly increasing the amount of hydrate. Where gas trapped below the GHSZ does rise, if it does not do so catastrophically (Dillon et al., 2001), it tends to rise through the GHSZ in either chimneys or vents (4.7.1 &.4.7.2) and some escapes into the sea through seafloor vents (Tréhu et al., 2004).

Hydrate growth experiments demonstrate that where free gas and seawater are brought together, hydrate forms only at their immediate boundary, which creates only thin hydrate, shelled bubbles (2.7.6). Further hydrate growth, which may cause aggregation, depends almost entirely on slow solid diffusion mechanisms. The movement of gas through the gas hydrate stability zone (HSZ) in a hydrate-armored channel, or chimney, will not allow time for further hydrate growth, and much gas may reach the seafloor and escape into the bottom water. Formation of large amounts of solid hydrate appears to be best accomplished, not where free gas and water are brought together, but where dissolved HFG is brought into the vicinity of hydrate, which can then grow into the water-filled pore space. The presence of free gas, therefore, may be less important than the nature of the local groundwater system, which is the best means for dissolved HFG to be transported.

A gas hydrate concentration cycle, however, can operate in oceanic hydrate deposits but not as originally envisaged as a direct result of the formation of gas deposits by the dissociation of hydrate that would provide concentrated gas to the receding GHSZ (Max and Lowrie, 1993). It is more likely that the main phase of concentrated hydrate formation would take place during phases of environmental change that favor the formation of hydrate, rather than during dissociation, as originally proposed. Formation of hydrate within zones in which gas had already been concentrated by migration upward from depth would also produce more concentrated hydrate.. Concentrated hydrate deposits could form near the base of an oceanic hydrate deposit in a manner very similar to that which would form concentrated hydrate in an existing gas deposit in permafrost terrane through the process of vapor diffusion.

Where the BGHSZ is depressed owing to increased pressure (from higher sealevels or seafloor subsidence) or where seafloor water cooled, conditions of hydrate stability would slowly progress downward into the free gas zone. Water within the gas zone initially would be available to form hydrate with the already concentrated gas. Where the gas zone would have an interface with groundwater outside of the descending conditions of hydrate stability, water would be available to diffuse into the gas as water vapor was removed from the gas at a higher level in the gas reservoir by incorporation into the hydrate. Thus, concentrated hydrate could be expected to be formed during depression of the BGHS rather than during its elevation, in response to changing environmental conditions.

Where conditions of hydrate stability progressed downward over a gas deposit that did not have an interface with groundwater, in either permafrost or oceanic terrane however, a free gas pocket could be formed. Gas pockets encased by hydrate that formed early in the establishment of a GHSZ might remain stable for considerable periods of time.

4.5. GAS HYDRATE: A DIAGENETIC ECONOMIC MINERAL DEPOSIT

Gas hydrate can be considered to be an unconventional type of gas concentration. But it is more unconventional than tight gas (gas in low porosity sands and shales), coalbed methane, or any other type of unconventional gas deposit in which the gas is actually present as gas phase concentrated in a geological reservoir of some type. Gas hydrate is a sold crystalline mineral substance that is similar in a number of ways to other solid crystalline economic minerals. Like other economic minerals, the hydrate must be processed to remove the valuable component, in this case, natural gas. This fundamental difference between gas hydrate and all other gas deposits has rendered hydrate difficult to characterize in conventional gas terms and requirements for successful exploration, identification, valuing, and recovery. Hydrate deposits are also very different from oil shale (which has to be mined to be processed), coal and other hydrocarbon deposits (which also have to be mined and chemically converted), and tar sands, which, although they may have to be processed in place to be recovered, do not have the potential for recovery offered by hydrate, whose phase transition is sensitive to environmental change.

4.5.1. Contrasts between Conventional Gas and Gas Hydrate Deposits

A conventional gas deposit is formed by impermeable cap or marginal rock or strata that overlies a reservoir rock or sediment holding gas in its porosity. The gas has migrated into the trap from source beds elsewhere. Conventional geological traps are usually mechanically very strong and prevent the gas from migrating or escaping to the surface because of the strength of the geological materials. The gas is usually held in place and pressurized from below by pore water (Fig. 4.11). An oil zone may lie between the water and the gas and may contain considerable dissolved gas. When gas or oil is removed from a conventional hydrocarbon reservoir, there is seldom any danger of reservoir collapse or distortion of the reservoir materials. As most of the conventional oil and gas fields that are currently being discovered are increasingly smaller, exploration methods have become more sophisticated. This enhanced technical capability is probably suitable for delineation of hydrate economic deposits.

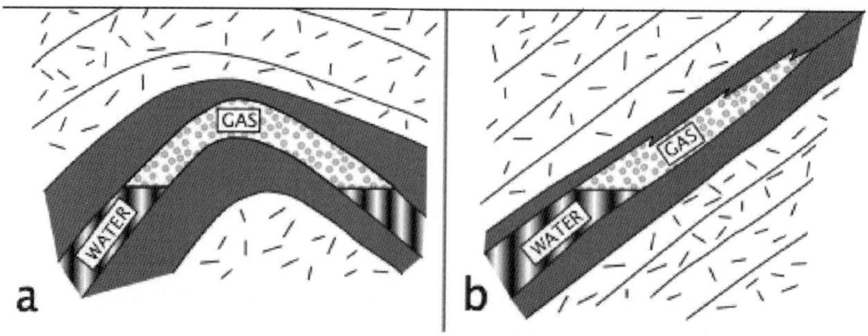

Figure 4.11. Conventional geological gas traps with porous beds (containing gas and oil) surrounded by impermeable strata (shaded). a. Anticlinal structural trap. b. Stratigraphic trap. Hundreds of different variations of these simple diagrams of traps (structural, fault, stratigraphic, hydraulic, etc.) exist, but are not discussed here.

Gas hydrate deposits are substantially different from conventional gas deposits primarily because of the difference in their geological hosts and their manner of entrapment. The major differences, from formation to extraction, are:

1. Gas hydrate deposits usually occur in virtually unlithified marine sediments in relatively simple structural and stratigraphic situations that are often not much different from their characteristics immediately following deposition.

2. The structural and mechanical characteristics of conventional gas deposits do not apply to oceanic gas hydrate and subjacent-trapped gas deposits, although they may apply at least in part for permafrost deposits.

3. Hydrate deposits comprise both trap and reservoir in that they contain the gas and are responsible for its being fixed as a crystalline solid within the GHSZ. Additionally, hydrate may trap free gas by effectively sealing sediment porosity, mainly below the BGHS, but often within it.

4. Gas hydrate has the potential to form part of the sediment framework, increasing its mechanical strength and dramatically altering its acoustic response in a manner completely different from gas in a conventional reservoir.

5. Hydrate concentrations are often underlain by or lie within a mobile water zone (Class 2 deposit, Chapter 7) and are unbounded above. That is, they do not reside in a geological trap but rather have concentrated at a particular place because of a combination of chemical and pressure / temperature conditions that favor formation of hydrate (Fig. 4.12).

6. Hydrate is metastable and responsive to its environment whereas conventional hydrocarbons are not. Conventional hydrocarbon deposits are isolated by the geological trap in which they reside. In a hydrate deposit, HFG in hydrate is usually in equilibrium with its groundwater environment. Where gas flux is above a certain level, more hydrate will form. Where gas flux falls, hydrate will dissolve into undersaturated groundwater.

7. Hydrate concentrations are localized within GHSZs both in permafrost and oceanic areas, in locations where gas is already present or where gas flux is sufficient, and only within a very limited distance from the ground surface or seafloor that is determined by burial pressure and temperature. Conventional gas deposits are found in geological traps often at great depths, and dry gas deposits can be found where temperatures are quite high. This implies that heavy seagoing drilling capabilities may not be required for gas hydrates as they are for deepwater conventional hydrocarbon deposits.

8. Especially in oceanic hydrate, the host sediments worldwide display a similar range of lithic and mechanical properties. This is particularly important for drilling because it implies that drilling conditions are similar for all potential oceanic hydrate deposits worldwide.

9. Pressurization in a conventional reservoir is usually provided by groundwater. In drawing down a conventional gas reservoir, replacement of extracted gas by water may restore the gas to the original pressure but not raise

pressure above its original pressure. Where hydrate is dissociated, it has the potential, depending on temperature, to raise the reservoir pressure well above the ambient pressure. This has important implications for safety as well as gas flow maintenance during dissociation and extraction.

10. Gas produced from hydrate has the potential to pump water against the natural water head. Gas in a conventional reservoir does not, unless an unusual (faulting) or artificial breach (drilling) is made to the groundwater system.

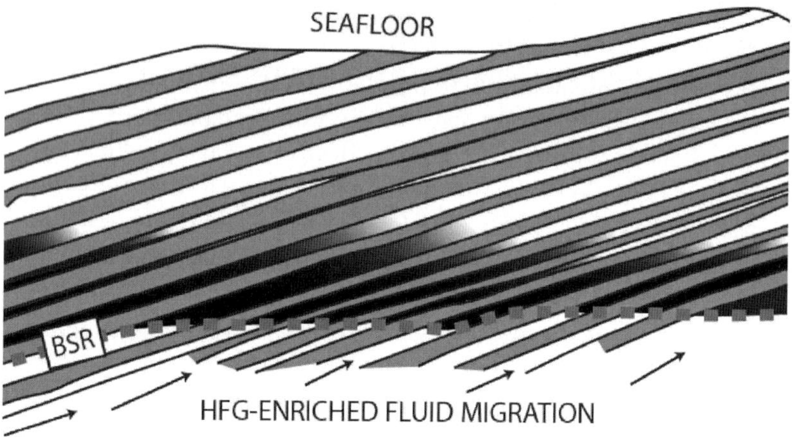

Figure 4.12. Diagrammatic representation of porous beds (white) containing hydrate (black) in a dipping stratagraphic sequence intersperced with impermeable beds (shaded throughout) caused by upward transport of HFG into GHSZ with BSR at its base. Diffuse upper margin of hydrate indicates gradational margin to hydrate concentration. Hydrate-enriched porous strata may consist of only one or a few beds where mineralization is confined to particular stratigraphic units (Figs. 1.17, 1.18, 1.19, 1.20). Exploration for hydrate payzones in individual porous strata will be similar to many current petroleum exploration projects using timeslice and acoustic analysis of seismic data to identify petroleum and gas enriched porous sands. Where beds are parallel to the BGHSZ, movement of HFG into the GHSZ may be more difficult and might depend on transport along joint or fault systems across low permeability sediments.

The thermodynamics of growth from enriched groundwate are primarily determined by temperature and strongly favor the development of a wide zone of hydrate formation because the temperature decreases upward. This temperature decrease has the effect of maintaining metastable zone conditions (Fig 2.9) for some distance above the BGHS to a point where substantial depletion of HFG from the ascending groundwater / diffusion is completed. The general tendency, however, is for hydrate to be higher grade nearer the source of HFG supply and lower grade further up structure.

As the HFG-enriched groundwater migrates upward into the GHSZ, hydrate will precipitate when saturation reaches the point where homogeneous nucleation takes place. Once hydrate nucleii are present, growth follows rapidly, considerably reducing HFG concentration. The initial nucleation and growth of hydrate may not take place immediately at the BGHS unless the HFG saturation is high enough. Because the temperature decreases upwards in the oceanic GHSZ, upward-migrating groundwater will become increasingly saturated and subsequent hydrate nucleation and growth may take place at a shallower levels. Thus, a number of stratigraphic horizons, or a number of zones within a dipping porous horizon, can be mineralized by the same upward migrating HFG-enriched groundwater.

In many respects, the hydrate reservoir may form a conventional gas trap similar to that formed by unconformities. In an unconformity trap, a dipping stratagraphic sequence having permeable and porous beds is cut across above by an unconformity. Impermeable 'trap' strata provides a seal. Gas and oil may pool in the porous strata beneath unconformities. Seafloor parallel hydrate also may provide a seal to dipping porous beds beneath the BGHSZ. The gas that ponds in porous strata beneath the BGHS is equivalent to conventional gas trapped by unconformities (Cameron and Ziegler, 1997; Quirk and Aitken, 1997).

4.5.2. Hydrate Mineralization

In many respects, oceanic hydrate deposits closely resemble stratabound metalliferous and non-metalliferous mineral deposits in both their form and manner of development. Soloviev (2003), for instance, who is familiar with seabed metaliferous mineralization as well as hydrate in the field, refers to hydrate as 'mineral' wealth. A mineral deposit is a volume of rock or sediment enriched in one or more (crystalline) materials. Oceanic hydrates are, in fact, economic mineral deposits, being composed of an economic mineral species that binds the economically valuable molecules in a crystal lattice. The methane can be converted or extracted from its mineral form much more easily than most economic minerals, however, and this can be accomplised in place rather than following extraction from its host sediments. Oceanic natural gas hydrate might even be considered to be a solid ore of methane, although the term ore is often reseerved for metallic mineral deposits. In practice, however an ore is a mineral deposit from which one or more valuable substances can be economically extracted, and methane satisfies this definition.

Hydrate deposits (Table 4.1), have been divided into four major classes (Chapter 7) on the basis of their concentrated or dispersed nature and whether the groundwater system is open or closed. In addition, where there is gas in contact with hydrate at the BGHS (Class 1), the optimum conditions for hydrate conversion by artificial means exist. It is anticipated, however, that many high-grade hydrate deposits will be of Class 2 character, essentially similar to the hydrate deposits at Nankai (5.3). Some characteristics of oceanic hydrate are:

<table>
<tr><th>Class</th><th>Hydrate</th><th>Bounded</th><th>Materials in contact[1]</th><th>Geological Situation</th><th>Groundwater System</th></tr>
<tr><td rowspan="2">1</td><td rowspan="4">Concentrated</td><td rowspan="4">Permeability boundaries / geological strata/ faults</td><td rowspan="2">Gas over water</td><td>Oceanic</td><td>Unconfined</td></tr>
<tr><td>Permafrost</td><td>Confined</td></tr>
<tr><td>2</td><td>Mobile water</td><td>Oceanic, Normal (2-OU)</td><td>Unconfined</td></tr>
<tr><td>3</td><td>No gas or water</td><td>Trap</td><td>No mobile fluids</td></tr>
<tr><td>4</td><td>Dispersed</td><td>No effective permeability boundaries</td><td>Pore water</td><td>Undifferentiated Marine Sediments</td><td>Unconfined</td></tr>
</table>

Table 4.1. Classes of hydrate deposits (See Chapter 7) that describe the general geological and groundwater relationships fundamental for the growth and dissociation of hydrate. [1]Beneath hydrate.

1. Oceanic hydrates occur within the GHSZ, which tends to occur about parallel to the seafloor. It is stratabound (ESA, 2004) to sequences of sediments where sediments are parallel with the seafloor. Where the sediments are not parallel to the seafloor and the BGHSZ cross cuts them, the hydrate forms selectively in more porous strata upward from the BGHS and is stratabound on a bed-by-bed scale. In dispersed hydrate, where few porous strata hosts are open, hydrate may be formed preferentially in secondary porosity or distributed in the relatively impermeable sediments in a random-appearing manner.

2. Hydrate is most likely to be of economic grade near the base of the GHSZ, where there is a direct upwelling supply of mineralizing solution.

3. Hydrate forms preferentially in coarse-grained sediments that allows groundwater movement or diffusion.

4. Hydrate also forms in secondary porosity joints and veins.

5. The BGHSZ can be locally pushed up where local heat flow sources, such as diapirs or vents, occur. In this case, mineralizing solutions may migrate laterally, away from the upward flow. These GHSZ thickness anomalies may form gas traps beneath them.

6. Oceanic hydrate is similar to low-temperature, diagenetic mineral deposits in that they are also deposited slowly from low-temperature groundwater fluids carrying the economic material in small quantities to the site of mineralization. Hydrate formation, especially in the densest hydrate deposits, appears to be the result of a long-term process rather than some quasi-instantaneous mineralizing process, such as observed at metal sulfide deposits around high temperature black smoker seafloor vents.

7. Hydrate is dissimilar to stable metalliferous mineral deposits in their interaction with the environment. Without an adequate gas flux and without an adequate supply of dissolved natural gas in the groundwater, the hydrate will dissolve and dissipate.

4.6. CLASSIFICATION OF OCEANIC GAS HYDRATE DEPOSITS

Oceanic hydrate deposits can be divided into high grade (Class 1, 2, and possibly 3) and low grade (Class 4) deposits that have many of the features of both metalliferous and non-metalliferous deposits. The nature of the supply of dissolved mineralizing solutions, the manner and rate of growth of the hydrate, and the resulting grade and value of the hydrate concentrations are controlled by the geology of the rock or strata that is available to host hydrate where sufficient gas flows occur. The geological framework of Class 2 (Class 2-OU, Chapter 7), which consists of highly porous sediments exposed to an unconfined groundwater system, and Class 4 deposits, which are developed in finer grained strata without the same excellent host beds that are found in Class 2 deposits, are significantly different. Thus, deposits of intermediate hydrate values can more properly be described as an immature high-grade deposit or an abnormally hydrate-enriched low-grade deposit.

4.6.1. High Grade Deposits

The principal attribute of a high-grade deposit is that there is a high concentration of solid hydrate in a relatively small physical volume. The hydrate interval (vertical section) tends to be relatively thin, the porosity tends to be high, the degree of pore filling (degree of saturation) by solid hydrate tends to be high, and gas volumes as a function of reservoir volume also tend to be high (Table 4.2). These are sweet spot deposits wherever they are best developed. The best-known example to date is the Nankai deposit of the outer, deep southeastern continental shelf of Japan (5.3).

In the Nankai type of deposit, hydrate mineralization is essentially confined to porous beds in a well-stratified sequence including sands, gravels and silts. The essentially impermeable silty interbeds appear to act as groundwater guides. Gas flow to below the base of the GHSZ may be focused or diffuse but the impermeable sediments focus the movement of dissolved and gaseous HFG into the porous beds yielding relatively rapid delivery of HFG to the zone of hydrate concentration in the GHSZ. At the Nankai site, the strata are dipping, which is the best case for allowing a maximum infusion of HFG-enriched groundwater into suitable host strata. This angular condition allows the BGHSZ to pass across the strata (Fig. 4.13). In this configuration, each of the stratigraphic horizons passes upward from a region where groundwater is being enriched with hydrate forming gas rising from below the GHSZ. As another good example of a best case, the BSR is relatively close to the seafloor. At the near seafloor level, groundwater could interchange with seawater and methane may diffuse into the seawater or be consumed by bacteria.

Site/Well	Hydrate Depth (m)	Hydrate Thickness (m)	Porosity (%)	Hydrate Saturation (%)	Gas Volume (km^2
Northwest Eileen State 2					
Unit C	651-681	29	37	61	1,031,000,000
Unit D	603-610	7	36	34	132,000,000
Unit E	564	17	39	33	347,000,000
Mallik L-38					
Units 1-10	810-1102	111	35	67	4,284,000,000

Table 4.2. Examples of high-grade reservoirs (clastic dominated sediment). Numerical values from T. Collett (2003) (rounded to nearest million).

Thus, there is a good possibility that the groundwater system is open to diffusion or percolation into the nearest seabed sediments up dip. Where deep groundwater systems in which HFG can be infused to near saturation conditions and has the potential to rise through the GHSZ and pass into or exchange solutes with the seawater, a completely open system exists. This situation is the 'unconfined' condition for groundwater interchange (Chapter 7). There will be many particular examples where some attribute of the geology prevents a fully open groundwater system from being established, but even less than ideal case situations may still yield high grade-type hydrate concentrations of economic interest (Fig. 4.13).

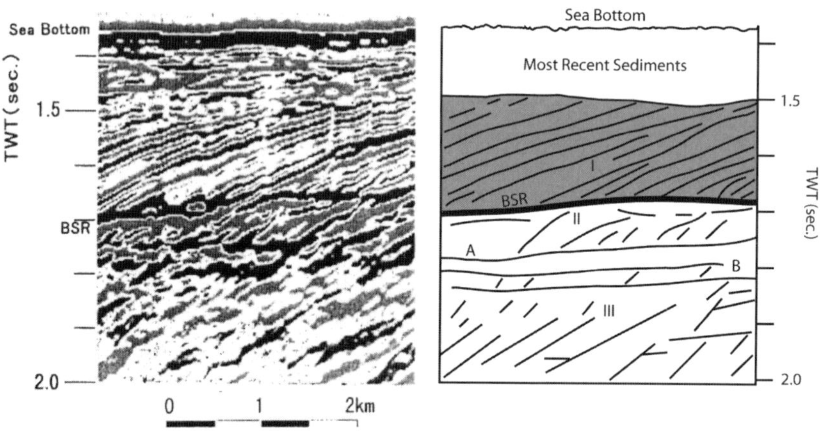

Figure 4.13. Reflection seismics from Figure 3 of Tsuji et al. (1998), with a diagram showing near-horizontal features associated with the existing BSR. Note horizontal features below the BSR, which likely mark the level of older BGHSZs.

The mineralized horizons generally are mechanically strong in that the sediment grains are usually impressing upon one another in the clastic tectosilicate-dominated porous beds in which the porosity is held open by the sediment grains. The host sediments are capable of maintaining their integrity during dissociation, although the apparent relic BGHS may reflect sediment collapse and redistribution from the dipping plane common to the sediments with a resedimented plane parallel with the BGHS (Fig. 4.13). Although sediment redistribution may have taken place, there is no evidence of collapse of overlying strata.

Where cycles of hydrate formation take place, gas may not be concentrated in the hydrate-enriched lower portion of the GHSZ. This is because the openness of the groundwater system may allow dissociation to molecularly infuse the groundwater and raise the saturation of HFG, which will then diffuse up dip into the sea down a methane saturation gradient (2.7.2). Alternatively, a gas phase could form at the BGHSZ that could then escape up dip in the porous beds, which have no natural blockage up dip, except hydrate filling of pore space. Little gas can be anticipated beneath the BGHS and there are likely no gas deposits that might be exploration targets in their own right associated with high-grade hydrate deposits.

4.6.2. Low Grade Deposits

The principal attribute of a low-grade deposit is that there is a low concentration of solid hydrate in any particular volume. The hydrate interval (vertical section) tends to be relatively thick, the porosity tends to be low, the degree of pore filling (degree of saturation) by solid hydrate tends to be low, and gas volumes as a function of reservoir volume also tend to be low (Table 4.3). This is typical of a Class 4 deposit (Table 4.2) (Chapter 7). The best known example of a low-grade deposit is the Blake Outer Ridge (3.3.2; Table 3.1). The hydrate tends to be disseminated as spots or blebs or in veins and irregular secondary porosity zones that could be related to dewatering, faulting, or the expansion due to hydrate formation. Because the sediments are poorly bed-differentiated and are dominated by fine grained sediments, permeability is much lower than in high-grade deposits. The location of hydrate sweetspots is problematical as they are likely related to some local original attribute of the sediment, or to a structurally imposed secondary cause for which general cases cannot be established in the absence of any good examples.

Although there is an 'open' groundwater system, communication and rates of groundwater movement are much lower than in high-grade deposits. Dissolved HFG probably moves through much of the lower part of the GHSZ by diffusion processes or, where packets of groundwater move irregularly through the sediments, as secondary porosity is created. Faults and chimneys may be the dominant mechanism by which HFG-enriched groundwater and gas pass into the GHSZ or through it to vent into the sea. Identifying sweet spot concentrations in these deposits is inherently difficult because concentration will depend on some secondary structural or local geological attribute that may be difficult to predict from a general geological analysis. In other words, a second or third order economic analysis may be required, which demands a much more complete dataset (6.3), rather than a first order analysis.

The mineralized horizons are generally mechanically weak in that the sediment grains are often not pressing on one another in a way that a structural framework exists for the sediment as a whole. These host sediments are capable of segregation of sediment and gas/water where dissociation might create overpressured gas deposits. They are also capable of forming thixotropic muds that will flow. They are also capable of collapse, especially where gas and water produced through dissociating of hydrate removes mass. Dillon et al. (2001) conclude that major sediment collapse associated with gas blow out from the vicinity of a collapse structure on the Blake Outer Ridge occurred on a very large scale. Mass flow deposits and profound sedimentary disruption within these sediments may be created relatively easily through hydrate dissociation, as there is no structurally suitable guide to allow overpressured gas to dissipate.

Where cycles of hydrate formation take place, gas is likely to be concentrated below the hydrate-enriched lower portion of the GHSZ. This results because of the absence of an open groundwater system within the GHSZ. Dissociating gas will form a gas phase at the BGHSZ that tends to be in physical continuity with the hydrate-enriched zone, which can only escape along some secondary porosity feature such as a fault, a vent, or a chimney (4.7). Thus, overpressured gas deposits (Max and Dillon, 1998) will tend to form beneath the sediment that has been rendered relatively impermeable by formation of hydrate in pore space. The gas itself may be an exploration target capable of extraction.

Site/Well	Hydrate Depth (m)	Hydrate Thickness (m)	Porosity (%)	Hydrate Saturation (%)	Gas Volume (km^2
Site 994	212-429	217	57	3.3	670,000,000
Site 995	193-450	257	58	5.2	1,268,000,000
Site 997	186-451	265	58	5.8	1,450,0,000
Site 889	128-228	100	52	5.4	447,000,000

Table 4.3. Examples of low grade reservoirs (clay dominated sediment). Numerical values from T. Collett (2003). Rounded to nearest million.

4.7. Migration of Hydrate-Forming Gas Into and Through the HSZ

There are no accurate measurements of methane and other hydrocarbon gases produced by thermogenic processes of the deep biosphere. Until recently, the scope of production of hydrocarbon gases beneath the sea has not been apparent. Although climate modelers once assumed that little or no methane reached the atmosphere from the ocean, recent surveys using combined side-scan sonar seafloor imaging and reflection seismics have shown considerable natural gas venting through the seafloor through gas chimneys (Petcher, 2002), and estimates of the quantity of gas produced beneath GHSZs are much larger (Sassen et al., 2003a). Deep biosphere and thermogenic production of methane and other hydrocarbon gases is prodigious (Soloviev 2002a, 2002b), and is now known to provide a large enough gas flux to generate the very large estimates for natural gas sequestered in hydrate (Kvenvolden, 1993). Petcher et al. (2004)

have also identified gas conduits and vents beneath oceanic hydrate. Ginsburg (1993) and Shoji et al. (2005), for instance, show that in the area of the Sea of Okhotsk, (eastern Russia, northern Japan Sea area) active venting is associated with large subjacent gas (estimated at about $8x10^9$ m^3 and between 10-40% hydrate in near seafloor sediments. Gas associated with active venting and seafloor hydrate in the southern Hydrate Ridge of the Cascadia margin of western North America is estimated at about $1.5\text{-}2x10^8$ m^3) by Trehu et al. (2004). The single Milano vent structure in the Mediterranean Sea may be associated with up to $2.5x10^9$ m^3 of gas (Soloviev, 2002).

Where HFG can be transported into the GHSZ in migrating groundwater, hydrate is likely to form in the most porous beds in the lower part of the GHSZ (Figs 4.12, 4.13), and little free gas will pond below the GHSZ because of the open nature of the groundwater system, unless pore fill becomes very tight. Ponding gas will tend to be confined to more porous beds where hydrate is hydrate seals pore throats. Fiiner grained sediments, which have little HFG-enriched water flowing through them and pore throats that will seal more quickly with hydrate, will tend to have less hydrate. During periods of dissociation, or where gas from beneath rises adiabatically into the gas concentration zone beneath the BGHSZ, considerable gas deposits may form.

Not all the deep gas rising toward the GHSZ will pass into it and form hydrate. Based on the relatively recent observations of widespread natural venting of gas from the seafloor, it appears that a great deal of gas rises through the GHSZ and into the sea. Published terminology for describing the manner of passage through the GHSZ is conflicting, in that the distinction between the composition of the venting gas and fluids does not appear to have been cosidered. Although the focus is on the venting of gas, the composition of the water-gas mixture that passes through the GHSZ is important because of the implications for the formation of economic deposits of hydrate.

There appears to be a considerable difference between the thermal properties of gas-dominated and water-dominated effusions through the GHSZ (Appendix B1). This essential thermal difference, when considered along with other variables such as the volume of warm gas and groundwater that pass through the GHSZ and the rate and nature of the rise of the gas and water, allow us to speculate on the mechanisms of hydrate mineralization within the GHSZ.

For the purpose of our discussion, we identify two end members for the materials penetrating the GHSZ that are responsible for natural gas being able to transit through the GHSZ and into the seaf from below. Chimneys are identified as being dominated by gas with little water while vents, are identified as being dominated by water flow. Vents appear to be more typical of high flux, focused flow gas source areas where while chimneys appear to be associated more with dispersed flow, low gas flux areas where gas has concentrated below a broad hydrate seal (such as in the Blake Ridge, SE US). Although the term venting may be applied to gas escaping from the seafloor, for instance on Hydrate Ridge where a specific gas-rich horizon has been identified as the source of venting gas (Tréhu et al., 2004), whereas the vented material is primarily gas rather than fluid, we would regard the nature of the passage through the GHSZ to be characteristic of a chimney because of the thermal effect within the GHSZ.

4.7.1. Chimneys

Free gas deposits tend to either remain trapped below the GHSZ or to pass through it and vent into the sea through gas chimneys that often pass along faults (Dillon et al., 1998). More rarely, seafloor overlying large deposits of trapped gas can collapse, allowing very large amounts of gas to reach the atmosphere (Dillon et al., 2001). Chimneys can be modeled as relatively narrow, open channels lined with hydrate which separates the internal gas from the groundwater in the surrounding strata. They can be identified on reflection seismic profiles where BGHSZ push-up has also been noted on the Cascadia margin (Riedel et al., 2005; Zykov & Chapman, 2005). These chimney-like profiles have vertical margins to the broad chimney-like structure, but are much broader than narrow, secondary-porosity-type gas chimneys. The gas within them may moves rapidly, on the order of m/sec or greater. The hydrate rim on the chimney occupies a double diffusion position (2.7.6) where free gas and water have come together in the GHSZ and a hydrate barrier has formed between them. Theoretically, all the gas could be consumed if it migrated or diffused into the water space or the water migrated or diffused into the gas space. Gas not mixed beneath the BGHS, however, has a very limited potential to mix with water within the GHSZ.

Chimneys in the Blake ridge appear to be dominated by gas and contain little water (Flemings et al, 2003). Hornbach et al. (2004) suggest that the free-gas concentrations beneath the Blake Ridge are highly variable, particularly when compared with the superjacent crest of the Blake Ridge. Little gas exists along the eastern erosional flank and within the Blake Ridge Depression (Dillon et al., 2001) where faults extend continuously from the BSR to the seafloor (Dillon et al., 1998). It would appear that both the near vertical faults that penetrate the GHSZ and outcrop at the seafloor act as secondary (porosity) conduits, and the more porous sediments that dip toward and intersect the seafloor may allow gas migration into the water column along primary porosity.

Gaseous methane migrates when buoyancy exceeds local capillary pressure. The relationship of overpressured gas to the formation of faults is uncertain, but if a volume of gas begins to ascend, it may have the potential to rise through a crack where the lateral gas pressure allows an upright fracture to propagate because the limit of brittle failure is exceeded. Pre-existing fractures may be used as passages of opportunity also. Regardless of the process, it is known that substantial gas may reach the seafloor with little attendant fluid.

Because gas can be cooled relatively rapidly by surrounding groundwater (Appendix B1), the gas chimneys will tend to remain cold and narrow. Although the BSR may be faulted or made irregular in the vicinity of the fault roots just below the BGHS (Fig. 4.6), there appears to be little hindrance to free gas percolation and little adjustment of the BSR in the vicinity of this type of fault. Thus, there is little opportunity for the HFG to dissolve in the groundwater and be able to enter into permeable sediments in the GHSZ where high-grade hydrate deposits could form.

Xun et al (2004) identify two gas rich stratigraphic passages through the HSZ at Hydrate Ridge. Their AVO analysis indicates that the gas is distributed in at least one of the stratigraphic units in isolated pockets rather than in a hydraulically continuous primary or secondary porosity. Free gas is more prominent in the units up dip, which is the general direction that gas would be expected to migrate if it were free to do so. Similar to the perched BSR associated with upwelling gas-rich fluids in the northern Gulf of Mexico, the highest identifiable hydrate in the HSZ occurs immediately above the highest identifiable free gas.

4.7.2. Vents

Vents are temporary, impermanent features that are dominated by fluid transfer in which gas may either be carried as microbubbles or dissolved in solution. Vents may have either a central conduit or a vertical porous zone, or a combination of both. Vents are associated with both pockmarks and mud volcanoes, as well as venting of large volumes of gas which exsolves from solution as the fluid mass rises adiabatically and coalesces. Mounds and mud volcanoes cannot form without the lifting power of ascending water, and the hydrate commonly found associated with them is related to HFG that forms hydrate on the seafloor owing to rapid chilling at the seawater/seafloor interface.

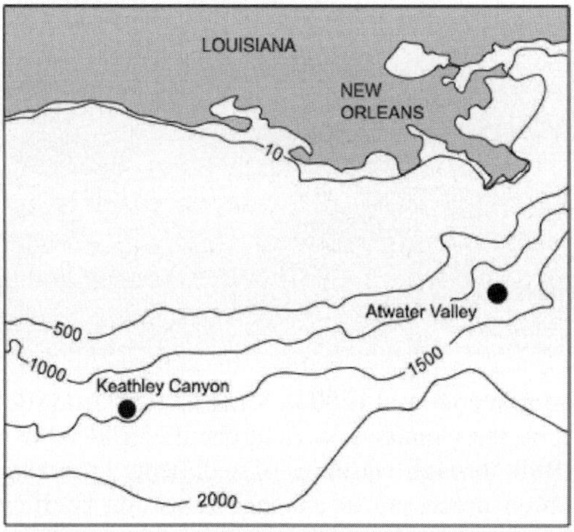

Figure 4.14. Location map of the two study areas of the Joint Industy Partnership (JIP) of the Department of Energy Gas Hydrate Research Program in the northern Gulf of Mexico.

Vents may also be associated with the formation of unusual seafloor minerals and with bio-communities that are dependent on the vents for their basic nourishment (Roberts, 2001). Venting has been particularly well studied in the northern Gulf of Mexico as part of the US DOE hydrate research program.

In vents, HFG-rich fluid from beneath the GHSZ will form gas bubbles as it traverses to the surface, but the asscoiated cooling effect (Apendix B1) is not significant. Water-dominated vents contain considerable heat and thermal capacity, in contrast to gas-dominated chimneys, and the latent heat has the potential to warm the sediments near the axis of the vent. Vents have the capacity to 'push up' the BGHS, creating a temporary BSR cone (Fig. 4.15) and to increase the area over which groundwater that is enriched with HFG may pass into the sediments. This process may be particularly important to the possibility of formation of economic deposits of hydrate within the GHSZ where sediment bedding is parallel to the seafloor. Where HFG-enriched groundwater has the ability to pass laterally into suitable high porosity sediments, the potential persists for the formation of high-grade hydrate deposits, even if this potential exists for only short periods of geological time. Once hydrate has formed in the GHSZ, the level of HFG saturation of groundwater in the GHSZ will probably be close to equilibrium with the hydrate. The hydrate will tend to persist in its

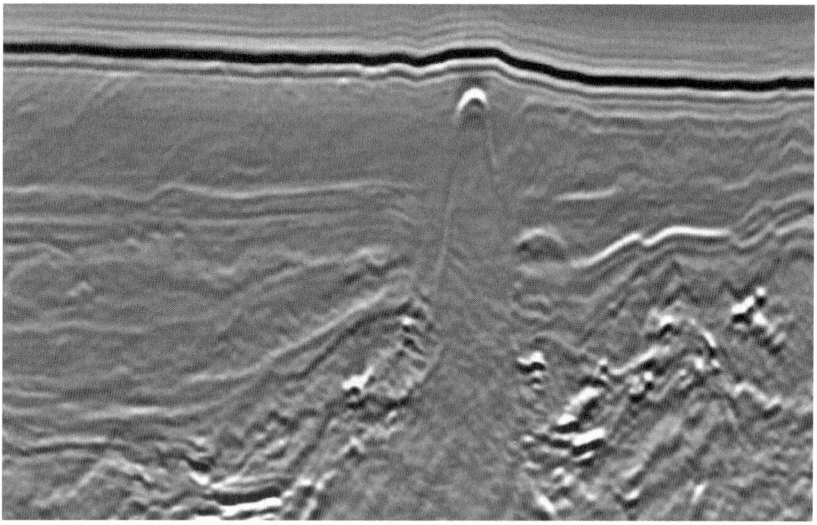

Figure 4.15. From Wood et al (2004). Unscaled reflection seismics showing acoustic response in the vicinity of a vent where considerable water is causing push up of the BSR through warming of sediments lateral to the main seep. The record has been processed to enhance reflection coefficients. Note the bright spot at the apex, thought to be caused by free gas beneath concentrated shallow hydrate. Courtesy of Warren T. Wood and Clifford I. Voss, Naval Research Laboratory.

concentrated form, even though it is liable to be cut off from its source of mineralizing solutions when the vent shuts down.

Two seafloor mounds formed by upwelling fluids, gas, and sediment lie on the floor of the Mississippi Canyon in Atwater lease blocks 13 and 14 in about 1300 m water depth (Figs. 4.15, 4.16, 4.17). These features were studied to discern subsurface thermal profiles and their relationship to upwelling methane-enriched groundwater (Wood et al., 2004). Fluid advection appears to have been minimal for at least 20 years. One mound is about 10 m high and 500 m in diameter, whereas the other is about 6 m high and 200 m in diameter. Both exhibited a several meter deep circumferential moat. Although Chen and Calthles (2003) suggests that the amount of natural gas stored in seafloor hydrate mounds (Chapman et al., 2004) may be surprisingly large, their estimates of hydrate in subjacent sediments do not take into account the likelihood that the BGHS may have been upwarped substantially and the conditions for hydrate stability may not continue downward for the anticipated full GHSZ thickness.

Finite element modeling of fluid and heat flux associated with bathymetric mounds in the Gulf of Mexico using high resolution seafloor thermometry has suggested that fluid flux is likely constrained to the immediate vicinity of a quiescent mound (where there is no active venting). When active, the fluid flux through the mound was at least two orders of magnitude higher than in the surrounding sediments and was associated with venting of methane. Thermal modeling of the thermal profiles results in a BGHS consistent with the top of gas estimated from seismic data, i.e. about 60-70 mbsf at the mounds. Most, if not all, of the apparent push up of the BGHS can be attributed to elevated temperatures associated with fluid advection. Elevated chloride concentrations (up to 2x seawater) indicate that hydrate formation is associated with the advected fluid. Coffin et al. (2005), in nearby Keathley Canyon (Fig. 4.14), also found both a greater vertical methane flux and thermal anomalies near both a fault and a seafloor sediment mound. Pore water chemistry also indicates abnormally high methane flux. In the vicinity of surface methane groundwater anomalies, the BSR is shallower, weaker, and discontinuous, indicating warm upwelling groundwater in excess of that which can be chilled by absorption of heat from the seafloor heat sink. Comparison of pore water chloride and methane concentrations indicated hydrate destabilization. Henriet et al. (2001), Vaneste et al. (2002) and van Rensbergen et al (2003) also find abnormal heat flow associated with vents elsewhere.

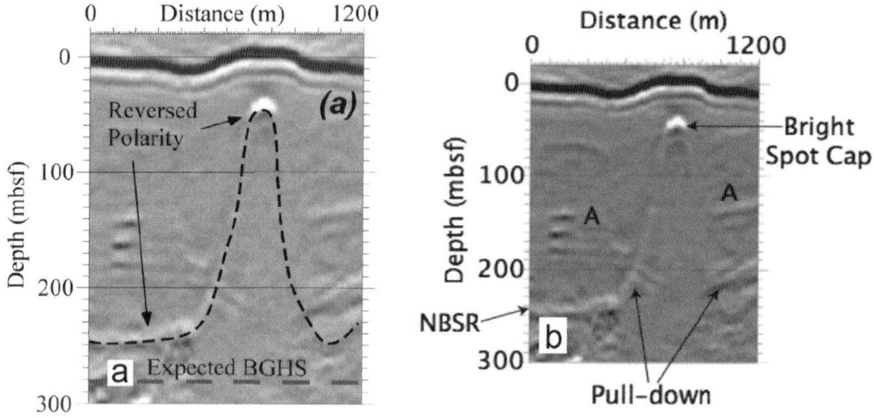

Figure. 4.16. Upward push-up of BHSZ in straddling a vent seep penetrating the HSZ, which brings deeper groundwater and gas from beneath the HSZ to the seafloor. In the Atwater Canyon area, a bright spot correlates with free gas at the top of the HSZ in figure 4.15 under mound F (Fig 4.17). Reflection seismic image produced by Fred Snyder, WesternGeco (a Schlumberger/Baker Hughes company) and kindly provided by W.Wood. a, dashed line shows interpreted BGHS cone merging with BSR at the BGHS. A, accentuated impedance contrast areas within GHSZ. b. Pull-down of reflections within the BSR cone indicates slower acoustic velocities than outside the cone within the area of hydrate stability.

Any upwelling of groundwater that is too warm to allow the spontaneous growth of hydrate into the GHSZ has the potential to dramatically alter the position of the BSR, which follows the BGHS.

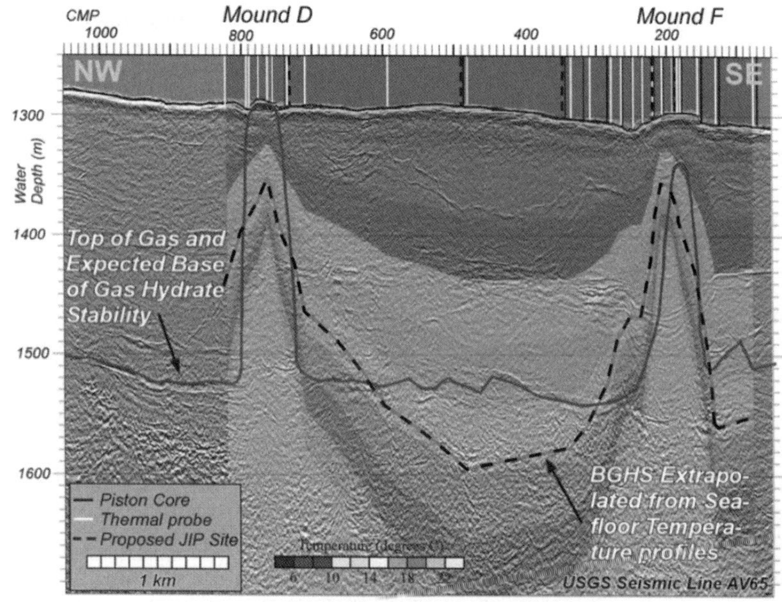

Figure 4.17. Thermal structure beneath seafloor vents in the Atwater Area, northern Gulf of Mexico. Used with kind permission Dr. R. Hagan (Hagan et al., 2004). Mound D is to the left of the section on figure 4.15. The figure is from a color image and does not show the temperature scale well. BGHSZ calculated in two manners noted with arrows. Top of the gas and expected BGHS coincides with the BSR beneath the hydrate bright spot beneath mound F.

Shallow sediment temperature profiles and multichannel seismic profiles across both mounds show a direct relationship between seismic features that indicate both the presence and absence of hydrate in appropriate relation to the position of the temporarily altered BSR (Fig 4.17). The seismic reflection believed to mark the base of the BGHSZ, is indicated about coincident with the top of the sub-BSR. Sediment temperatures are extrapolated to depth (using conduction only) and superimposed on the multichannel seismic profile (Wood et al., 2004). Lines tracing the seismic reflector believed to mark the base of gas hydrate stability as modeled by the temperature data is very similar (Fig 4.17). Mound D has no bright spot, and the conditions of hydrate instability found beneath the BGHS reach the seafloor.

Examples of seafloor venting that do not dramatically upwarp the BGHS also occur. Gorman et al. (2002), show seismic profiles with a BSR only slightly disrupted below seafloor vents. Reflections in sediments intermediate between the BSR and the seafloor vents occur at a number of levels associated with apparent bright spots and wipe-out that they regard as indicating the presence of free gas in the chimney system. This indicates that gas pathways to the surface are not continuously open. It is likely that the gas in the chimney structure formed hydrate in excess to that needed to line the chimney and blocked the pathway. Further ascending gas became trapped within the HSZ, unable to form hydrate because it was separated from groundwater by solid hydrate, through which two-way diffusion that would consume the gas is a very slow process.

4.8. Implications for Hydrate Concentrations not Directly Associated with a Seafloor-simulating BGHS

The geology of sedimentary strata and the presence of a BGHSZ control the rate at which gas and HFG-enriched groundwater can migrate in porous sediments. Where groundwater recharge rates in GHSZs are high, economic hydrate deposits have the potential to form. Where the BGHS is roughly parallel with the surface, a dipping attitude of the sediments exposes them to infiltration by mineralizing groundwater (Fig. 4.12). Where porous sediments above the base of a normal BGHSZ that have the potential to host high grade hydrate deposits are either not in a position to be exposed to the subjacent mineralizing solutions or are sealed at a lower elevation by hydrate. This situation causes a push-up of the BGHS by establishment of a thermal structure related to upwelling vent fluids. The result is a potential to cause hydrate mineralization throughout the GHSZ, depending on the availability of suitable porous horizons and groundwater response.

There are important ramifications of the temporary alteration of the groundwater plumbing system that may be introduced by fluid venting upward through a GHSZ. The primary effect of BGHS push-up is dramatic increase of the BGHS surface area. Instead of a normal near-horizontal BGHS (NBGHS, Fig. 4.1-2), a cone of sediment beneath the pushed-up BGHS is rendered unstable for hydrate because of heating (Fig 4.16) caused by the upwelling warm fluids. The upwelling warm HFG-saturated fluids and gas (Fig. 4.18, F&G) establish a thermal equilibrium boundary, probably of both a conductive and convective nature, wherein the temperatures below the boundary are not suitable for hydrate stability. Above the boundary the conditions are relatively normal for a normal BGHSZ, with the exception that the thermal gradient between the BGHS and the seafloor is steeper. In addition, the steeper-dipping, temporarily pushed-up BGHS (Fig 4.18-1) becomes much more cross-cutting to bedding compared with a normally disposed BGHS. This mechanism for producing a temporary cross cutting situation between strata and BGHSZ is particularly important where sedimentary strata lie about parallel with the seafloor (which would be the common case for most near-seafloor sediments). This geometry of sediments and the temporary BGHS is thus suitable to introduce groundwater HFG fluids laterally into porous sediment that may occur within the GHSZ. Whereas chimneys essentially insulate the escaping HFG from the groundwater in surrounding sediments within the GHSZ, the vent structures offer the main opportunity to allow HFG-enriched groundwater (with/without fine gas bubbles) to percolate through porous strata where hydrate growth would be promoted.

These strata would otherwise be inaccessible to mineralizing groundwater solutions located beneath a normal BSR.

Venting consists of an active phase during which the warm fluids beneath the BGHSZ penetrate through the zone more rapidly than they can be cooled. The rate of fluid and gas venting can vary considerabley (Chen and Cathles, 2003). In certain circumstances, the rate of fluid ascent is such that temperature conditions unsuitable for hydrate formation reach the seafloor within the vent. Venting may persist or be periodic, depending on local groundwater conditions and the rate of ascent of subjacent gas and fluids from depth (for instance, focused flow from a subduction complex). At some point, however, active venting either ceases or jumps to another location, and the venting enters a recessive phase where upwelling of fluids diminishes and the normal temperature structure related to cooling of the seafloor by the cold bottom waters is reestablished.

Active venting (Fig. 4.18, 1) has the potential to establish an cone-shaped, pushed-up BGHS that may reach to the seafloor (Fig 4.17, mound D). Where the apex of the BGHS does not reach the seafloor, ascending vent waters cooled by heat exchange with the seafloor, fall below the temperature where hydrate is stable. Pockmarks (Fig 4.18-1, P) will almost certainly form above vents whether the apex reaches the seafloor or not as surplus HFG is released into the sea. Hydrate is likely to form immediately above the phase boundary in the thin GHSZ apex (Fig 4.18-1, a) and laterally, along the flanks (Fig 4.18-1, b). Hydrate formed in the GHSZ in this manner has the potential to produce seismic bright spots denoting hydrate concentration (4.18-2, T) that are well away from the normal BGHS (Figs 4.16, 4.17), where most of the mineralization would be anticipated (Fig. 4.12).

Apex or terminal hydrate enrichment may occur irrespective of lithology as the gas flux may be high enough to indurate the sediment through the formation of innumerable coalescing secondary porosity fractures that fill with hydrate owing to the pressure of the ascending gas. Solid hydrate, however, will continue to be grown from the dissolved HFG in the ascending fluid. This apex zone could migrate laterally, as suggested by Milkov and Sassen (2002, Fig. 3), so long as gas flux (HFG in vent groundwater) is maintained. Where hydrate is developed on or near the seafloor, however, it will dissolve through diffusion processes (2.6.3) very rapidly following cessation of vent activity. Seawater will return to its strongly undersaturated natural state.

A further possibility is that venting may be cyclical, related for instance to fluids being driven out of a long-standing crustal structure by large scale tectonic activity that waxes and wanes in their gas saturation and the volume of water venting over time. Where vent activity ceases in a series of pulses such that a new and recessional thermal equilibrium boundary is formed at a lower elevation, while HFG venting through the BGHS continued, recessional apex hydrate deposits (Fig 4.18-2-R) could form. Hydrate apex deposits could also form stable compound hydrates (2.5.2) below the position of a methane hydrate apex deposit if the gas mixture changed to include higher proportions of higher density hydrocarbon gas.

Of greater importance is the potential for mineralization of high porosity sediments that occur laterally to a push-up BGHS cone (Fig 4.18-1, b) in which the sides of the cone are transitional passages for HFG-enriched fluid to migrate into suitable host strata. This geometry has the potential to bring HFG-charged groundwater into direct contact with essentially the whole of the stratigraphic succession within a GHSZ in a manner that could cause the HFG groundwater to pass into the porous strata and traverse into pressure-temperature conditions of hydrate stability and form isolated lateral deposits (Fig 4.18-2, L). Porous strata that dip upward toward the seafloor that could host groundwater interchange between the GHSZ vent and the seafloor would be particularly likely to be mineralized.

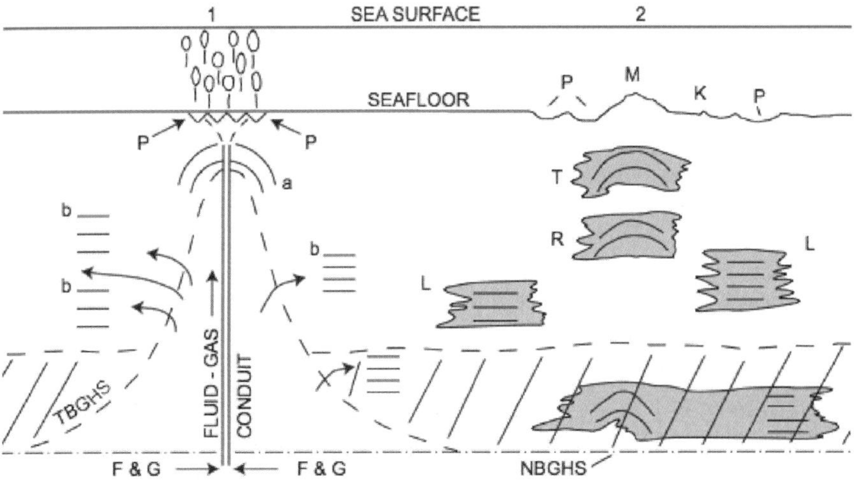

Figure 4.18. Vent model for hydrate deposits formation from circulating groundwater. 1. Active vent (see Fig. 4.15). Gas/fluid conduit warms adjacent strata, causing upwarp of the base of normal gas hydrate stability (NBGHS) into a temporarily pushed-up cone (TBGHS). Arrows indicate movement of dissolved HFG-enriched fluids upward and laterally in GHSZ. a, apex hydrate where BGHS does not reach the seafloor, b, lateral hydrate. P, pockmarks. 2. Potential for hydrate deposits following cessation of vent activity and reestablishment of NBGHS. Shaded areas with lines are potential hydrate deposit locations. P, pockmarks, M, mounds, K, rough ground. T, terminal hydrate deposit, R, recessional hydrate deposit, L, Lateral hydrate deposit. Lowermost in hachured zone is the likely position of a deposit whose formation would not necessarily depend on vent activity.

Thermal modeling of the vents is represented as being essentially symmetrical by Wood et al. (2004) and Hagen et al. (2004), because that is the simplest thermodynamic model. It fairly accurately describes an actual case revealed by the reflection seismics (Figs 4.16, 4.17). Where passageways exist, however, for instance along dipping thrust faults in a subduction complex such as is seen on the Cascadia margin (Trehu et al., 2003), or in upright faults in the Blake Ridge (Dillon et al., 2001), vent passages may follow these features. Upward flow can be expected to jump from one course to another where buoyant force and the mechanical framework of the sedimentary succession allow.

Mineralization of porous strata at different levels in the GHSZ, associated with hydrocarbon gas and HFG-enriched vents was suggested by Milkov and Sassen (2002), but without a distinct mineralizing mechanism (They refer to gas transport, which does not have the potential, without first being dissolved in groundwater, to promote large scale crystallization of solid hydrate). Their basic suggestion, however, that high-grade hydrate deposits could be found in association with vents appears to be physically achievable.

BGHSZ push-up in association with upwelling HFG-rich fluids also may occur on a much larger scale than has been seen to date in in the northern Gulf of Mexico. Multiple push ups of BSRs are seen on seismic sections of the Atlantic margin of Canada off Nova Scotia, where the seafloor espression of the vents are sediment or vent mounds (Mosher et al., 2005, Fig. 4). Asymmetrical pushed-up BGHSZ immediately beneath a mud volcano in SW Asia (Fig. 4.19, MV) can be traced laterally into conventional BSR on either side of the vent structure. The mud volcano appears to be relatively young because of its conical form, and not to have suffered any significant subsea erosion. Second, the push-up does not reach far toward the seafloor now. It is not clear as to whether the BGHS is in the process of migrating upward, has reached a steady-state, or is in the process of receding toward its original position that may be marked by discontinuous relics of a BSR (Fig 4.19, R). The discontinuous reflections (R) are about level with the BSR, discounting a general but gentle upwarping over the anticlinal structure that appears to have slightly affected the local geothermal structural. These could be relic hydrate deposits, a chemical precipitate associated with the BSR, or zones of resedimentation along the BGHS caused by hydrate dissociation and sediment collapse into the plane of the BSR. Accentuation of the impedance contrasts (G), beneath the BSR, reveals the location of gas trapped in strata beneath sediment whose porosity is sealed with hydrate. If large enough amounts of gas were trapped in this fashion, they would constitute conventional gas traps, even though their trap is related to hydrate sealing of porosity.

Acoustic accentuation within the GHSZ is commonly seen to the right of the assumed central fluid-gas conduit and blanking (L) occurs to the left (Fig. 4.19). Both of these features might indicate the presence of hydrate in a concentrated form, with the different seismic expression caused by some subtle factor controlling hydrate growth for which there is no ready explanation. It is unlikely that some sedimentary or stratigraphic factor would be responsible for the different seismic response. The anticline is symmetrical (with folding apparently continuing as sediment was being deposited). It is unlikely that any systematic difference in depositional environment existed on either side of the anticline. Apparently sedimentation affected the thickness on either side and marks a knick line along which the seafloor somewhat deepens. It is also unlikely that the difference in seismic response could be an acoustic artifact, so long as no changes were made in the data acquisition or processing set-up. There is a strong possibility, however, that the poor signal return areas are an artifact. They may have been caused by an automatic gain control (AGC) response to the change in bottom-bounce signal as the array began to pass over the mud volcano, or some other seafloor-induced interference. Acquisition could have been biased toward a higher frequency range, for instance, which would could cause significant differences in apparent response. If the difference in seismic response is real, the blanked zone might be correlated with potential hydrate development. In the accentuated zone, individual strata also could be idenntified with preferential hydrate mineralization. Speculative shallow drilling targets can be picked. Note that the existence of a BSR alone, or its persistence, cannot be regarded as a critical factor in defining potential economic hydrate deposits.

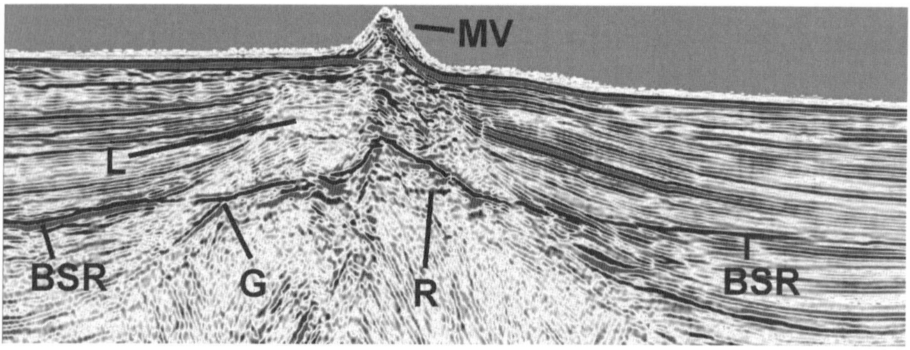

Figure 4.19. Unscaled diagram of reflection seismic lines showing BSR passing across an anticline but being pushed up beneath a mud volcano (MV). Data is 2D seismic record from the Makassar Straits, Indonesia. Seismic data provided courtesy of TGS-NOPEC Geophysical Company. G, gas in sediments below BSR terminated at BSR. R, Possible relic of BSR prior to push up and establishment of BSR 'cone'. L, possible lateral hydrate-enriched blanked zone.

In addition to the temporary dislocation of BSR where a vent pushes up the BGHS, BSR may move up or down in response to environmental changes. In the Nankai area of the southeast Japanese continental shelf, there are several seafloor parallel discontinuous reflectors (Fig. 4.13) which may reflect the position of older BSRs. These BSRs may have been abandoned owing to continued sedimentation or crustal uplift that has caused the GHSZ to migrate upwards (Lowrie and Max, 1993). A number of discontinuous reflectors also occur on other seismic lines in the Makassar Straits region of Indonesia between

Sulawesi and Kalimantan, which appear to reflect a general upward movement of the BSR (Fig. 6.20). Ideally, correlation of these features would be able to show the effect of combined uplift / relative sedimentation rate from place to place. No vents appear to have been active here, although some faulting of reflectors appears to exist and the seafloor appears to be affected by growth of the asymmetrical upright anticline. Displacement of the BSR among the steeper dipping strata of the anticline may be due to differential groundwater flow into the GHSZ in more porous beds.

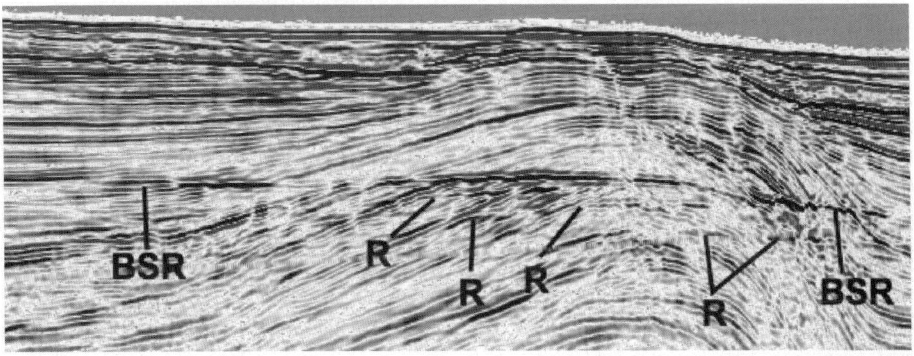

Figure 4.20. Unscaled diagram of reflection seismic lines showing BSR passing across an anticline. Data is 2D seismic inforamation from the Makassar Straits, Indonesia. Seismic data provided courtesy of TGS-NOPEC Geophysical Company. R, Possible relic of BSRs.

Prospecting for one-time vent structures that may have formed economic concentrations of hydrate deep enough into the GHSZ so as to inhibit rapid dissolution, can best be done through the use of side-scan sonar imagery (Miles, 2003). This technique can identify palaeostructructural zones of pockmarks, mounds, and rough ground (Fig 4.18-2, K, M. P) formed by coalescing or interfering pockmarks and mounds. Whereas active venting is liable to be related to pushed up BGHSZ and immediately subjacent conditions which are not generally suitable for the formation of hydrate deposits. Inactive features may indicate the general vicinity in which a groundwater system had been active. Within these systems, or adjacent to them, significant hydrate mineralization may have developed during the recessive phase of venting.

Reflection seismics, however, are required to better assess the subsurface situation for the likelihood of hydrate concentrations. It is not known whether side-scan imagery of the seafloor exists in the vicinity of the trace of the anticlinal hinge whose cross section is shown in Figure 6.20. Thus, it is not possible to rule out minor venting that did not push up the BSR in the same manner seen in Figure 6.19.

4.9. EXAMPLES OF STRATABOUND MINERAL DEPOSITS

Methane hydrate deposits are physically similar in many ways to metaliferous strata-bound mineral deposits, in particular, Mississippi Valley Pb-Zn deposits. Mississippi Valley type mineralization displays a consistent history transport of mineralizing reactants by movement of groundwater and by diffusion of mineral components in the groundwater matrix, and deposition caused by pressure-temperature and saturation changes that cause and localize mineralization. Some examples are:

The Viburnum New Lead Belt in southeast Mississippi, which is the world's largest lead producing province and was deposited by the action of sediment-hosted connate fluids in porous Cambrian dolostones over 200 million years later during Carboniferous times (Robb, 2005). Mississippi Valley type sediment-hosted mineralization has also been deposited through the action of meteoric groundwater and hydrothermal solutions that may have a high degree of connate water lineage.

The George Fisher Zn-Pb-Ag deposit, situated 25 km north of Mount Isa, Australia was precipitated from groundwater carrying dissolved metals in sediment porosity nodular carbonates (Chapman, 2004). The deposit is hosted by the ~1655 Ma Urquhart Shale in the Western fold belt of the Mount Isa inlier and contains eleven mineralized stratigraphic intervals. Transgression of the stratigraphy by mineralization may represent a surface parallel change of chemical potential with depth that would be analogous to a bottom-parallel chemical potential feature similar to that of the gas hydrate system.

Lee and Wilkinson (2002) identify a paragenetic series of dolomitization and sulfide mineralization as part of a diachronous, prograding mineralizing system that formed in the late Courceyan, ~352 to 355 Ma Waulsortian Limestone micrite. The mineralized zone was part of a groundwater system that was open to circulating subjacent and lateral groundwater and also venting from the seafloor.

The Kuh-e-Surmeh carbonate-hosted zinc-lead deposit, located within the Zagros Mountains in southwestern Iran, is an orogen-related Mississippi Valley type deposit (Liaghat et al., 2000). Ore deposition took place as fillings in the porosity of carbonate sediments. The metals precipitated from basinal brines (15 wt% equiv. NaCl) at low temperatures (less than 200 °C), typically within strata of a Late Paleozoic carbonate platform. The sediments with greater porosity are preferentially mineralized, indicating a groundwater transport mechanism.

The Rammelsberg polymetallic massive sulphide deposit has been the object of mining activity for nearly 1000 years before finally closing in 1988. The deposit is hosted by Middle Devonian pelitic sediments in the Rhenohercynian terrane of the Variscan Orogen in north-central Europe. The deposit consists of two main orebodies that have characteristics of sediment-hosted massive sulfide class of ore deposits (Large and Walcher, 1999).

Mineralization from migrating groundwater fluids may take place even where no significant primary porosity remains. For example, the Nanisivik mine, which is located on Borden Peninsula, northern Baffin Island, about 750 km north of the Arctic Circle is developed in Mesoproterozoic strata. Mineralization took place from groundwater migrating in secondary porosity related to faulting. Sulfides developed in relation to a surface-parallel gas-water interface that controlled precipitation of the sulfides.

Jassim et al. (1999) describe the formation of a stratabound sulfur deposit that was formed by gypsum dissolution followed by sulfate reduction in the presence of hydrocarbons. The process was driven by natural gas-rich groundwater brines rising from underlying hydrocarbon reservoirs that mixed laterally with descending meteoric waters and produced the peculiar oscillating reduction conditions required for sulfide deposition.

4.10. CONCLUSIONS

Geological conditions for the formation of gas hydrate deposits are similar to the conditions for formation of low temperature strata bound metaliferous and non-metaliferous mineral deposition. Permeable strata are preferentially mineralized by reactants brought into the permeable strata by groundwater that may be of connate, meteoric or hydrothermal sources. Both physical movement of water and diffusion of reactants within the water matrix are active mineralizing agents. Minerals are preferentially deposited in coarser grained sediments having superior porosity whereas little mineralization is found in associated sediments in the sequences that had low original porosity. This indicates that actual water flow took place during the mineralizing events.

The physical chemical parameters and constraints governing these stratabound mineral deposits and hydrate formation are analogous in many respects. Marine sediments are rich in connate water. During compaction, which is a normal feature of marine sediments through time, this water is expelled, most of it upward into the sea both through movement of the water through porosity without disturbing the strata and by venting, where considerable disruption of the original stratigraphic character may take place. Also, the formation of hydrate (hydrate expansion) and especially the dissociation of hydrate provide mechanisms for the pumping of groundwater (Max and Chandra, 1998). In addition, especially in the sedimentary prisms of active continental margins, large volumes of natural gas-rich water derived by the tectonic compression of connate water in the marine sediments are pumped into the sea through the GHSZ. Water in marine sediments is both the transport media and one of the principal reactants for hydrate mineralization.

The body of economic geological knowledge available for much sediment-hosted mineralization can be applied to hydrate mineralization in order to assist identification of the location of economic deposits of gas hydrate.

Appendix B1. Thermal Character of Gas and Fluid Mixtures Rising Through the GHSZ.

Considerable volumes of natural gas and water vent through the seafloor into oceans and lakes. In some cases, the venting is dominated by gas, in other cases the venting is dominated by water. There has been little study of the composition of the fluids and the relative volume of gas and fluid emitting from the seafloor and attention has tended to focus on the gas and its composition. Imaging and study of the fluids is much more difficult. Knowledge of the composition of the gases is important to hydrate provincing (6.2) but the alteration of the thermal structure in the GHSZ by transiting gas and water may be of primary importance to the likelihood of formation of concentrated hydrate deposits (4.7). Thus, this appendix considers the relative thermal impact upon the sediments (and hydrate) in an oceanic GHSZ of gas and water rising to the seafloor from below an oceanic BGHSZ to examine the hypothesis that the relative transiting gas-water composition may be an important factor in hydrate formation and disposition.

Observation and examination of reflection seismic data (4.7) suggests that gas-rich GHSZ transiting material may pass through the GHSZ with little thermal perturbation; whereas water dominated material has the potential to alter the thermal structure within the immediate vicinity of the transit course (usually a fault). These are the two end members, each of which is unlikely to be common in nature where gas and fluids are likely to be mixed to some degree, especially during a natural depressurization event (such as the case where overpressured gas blows out along a fault). This modeling is intended to establish whether there is a significant difference between transiting gas and water.

Our model reports the heat transfer from an upwelling mass upwelling to successively lower depths in the sediment. The model assumes that the mass travels to the specified depth and then comes to thermal equilibrium with an infinite thermal reservoir (the sediment). There is no material exchange between the advecting, transiting material and the material it is passing through. Other factors, such as the ability of the particular material within the GHSZ to exchange and transfer heat and the preferential movement of groundwater within porous beds in the GHSZ are not considered, although they may be important to the formation of hydrate away from the thermal effects of transiting material. Although these assumptions are patently unrealistic, for instance, it is known that the temperature of water emitting from a vent is usually colder than the bottom water, which indicates that equilibration has not been reached, the model can be used to estimate end member performance. Actual performance will lie between the end-members and will be case-dependent.

The latent heat of gas and water, in terms of the potential to elevate a certain volume of material within the GHSZ, is different for different examples of gas and water. These differences are factored into the calculations rather than being simply expressed as potential.

The heat that a given volume of material can transfer to the surrounding environment is related to the mass of the material, the heat capacity of the material, the temperature difference between the material and the environment, and any phase changes the material may experience. As material rises from the BGHSZ into cooler sediments it carries with it thermal energy. To a first approximation this thermal energy can be calculated as:

$$Q_{released} = -\sum_{i} m_i * C_{v,i} * (T_{\text{sediment}} - T_{\text{BGHS}})$$

Where $Q_{released}$ is the heat released by the rising mass of material, m_i is the mass of component *i* of the rising material, $C_{v,i}$ is the constant volume heat capacity of component *i*, $T_{sediment}$is the temperature of the sediment at some depth above the BGHSZ, and T_{BGHS} is the temperature at the BGHSZ, and the system performs no work. This is a linear function that increases with decreasing depth as long as material is not changing in amount or state. Since the C_v of water is greater than gas, the contribution of water outweighs the gas contribution. However, at great depth the gas becomes so compressed that it contains considerable thermal energy per unit volume.

As a saturated solution of water rises some gas may evolve depending on the geothermal gradient. The amount of gas evolved will consume heat. Again, to a first approximation this process can be modeled as:

$$Q_{\text{consumed}} = n_{evolved} * \Delta H_{vap}$$

Where $Q_{consumed}$ is the amount of heat consumed while vaporizing the gas, $n_{evolved}$ is the number of mols gas evolved, and ΔH_{vap} is the heat of vaporization of the gas. The total heat flow, $Q_{transfered}$, is:

$$Q_{\text{transfered}} = Q_{\text{released}} - Q_{\text{consumed}}$$

By combining the heat evolved from warm water rising into cooler sediments, evolving gas as a function of Henry's Law, and consuming heat during vaporization, the heat flow can be estimated (Table B1).

At the P/T conditions explored, the effect of the gas can be considered a perturbation on the heat flow due to the rising water. At the greatest depth the gas accounts for 2-4% of the heat flow. As the depth within the BGHS increases, the dissolved gas in the rising fluid causes a decrease the heat evolved because of the heat of vaporization (Table B2). This is a slight underestimate because the total gas evolved at each depth is considered to have carried heat, as free gas, all the way from the BGHSZ while the all of the gas evolves after the fluid rises above the BGHSZ.

Seafloor (km)	BGHS (mbsf)	Temperature at BGHS[1] (°C)	Heat evolved, 1 L rising water[2] (kJ)	Heat evolved, 1 L rising gas[3] (kJ)	Heat evolved, 1 L rising solution[4] (kJ)
1	200	23	29.2	0.95	28.7
2	400	25	58.2	3.76	57.1
3	600	28	87.8	8.4	85.06
4	800	30	117	14.8	112

Table B1. Heat evolved at the seafloor from material rising from the BGHS. [1]Heat capacities, heat of vaporization, and geothermal gradient assumed constant. Geothermal gradient 35 K/km. Heating due to expansion of the gas not calculated. [2] Pure water. [3] Pure methane gas. [4] Methane gas saturated in otherwise pure water at the BGHS pressure and temperature (Appendix A2).

Seafloor (km)	BGHS (mbsf)	Gas process contribution[1]
1	200	-1.8
2	400	-2.5
3	600	-3.2
4	800	-4.2

Table B2. Contribution of gas evolution and heat capacity from a solution relative to rising water. [1]$(Q_{solution\ process}-Q_{rising\ water})/Q_{rising\ water}$. The free gas consumes and exchanges heat.

Figures B1-B4 show the heat flow for several systems as a function of mbsf. Each line represents the heat produced if fluid at the BGHS rose to that depth and then came to thermal equilibrium with an infinite surrounding. In all cases the trend from lowest $Q_{transfered}$ to highest is 1 liter (L) of gas rising, 1 L of methane in water solution rising, 1 L of water rising, 1 L of water plus 1 L of gas rising. All of the systems are linear except the rising methane in water solution, which has a slight upward trend.

There is a significant difference between the rising pure gas and any of the rising water bearing materials, with the pure gas small compared to the water including processes. However, as the total depth increases, the pure gas heat flow increases faster from 3% (1.2 km total depth) to 12 % (4.8 km total depth) relative to rising pure water (Table B3).

Total depth (km)	Heat flow gas/heat flow water[1] (%)
1.2	3.2
2.4	6.4
3.6	9.5
4.8	12.6

Table B3. Heat flow from rising gas increases faster than rising water. Heat flows are for pure rising gas and pure rising water. No phase changes occur.

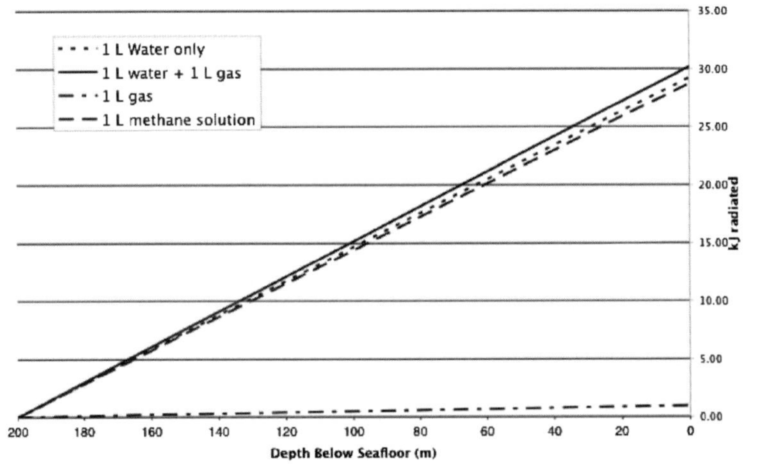

Figure B1. Heat radiated. 1 km seafloor, 200 m BGHSZ, 23 °C.

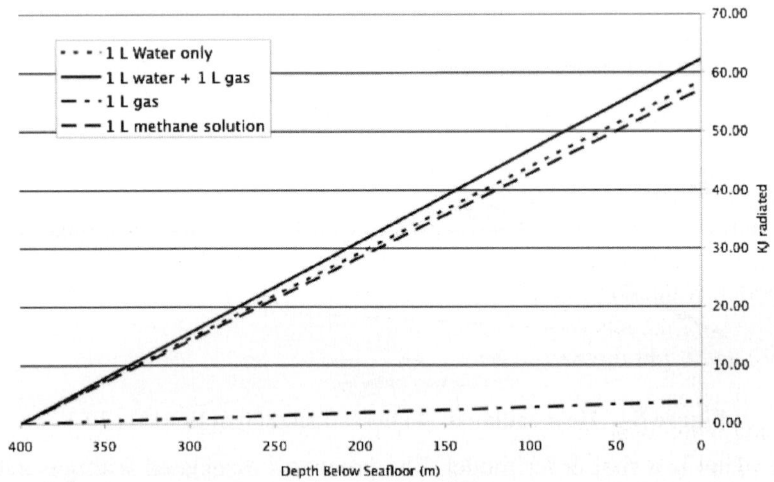

Figure B2. Heat radiated. 2 km seafloor, 400 m BGHSZ, 25 °C.

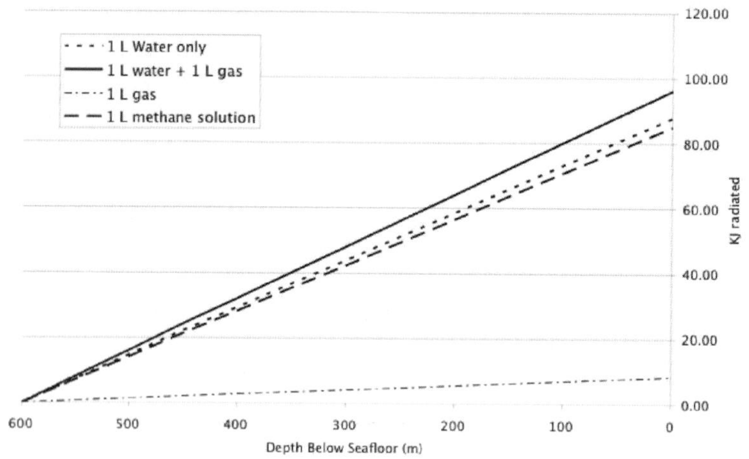

Figure B3. Heat radiated. 3 km seafloor, 600 m BGHSZ, 28 °C.

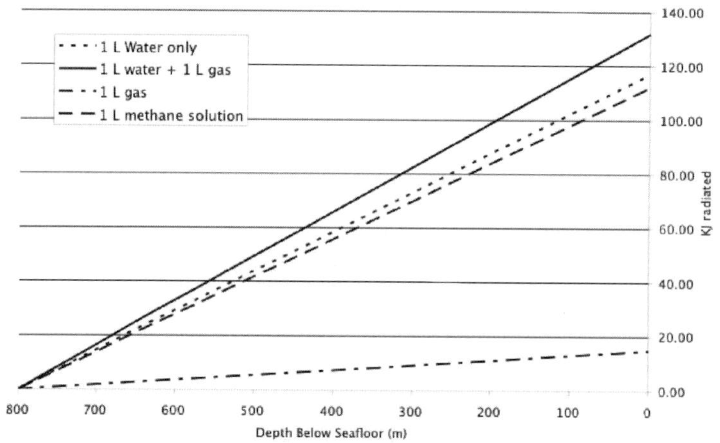

Figure B4. Heat radiated. 4 km seafloor, 800 m BGHSZ, 30 °C.

This is a first order model. The processes associated with gas evolving from solution have been described by linear terms when differential equations are more appropriate. Additionally, the rate of heat transfer has not been taken into account and kinetics are likely to play a significant role in any real system. In other words, the rate of ascent of the material rising is liable to be very important. If the rate of rise is too fast to allow for thermal equilibration, heating will be diminished.

Heating of GHSZ Material by Rising Fluid and Gas

The heating of sediment due to material rising from the BGHS can be calculated using a simple model. A given volume of material, either water or gas, rises from the P/T conditions at the BGHS to some location above the BGHS. The material then releases heat until it is at the same temperature as unheated sediment at that pressure/temperature. The 'extra' heat then is redistributed among the rise material and 1 L of water, which represents the sediment (Fig. B5). This physical path is possible since Q is a state function. The temperature of the sediment plus rise fluid will be:

$$T_{\text{equilbrium}} = \frac{Q_{\text{released}}}{(m_r C_r + m_l C_l)} + T_{\text{depth}}$$

Where m_r is the mass of the rising material, C_r is the heat capacity of the rising fluid, m_l is the mass of the fluid above the BGHS that the rising material is coming into thermal contact with, C_l is the heat capacity if the fluid above the BGHS, T_{depth} is the initial temperature of the sediment above the BGHS, and $T_{equilbirium}$ is the temperature that the rising material and the fluid above the BGHS will come to once heat is distributed evenly.

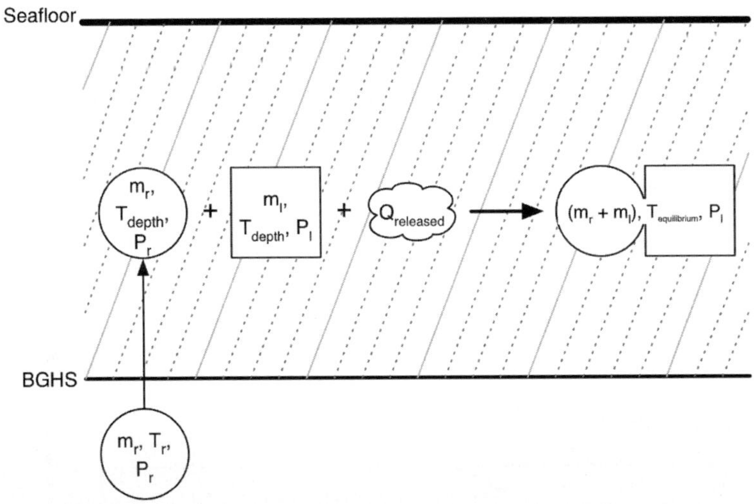

Figure B5. A mass of 1 L of rising material from at or below the BGHS to a location within the GHSZ gives off all of its excess heat. The excess heat is then redistributed to the mass of rising material plus 1 L of water. This calculation path is possible because the heat of a system is a state function.

This model does not include any heat loss of the rising fluid until it reaches the test depth. At the test depth the fluid and sediment come into equilibrium. Additionally, the model does not take into account any heat loss from the sediment heat flux. Phase changes, frictional heating, work performed by the system, amongst other variables, which can alter the results in relatively small manners, are not included.

The Temperature rise is dependent on how far above the BGHS a bleb of material raises before it comes to equilibrium and how much mass is in the bleb. For a rising plug of water, the mass is essentially constant regardless of start conditions. For a rising plug of gas, the mass depends on the P/T conditions at the BGHS. Figures B6-B9 show the temperature increase of 1 L of in-place solution, when exposed to the rising material.

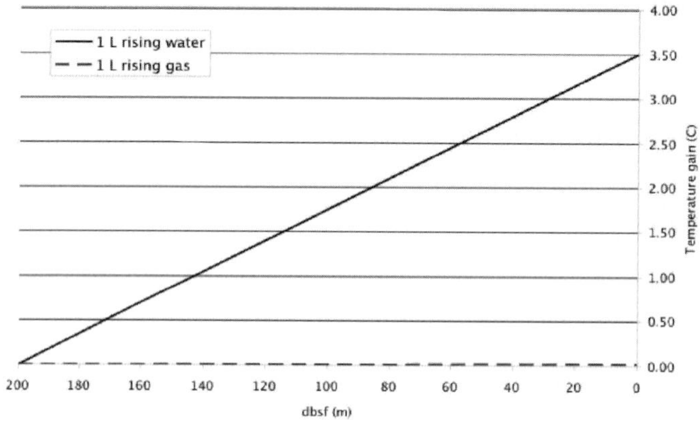

Figure B6. Equilibrium temperature. The rise in temperature of 1 L of water at a given pressure-temperature condition after exposure to 1 L of material that was at BGHS pressure/temperature conditions. The BGHS is 200 mbsf. The seafloor is 1 km msl.

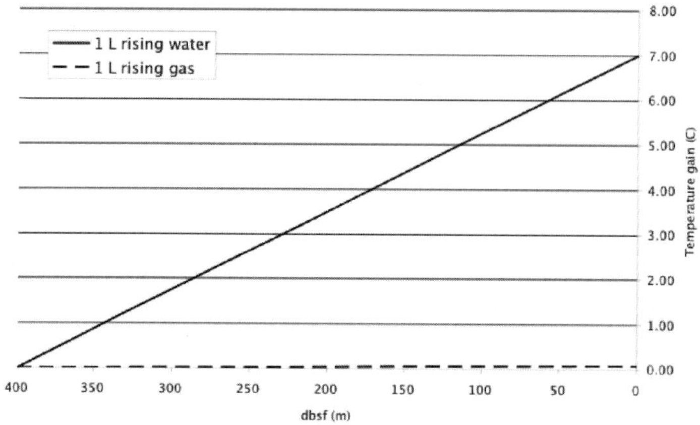

Figure B7. Equilibrium temperature. The rise in temperature of 1 L of water at a given pressure/temperature condition after exposure to 1 L of material that was at BGHS pressure/temperature conditions. The BGHS is 400 mbsf. The seafloor is 2 km bsl.

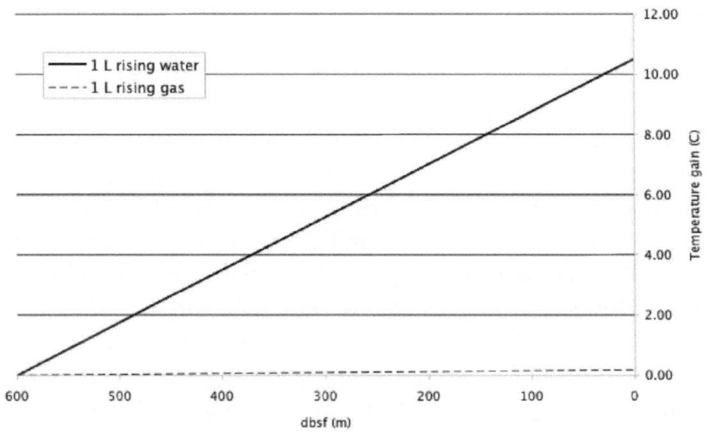

Figure B8. Equilibrium temperature. The rise in temperature of 1 L of water at a given pressure/temperature condition after exposure to 1 L of material that was at BGHS pressure/temperature conditions. The BGHS is 600 mbsf. The seafloor is 3 km.

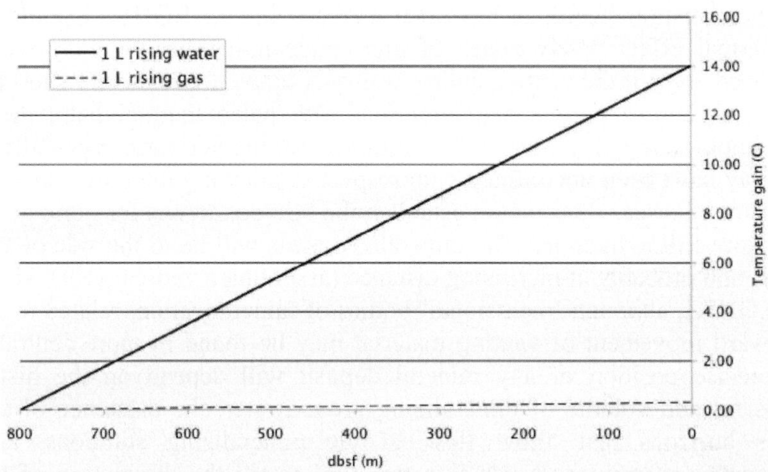

Figure B9. Equilibrium temperature. The rise in temperature of 1 L of water at a given pressure/temperature condition after exposure to 1 L of material that was at BGHS pressure/temperature conditions. The BGHS is 800 mbsf. The seafloor is 4 km.

There appears to be a much greater potential for disruption of the geothermal gradient in a GHSZ in the immediate vicinity of the transit courses where the rising material is dominated by water. In transit courses dominated by gas, the significance of the gas to heating, relative to water, is in the region of < 2%. The temperature change due to 1 L rising gas on 1 L of sediment water ranges from 0.6% (1 km seafloor) to 2.2% (4 km seafloor) compared to rising water. It would appear that where pressure-temperature conditions in the vicinity of a transit course through a GHSZ have been altered so that the BGHS has been effectively pushed up (4.7.2), water might be the dominating material for thermal modification of the local geothermal gradients, even though it might first appear that mainly gas is being emitted from the seafloor into the seafloor water. The water may also be bringing other dissolved material to the seafloor from deeper in the sediment column, and in larger volumes, than if the venting material was dominated by gas.

Hydrate mineralization within a GHSZ depends on delivery of dissolved HFG to porous beds through which the groundwater carrying the mineralizing solutions can pass. Because the ascending vent material is depressurizing as it rises, it is likely that the HFG in the groundwater will be very nearly saturated, even before it begins to cool at the margin to the thermally affected halo. This is an ideal case for rapid growth of solid hydrate and high pore fill of porous beds through which the mineralizing solutions can pass.

Where indications of a thermal halo or 'cone' are found to exist (on reflection seismic records or any other method) and the BGHSZ is pulled up by the thermal effect, likely zones of high-grade mineralization may well have developed, even if the venting course was only active for relatively short periods of time. Where a porous bed intersects with paleo thermal halo, the likely exploration zone will be generally up-dip in porous horizons, especially where they may have been unconfined with respect to allowing the pore water to seep up dip into the sea. This would establish the best conditions for rapid growth of solid, pore-filling hydrate. The mineralized areas will lie to the side of the vent course, and probably at increasing distance (assuming a vertical vent) with depth in the GHSZ, although 'recessional' bodies of mineralization, related to the end of upward movement of venting material may be found in more central areas. The precise position of any mineral deposit will depend on the history of movement and volume of mineralizing groundwater, the existence of suitable porous horizons that allow flow of the mineralizing solutions, and the maintenance of a positive gas flux that will retard the dissolution of hydrate concentrations.

Evidence of paleoventing on the seafloor, especially where this may have continued for longer periods of time, may provide a central point from which to explore outward, depending on the geology.

Chapter 5

State of Development of Gas Hydrate as an Economic Resource

5.1. INTRODUCTION

An important step along the path to developing a commercial hydrate natural gas resource is valuation of discrete, continuous deposits that would be the equivalent of conventional hydrocarbon fields. Where there are only limited drill hole data to provide ground truth, parts of a deposit can be valued on an aerial basis using assumptions for lateral continuation of mineralization for some distance away from drill holes, in the same manner used for valuation of metallic mineral deposits. Once a viable economic target has been identified, further work, including fine-scale drilling, is justified. Regardless of how interesting or important to basic science issues, such as the part hydrate plays in the Earth's carbon cycle and global climate change, or in geological seafloor stability, unless a resource potential can be proven, industrial hydrate science will remain focused on the issue of prevention and remediation of unwanted hydrate and on seafloor safety and drilling issues. Once a valuation can be established, the worldwide and regional estimates of available hydrate natural gas, and the discussions associated with the accuracy of these estimates (Collett, 2003, Table 2), can be left to speculation. Collett (2003, Table 3) shows substantial estimates for in-place gas and hydrate beneath an area (in resource per km^2) and potential basis for field development. These resource estimates justify further work on development of the hydrate resource.

There are a number of elements that will comprise production models for commercially recovering hydrate gas deposits. The first of these is achieving controlled conversion of hydrate to its constituents by dissociation of the hydrate and the formation of discrete gas concentration that can be recovered within the hydrate reservoir. Only one deposit has been the subject of verifiable hydrate conversion testing. In 2002, the Mallik permafrost hydrate was the subject of a short production test. This test confirmed that a sustained production of gas could be maintained. Other aspects of reservoir character were also determined during the testing. While only two wells have been used in hydrate research at Mallik (2L-38 and 5L-38), a total of 20 wells have been drilled on the Nankai offshore area of the SE Japanese continental margin in a hydrate program (Takahashi and Tsuji, 2005). Although the Mallik and Nankai locations are in different parts of the world and in very different geological settings, the Government of Japan has provided most of the funding for research at both sites.

Michael D. Max et al. (eds.), Economic Geology of Natural Gas Hydrate, 191–206.

The history of evaluation and testing at these two sites has been short but demonstrates the telescoped nature of rapid hydrate development.

Other locations, principally the Blake Ridge on the SE continental margin and the Hydrate Ridge locality off the NW continental margin of the contiguous United States have been drilled and subject to extensive scientific research. A number of other localities have been drilled (Kvenvolden and Lorenson, 2001) on continental margins. Other permafrost localities are also known. Nankai and Mallik are the only sites that have been selected for study on any commercial basis such as, proximity to infrastructure, markets, other shared resources, water pressure-depth, and hydrate growth models.

5.2. MALLIK

Mallik is the best known locality for permafrost hydrate. The locality is located in the Mackenzie delta of Arctic Canada (Fig. 5.1) amidst many other wells showing indications of definite and probable hydrate (Majorowicz and Osadetz, 2001; 2003; Osadetz and Chen, 2005). The site can only be occupied for drilling during the winter, when the ground is frozen.

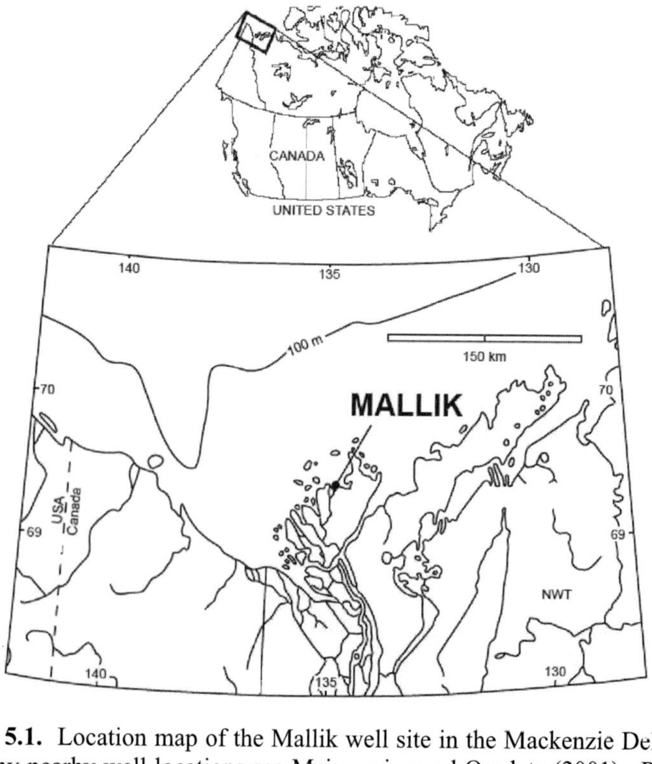

Figure 5.1. Location map of the Mallik well site in the Mackenzie Delta. For the many nearby well locations see Majorowicz and Osadetz (2001). Redrawn from Osadetz & Chen (2005).

5.2.1. Background

The Mallik structure is a four-way anticlinal feature located beneath Richards Island in the Mackenzie Delta of Northern Canada. The initial well on the structure (Mallik L-38) was drilled by Imperial Oil in 1972 and was believed to have encountered at least 10 hydrate-bearing units with a total of 361 feet (110 meters) of hydrate-bearing sands between depths of 2657 an 3615 feet (810 and 1102 meters) (Collett, T.S. and Dallimore, S.R., 2002). Analysis of wireline logs indicates that gas hydrate is also present in 25 additional wells in the Mackenzie Delta and adjacent Beaufort Sea (Smith and Judge, 1993) (figure 5.1).

For 25 years following the determination that large amounts of gas hydrate are present in the Mackenzie Delta and Beaufort Sea, little was done with this knowledge due to the discovery of over 1.5 billion barrels of oil and over 10 TCF (283 MMCM) of conventional gas in the region. Without pipelines, neither the conventional nor unconventional gas could be developed.

5.2.2. The 1998 Mallik Program

As interest grew in the 1990s regarding the large resource potential of gas hydrate (especially in Japan), the Mallik structure was a logical candidate for study because of its known high concentration of gas hydrate. For the Japanese, the Mallik location had the added advantage of bearing many similarities to offshore Japan in terms of stratigraphy and temperature-pressure conditions. A collaborative drilling and evaluation program was carried out in 1998 with the drilling of the Mallik 2L-38 well as a joint project of the Japan National Oil Company (JNOC) and the Geological Survey of Canada (GSC), with technical participation by Japan Petroleum Exploration Company LTD (JAPEX) and the U.S. Geological Survey (USGS).

Mallik 2L-38 was located 361 feet (100 meters) from the Mallik L-38 and was drilled to a total depth of 3773 feet (1150 meters). The field investigation program included the collection of permafrost and gas-hydrate-bearing core samples, downhole geophysical logging and a vertical seismic profile survey. The field studies were followed by laboratory and modeling research that evaluated the sedimentology and the geochemical, geophysical and petrophysical properties of the Mallik gas hydrate accumulation. The Mallik 2L-38 validated the presence of large amounts of natural gas stored in the structure as gas hydrate, with total gas in place of approximately 3.9 TCF (110 MMCM). A significant result of the program was the confirmation that the greatest concentration of natural gas occurs within sands and gravels. Some zones had 80% pore saturation by hydrate. In contrast, finer grained sediments contained little or no hydrate.

5.2.3. The 2002 Mallik Program

The 1998 program at Mallik successfully investigated the origin and occurrence of gas hydrate, and led to the development of new methods of drilling, coring, and evaluating hydrate-bearing strata; however, the producibility of gas hydrates was not evaluated. In 2001 the GSC and JNOC established a wider partnership, that included GSC, JNOC, USGS, U.S. Department of Energy, India Ministry of Petroleum and Natural Gas (MOPNG) with the Oil and Natural Gas Corporation Ltd. (ONGC) and Gas Authority of India, Limited (GAIL), GeoForschungsZentrum (GFZ), and the Chevron-British Petroleum-Burlington joint venture group. The program was carried out in collaboration with the International Scientific Continental Drilling Program and Imperial Oil Ltd.

Three wells were drilled under the "Mallik 2002 Research Well Program". Two were observation wells, designated Mallik 3L-38 and Mallik 4L-38. The third well in the program was a production well designated Mallik 5L-38. The 2002 Mallik program recovered over 480 feet (150 meters) of high quality cores. These cores contained hydrate-bearing sands and gravels from several intervals between 2887 and 3773 feet (880 and 1150 meters) depth. Preservation of the hydrate within the cores has allowed for detailed evaluation of the occurrence of hydrate within the sediment at both microscopic and macroscopic scales. The analyses that have been conducted or are ongoing include kinetics of gas hydrate dissociation, petrophysical properties, geotechnical properties, and molecular chemistry.

As with the Mallik 2L-38, the highest concentrations of hydrate occurred in coarse-grained sediment, with finer-grained section containing little or no hydrate even in silt units located between hydrate-bearing sands.

One of the goals of the 2002 program was to conduct geophysical studies that could quantify gas hydrate distribution. Downhole and cross-hole geophysical measurements were carried out along with surface studies. Of critical value were measurements of the permeability of hydrate-bearing units and the occurrence of natural fractures. These investigations have provided data for detailed geophysical models that will be of great value in determining future commercial drilling sites.

Of particular significance for consideration of the viability of gas hydrate as an energy resource was the production tests carried out by the Mallik 2002 Research Well Program. An extended production test might have been optimal from an industry standpoint, but logistical and budget constraints made that impractical. Instead, a series of carefully controlled production experiments were conducted that tested the response of hydrate-bearing sediments to heating and depressurization. The highly accurate measurements that were made allow for the calibration and refinement of reservoir simulation models. These, in turn, are capable of predicting the long-term response of reservoirs.

The three depressurization experiments showed that a reduction in formation pressure alone could yield gas production. Production was shown to be enhanced by artificially fracturing the reservoir. A significant additional observation was that hydrate-bearing formations are more permeable than previously thought (Mallik 2002 Partners, 2003).

In addition to the depressurization experiment, a five-day thermal heating experiment was conducted on a 58 foot (17 meter) interval of sediment with very high hydrate content. The heating was carried out by the circulation of warm drilling fluid in the wellbore past perforations in the casing. The bottomhole temperature was held constant in excess of 50° C. Gas was continuously produced during the test, with a maximum flow rate of 53,000 cubic feet (1500 m^3) per day. This volume is quite small relative to that expected from a conventional natural gas test; however this volume reflects the controlled, experimental nature of the Mallik test. With a higher heat source of longer duration, the production rate would certainly have been much greater. Some concerns have been raised regarding the degree to which natural and enhanced fractures may have acted as conduits that allowed dissociated gas to flow away from the well (Mallik 2002 Partners, 2003).

The Mallik 2002 Research Well Program proved that gas production from gas hydrate is technically feasible. The data collected have been utilized in models to show that the rate of gas production could be high enough to make gas hydrate a technically recoverable resource. Models based on the Mallik tests show that a combination of heating and depressurizing will yield the highest production rates (Moridis, et al., 2003).

5.2.4. Planned Follow-up and Options

At present there are no confirmed plans for additional drilling and production testing at the Mallik structure. An extended production test would be of great value, especially if lateral gas collector drill holes were carefully sited with respect to a geological reservoir model that took into account gas generation, coalescing and separating of gas from water, and concentration of gas following conversion of hydrate. The construction of a gas pipeline linking the Mackenzie Delta to North American gas markets, which is being done to carry abundant conventional gas resources, should provide an incentive for further development within a few years. Producing natural gas from hydrate would increase the overall value of this and other infrastructure investments. No plans have been made public for more widespread differentiation of natural gas hydrate within the compound ice-hydrate cryosphere (4.4.2; Fig. 4.9). It is seismically difficult to distinguish, ice from hydrate but this step is necessary to make a fuller assessment of Arctic natural gas resources.

5.3. NANKAI

This locality on the deep continental shelf to the SE of Nagoya and SW from Tokyo Bay (Fig. 5.2) is named after the nearby Nankai Trough, which is one of the major subduction zone trenches that form the eastern margin of the Japanese archipelago. The Nankai Trough marks the boundary at which the Philippine Sea Plate is being subducted beneath the Eurasia Plate. Although the term

'Nankai Trough' is currently used, the hydrate locations are not geographically located within the deep bathymetric trough.

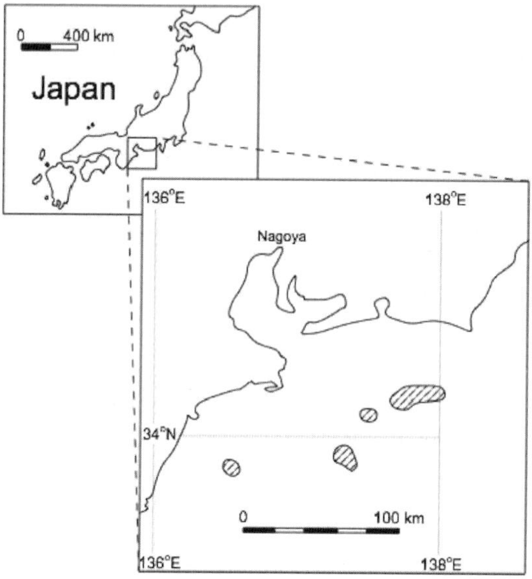

Figure 5.2. Drilling areas in the Nankai hydrate play. For locations of wells within the study areas shown, see Takahashi et al. (2005).

5.3.1. Background

Japan imports 99% of its primary oil and gas supply. As a result, natural gas only supplies 13% of Japan's energy supply compared with the OECD average of 23%. To enhance its energy security, Japan's Ministry of International Trade and Industry (MITI) established an aggressive gas hydrate program beginning in 1995 and has participated in investigations off its own coast as well as in Canada and the U.S. Gulf of Mexico. As noted in the previous section, Japan was a key participant in the 1998 and 2002 Mallik programs in the Mackenzie Delta of the Canadian Arctic. Between the 1998 and 2002 Mallik programs, Japan conducted a series of gas hydrate field evaluations and experiments at the Nankai Trough, and additional drilling in the Nankai Trough was carried out in 2004.

5.3.2. 1999-2000 Nankai Drilling Program

A drilling program was undertaken from November 1999 through March 2000 by JNOC with JAPEX as well operator. Six closely spaced wells were drilled, cored, and logged in a water depth of about 3100 feet (945 meters). The wells were designated as two pilot holes, the main hole, and three post survey wells (Tsuji, et al, 2003). The sites were chosen on the basis of both BSRs and the work of Tsuji et al. (1998) that identified a high probability of inclined stratigraphic layers that would include reservoir-quality sand units.

Core samples and well logs confirm the existence of four sand units containing gas hydrate at depths of from 679 to 869 feet (207 to 265 meters) below the seafloor. The sands are turbidite fan deposits, and the total thickness of the hydrate-rich sands is 39 to 46 feet (12 to 14 meter). Wireline logging data indicates that the hydrate saturation of the pore space in the sands is 20 to 80% (Tsuji, et al, 2003).

The 1999-2000 Nankai results have significance beyond Japanese energy program. First, the wells confirmed the occurrence of a highly concentrated gas hydrate in marine sediments for the first time. The capacity of gas hydrate to concentrate natural gas resulted in large in-place reserves, even though the sand thickness was not extremely large. More important, the Nankai results validate the concept that lithology exerts a critical control on gas hydrate occurrence, with the effect being as strong in marine sediments as in those in Arctic settings.

5.3.3. 2004 Nankai Drilling Program

In 2001, METI initiated "Japan's Methane Hydrate Exploitation Program", a 16-year program in which methane hydrate is defined as a future energy resource that is expected to exist in large amounts offshore around Japan. The success of the 2002 Mallik program supported additional drilling at the Nankai Trough. For FY2004 the Japanese program was budgeted at an estimated $65-100 million (U.S.) and included 250 people and 30 organizations (Tanaka, 2003). The 2004 program included a substantial drilling program and studies of resource assessment, production methods, and environmental impacts.

Sixteen wells were drilled in water depths of 2376 to 6670 feet (720 m to 2033 m) with drilling depths ranging from 538 feet (163 m) to 1888 feet (572 m) beginning in January 2004. The program lasted for a total of 122 days including 6 port calls (Takahashi and Tsuji, 2005). The wells targeted locations with and without BSRs (Tsuji and Nonaka, 2004). The objectives were to delineate hydrate occurrence and resource potential and to refine drilling techniques. Sixteen wells were logged while drilling (LWD), two wells were cored by conventional wire-line, 12 other wells were cored (6 using the Advance Piston Corer/Extended Core Barrel [APC/XCP] of ODP and 6 using the Japanese Pressure Temperature Core Sampler [PTCS]). APC/XCB recovery was 55.7% and PTCS core recovery was 79.3%. In addition, an important proof of concept was achieved when a 330 foot (100 m) lateral well was drilled for the first time in a hydrate-enriched sediment interval. It is considered that horizontal drilling in hydrate-enriched horizons will be one of the necessary conditions for hydrate natural gas recovery. One well was successfully monitored for its temperature profile over a 51 day period. The near linear temperature gradient of the drill hole was gradually superceded by a stepped profile caused by the thermal conductivity effect of hydrate in part of the hole.

Complete results of the drilling program have yet to be released, but initial results suggest that in some locations with well-developed BSRs, no sands were encountered. This observation has caused a reassessment of the potential gas hydrate resource base for Japan, and apparently a reassessment of the direction of the Japanese gas hydrate program. On the positive side, the 2004 Nankai Drilling Program confirmed the link between lithology and hydrate occurrence. While it is more difficult to conduct detailed stratigraphic analyses rather than to merely map BSRs, a more complete geological analysis for any potential gas hydrate location appears to be a critical component in successful gas hydrate exploration.

5.3.4. Future work

A conservative and carefully laid out plan to test drilling technology, production techniques, and commerciality has been established. More drilling is planned, and well-constrained resource evaluation will undoubtedly be carried out in a completely professional manner. In addition, we regard it as highly likely that further deposits of hydrate will be identified on the continental slope, where thicker GHSZs may host larger deposits of hydrate. Only the relatively shallow resource has been examined to date.

5.4. GULF OF MEXICO

Hydrate prospects are currently being investigated in the northern Gulf of Mexico off Louisiana and Texas are currently in the Atwater Valley and Keithley Canyon localities (Fig. 4.14 localities). Hydrate and seafloor seeps occur over a broad area (Fig. 5.3).

5.4.1. Background

Gas hydrate mounds are a common occurrence near deepwater vents in the Gulf of Mexico (Roberts et al., 2002) yet hydrate-bearing sands have not been reported. This observation is remarkable considering that over 1000 wells have been drilled by the oil industry in water depths where the GHSZ extends to a substantial depth beneath the seafloor. In addition the depositional systems active in the Gulf during Neogene and Pleistocene times are known to include abundant sands. The lack of observed hydrate-bearing sand is best explained by the lack of effective evaluation of the shallow interval. That interval is seldom logged with a full wireline suite, and the LWD (logging while drilling) tools typically employed for the shallow section are not designed for identification of hydrate-bearing sands. Fortunately LWD gamma ray logs are able to delineate sands and these logs are run often enough that a careful evaluation of all available logging measurements indicates that sands are common (although not universally present) within the GHSZ of the Gulf of Mexico.

Developing a valid model for gas hydrate occurrence in the Gulf is made challenging by the Gulf's complex geology. Relative to other areas of interest for gas hydrate researchers, the geology of the deepwater Gulf is far more complex involving ongoing salt diapirism and dissolution, recent sedimentation and faulting, and high volumes of migrating fluids. These factors result in a high lateral variability in temperature gradients, fluid flow, and water chemistry. This variability makes it unwise to form regional generalities from the sampling results of a limited number of locations. While high temperatures, low gas flux, and unfavorable water chemistry will prevent many locations in the deepwater Gulf from having gas hydrate accumulations, any predictions that the Gulf of Mexico will have minimal amounts of gas hydrate (Ruppel, et al, 2005) are premature. Gas hydrate research has often focused on the seafloor hydrate mounds associated with salt features where the flux of warm saline waters is highest. These sites should not be viewed as typical of the Gulf of Mexico as a whole. In addition, most sampling has been carried out with shallow piston cores that are not able to assess directly the resource potential of deeper sands that could be hydrate bearing.

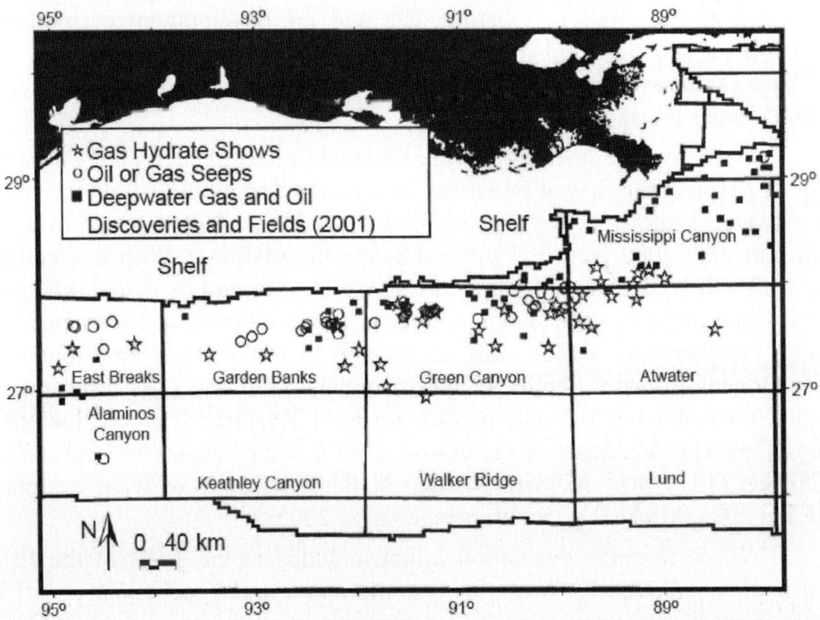

Figure 5.3. Hydrate locations and offshore lease blocks in the northern Gulf of Mexico. Redrawn from Sassen et al. (2001).

5.4.2. ChevronTexaco Joint Industry Program

For oil and gas companies with operations in the Gulf of Mexico, the primary interest in gas hydrates has been their potential as a geohazard, with interest in gas hydrate as a potential energy resource being a significantly lower priority. While the offshore industry has an excellent safety record with deepwater operations, the recognition that gas hydrate could pose a hazard led to the formation of a Joint Industry Program (JIP), headed by ChevronTexaco and funded jointly by the US DOE and member companies. The members of the JIP are ChevronTexaco, Schlumberger Oilfield Services, Halliburton Energy Services, ConocoPhillips, Total, JNOC, and Reliance Industries Ltd. (India). The U.S. Minerals Management Service (MMS) is a non-cost-share member of the JIP. The USGS provides technical support for the project.

The primary goal of the Gulf of Mexico JIP is to address deepwater safety issues related to conventional drilling and production operations by developing techniques for using seismic data to (1) quantify the extent of gas hydrate deposits, (2) develop protocols for safely drilling and coring through hydrate-bearing strata, and determine how the presence of gas hydrate in sediment affects wellbore stability. In addition, theoretical research and laboratory investigations are being carried out to provide a better understanding of the mechanical properties of hydrate-bearing sediments.

The JIP is a multi-phase, multi-year program. Phase I involved the collection and analysis of existing data and the development of models and protocols for detecting and characterizing hydrate-bearing sediment. The results of Phase I were used to design the drilling, coring, and logging program initiated in 2005 under Phase II.

In April 2005 operations were initiated in the Atwater Valley area of the Gulf of Mexico, the first of two areas of investigation by the JIP under Phase II. At Atwater Valley, the two pairs of wells will be drilled, each to a depth approximately 1000 feet (305 meters) below the seafloor. With the well pairs, one well will be drilled and logged using LWD tools and its paired well will be cored using pressure-core samplers over intervals of interest determined by the LWD. The Atwater Valley sites are located 110 miles (177 km) southeast of New Orleans in water depths of approximately 4300 feet (1311 meters). The second area that the JIP will investigate is in Keathley Canyon, located 160 miles (258 km) southeast of Galveston, also in water depths of approximately 4300 feet (1311 m). As with Atwater Valley, two pairs of wells are planned (NETL, 2005, Michael A. Smith pers. comm., 2005).

While resource evaluation is not included in the goals of the JIP, the information gathered on heat flow, stratigraphy, and geochemistry will have direct application for assessing the resource potential of marine gas hydrate. The locations were chosen in part because of active vents and mounds in each area. The heat flow, salinity, and lithology at these sites may prevent significant hydrate formation, although the results of Phase II - positive or negative - must be applied carefully to the Gulf of Mexico as a whole.

5.4.3. MMS Gulf of Mexico Gas Hydrate Assessment

The U.S. Minerals Management Service has undertaken a project to assess the gas hydrate resource potential on acreage under MMS jurisdiction in anticipation of eventual gas hydrate leasing and drilling in Federal waters. By law, the MMS must ensure that the Federal government receives fair market value for acreage offered in lease sales. Gas hydrate resources will be included in MMS resource assessment models in order to evaluate resources at a statistical level.

The MMS hydrate assessment process is integrating seismic data, well logs, temperature data, and geotechnical boreholes. Four potential gas hydrate settings are being considered (Collett, 2004): (1) vein-filled muds adjacent to vent areas, (2) hydrate-bearing sandstones completely within the GHSZ, (3) hydrate-bearing sands that straddle the GHSZ with free gas trapped below, and (4) hydrate-bearing caprock associated with salt domes. Economic models are being developed as part of the evaluations. Results are due in December 2005.

5.5. ALASKA

Alaska also has permafrost hydrate in its Arctic margin, where the deposits are associated with conventional gas and petroleum (Fig. 5.4).

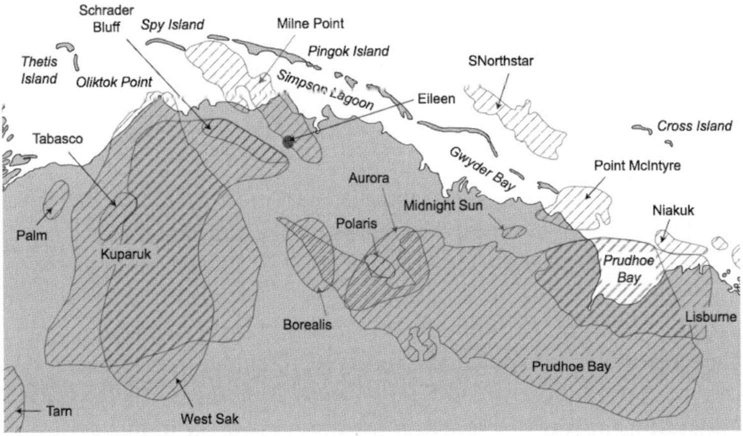

Figure 5.4. Major gas fields of the North Slope of Alaska. Redrawn from a figure supplied by T. Collett. Eileen (North Eileen State 2) dill site has been evaluated for hydrate (Collett, 2003). Also see Table 4.2.

5.5.1. Background

As noted in the previous book in this series, the Northwest Eileen State-2 well located in the northwest part of the Prudhoe Bay Field provided direct evidence of gas hydrate in 1972 when a core containing gas hydrate was recovered from the well. Since that time, well log analyses indicate the presence of gas hydrate in at least 50 wells.

Most of the well log inferred gas hydrates occur in six laterally continuous sandstone and conglomerate units that are restricted to the area overlying the eastern part of the Kuparuk River Field and the western part of the Prudhoe Bay Field. The total mapped area of all six gas hydrate occurrences is about 634 mi^2 (1,643 km^2). The volume of gas within the gas hydrates of the Prudhoe Bay-Kuparuk River area is estimated to be about 35 to 42 TCF (1.0 to

1.2 TCM), or about twice the volume of known conventional gas in the Prudhoe Bay Field (Collett, 1993).

5.5.2. BP Exploration Alaska

BP Exploration Alaska (BPXA) is one of the largest operators of oil and gas leases on the North Slope of Alaska. In partnership with the DOE, BPXA has undertaken a project to characterize, quantify, and determine the commercial viability of gas hydrates and associated free gas resources in the areas of the Kuparuk River, Prudhoe Bay, and Milne Point Fields. The project is utilizing the technical assistance of the USGS, the University of Alaska in Fairbanks, and the University of Arizona in Tucson (Collett, 2004).

During the initial phase of the project, reservoirs and fluids are being characterized in terms of stratigraphy, structure, continuity and various physical properties to determine their commercial potential. In addition, procedures are being developed for drilling, logging, completion, and production. Phases 2 and 3 will involve one or more production tests. As yet, the dates for a production test have not been set.

5.6. CASCADIA MARGIN

The main hydrate locality of the Cascadia margin (Fig 5.5) lies about 50 miles off the coast of Oregon, Washington, and British Columbia.

5.6.1. Background

The Cascadia Margin, a convergent margin that extends from northern California to British Columbia, has received intensive gas hydrate study since BSRs were identified there in 1985 (Davis and Hyndman, 1989). Early work on the Cascadia Margin was discussed in the previous book in this series, including the results of Ocean Drilling Program (ODP) Leg 146 in 1992. Additional data have been obtained from ODP Leg 204 in 2002, and Integrated Ocean Drilling Program (IODP) Expedition 311 is scheduled to investigate the Cascadia Margin in the late summer of 2005.

5.6.2. ODP Leg 204

ODP Leg 204 was a dedicated gas hydrate research cruise in the Cascadia region in 2002 designed to assess the gas hydrates at a location termed "Hydrate Ridge", offshore Oregon. Hydrate Ridge is part of an accretionary basin with water depths ranging from about 2625 to 3937 feet (800 to 1200 m). The crest of Hydrate Ridge exhibits very high gas flux with active surface seepage and seismically imaged gas chimneys (Tréhu. et al., 2003). Nine sites were cored and logged to (1) determine the distribution and concentration of gas hydrates in an accretionary ridge and adjacent slope basin, (2) investigate the mechanisms that transport methane and other gases into the GHSZ, and (3) determine constraints on physical properties of *in-situ* gas hydrates.

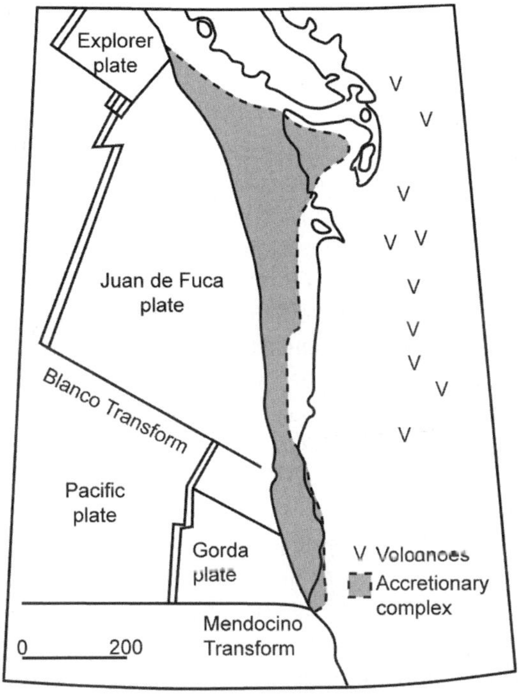

Figure 5.5. Generalized location of the Hydrate Ridge area in the Cascadia offshore region. Redrawn from Bangs et al. (2005).

Leg 204 obtained samples under pressure using the ODP pressure core sampler (PCS) system and the newly developed Hydrate Autoclave Coring Equipment (HYACE) system. HYACE includes a laboratory transfer chamber for maintaining pressure while making measurements of physical properties. In addition, extensive use was made of infrared (IR) cameras to identify rapidly potential hydrate-bearing samples that were preserved for detailed study. High-resolution measurements were made of the chemistry of interstitial waters. *In-situ* temperature and pressure measurements were frequently made, especially in zones where rapid changes in the physical properties of the sediments were detected by LWD. An extensive suite of downhole and two-ship seismic experiments were conducted.

The results of Leg 204 validated the concept that gas flux and lithology play a dominant role in determining where high concentrations of gas hydrate will occur. Away from areas of high gas flux, clay-rich sediments were observed to have low gas hydrate saturations, but gas hydrate occurred within distinct stratigraphic horizons and along faults. At the crest of Hydrate Ridge, very high gas flux results in active surface seepage, and a gas chimney was imaged on seismic lines. The clay-rich sediments on the crest of Hydrate Ridge contain very high gas hydrate content (about 90% saturation or about 70% bulk volume).

5.6.3. IODP Expedition 311

IODP Expedition 311, scheduled in the Cascadia region for autumn, 2005, is designed to constrain models for the formation of marine gas hydrate in an accretionary prism. The expedition will investigate the deep origin of the methane in the system, its upward transport, its incorporation in gas hydrate, and its subsequent loss to the seafloor. Of particular interest is a widespread layer of dispersed hydrate located just above the base of the stability field and parallel to the seafloor. Long-term monitoring in the boreholes is planned and, given the seismically active nature of the area, will assist in determining (1) the role of shaking in the sediment consolidation, (2) episodic upward fluid transport, and (3) hydrate formation (USSSP, 2005).

5.7. MESSOYAKHA

Many areas of Siberian Russia have sediments in petroleum and gas provinces that lie within the gas hydrate stability zone. The Messoyakha gas field in the West Siberian Basin was one of the first hydrate localities recognized for its commercial possibilities (Makogon, 1965), before hydrate was recognized in North America. Makogan et al. (2005) estimate that about 53% (230 billion m^3 of the gas was derived from hydrate through reservoir depressurization (Makogon et al., 1971; Makagon, 1997), but Collett and Ginsburg (1997, 1998) suggest that the volume of gas converted from hydrate is much smaller. The field was developed for conventional gas, but the additional presence of gas hydrate and its dissociation have been implied by analysis of time-pressure variations recorded in the field. If hydrate has been fortuitously converting through the simple depressurization of the gas reservoir, then at least some of the produced gas originally resided in hydrate rather than the free gas deposit. Gas production at Messoyakha would demonstrate the general feasibility of gas hydrate to rapidly become a commercial resource in those locations where similar hydrate-bearing sands are continuous with large down dip free gas reservoirs, and where there is a natural interface within porous sediments between natural gas hydrate and natural gas.

No additional research that would quantify the hydrate conversion effect through depressurization is currently planned for Messoyakha. However, if a large volume of gas hydrate has been dissociated over the past 30 years, as Russian scientists believe, significant information could be obtained through a scientific drilling program in the field.

5.8. INDIA

India appears to be committing resources on a national level and has formed international consortia for hydrate research with Russia (Introduction).

Discussions for joint research programs with the United States are continuing. Verifiable details about the nature and timing of the Indian gas hydrate research program, however, have proven difficult to pin down, although potential hydrate deposits have been identified (Max, 2003, Fig. 1). A drilling program may commence as early as late 2005.

5.9. COMMENT ON HYDRATE RESEARCH: OBJECTIVES AND PROGRESS

What kind of research is necessary to develop hydrate natural gas resources commercially? It is not the type of research where the answers may be predicted with accuracy from a background of experience. There are as yet no look-up tables or spreadsheets for reservoir engineering. The degree of uncertainty in the reservoir modeling has been demonstrated by the hydrate conversion and recovery performance experienced in testing to date, which did not meet predictions.

Considerable controversy surrounds the results of the Mallik natural gas production tests. Although little of the controversy is well documented for public dissemination, it would appear that some basic issues such as, how much gas was actually converted from hydrate, how much of the gas that was converted was recovered, and where did the gas go that 'got away', are far from being well understood and agreed to by the senior participants. What is known is that gas was converted and recovered in both tests, although not as much was recovered as originally predicted. In addition, the movement of the gas in the reservoir is also not well understood. The Mallik test at least released enough gas to produce a dramatic flare (cover figure). Some may conclude that because gas recovery did not reach predicted values, the tests were failures. A wide spectrum of views can be anticipated from the results of any research project, especially where little data were actually recovered from within the hydrate-enriched strata in which the gas was converted except in the immediate vicinity of the drill holes. It is important that scientists and more importantly the agencies and companies funding the development understand the nature of frontier research.

Hydrate natural gas development research presently, and probably for some time to come, must be characterized as high-risk. Most funding agencies, and most companies that are focused on their bottom lines, however, do not normally support high-risk research (Mervis, 2004), although they may publicly claim to do so. High-risk research is generally avoided because it is difficult to predict results in detail. This makes it difficult to assign the levels of success or failure to results. High-risk research is often left to governments, which in the field of hydrate natural gas recovery, is demonstrated by the preponderance of funding coming from national gas hydrate programs (Introduction). Small, innovative companies or independent researchers' bootstrapping breakthrough research, often without the cumbersome overhead and delay that are associated with structured research, often achieves the development of new paradigms and breakthrough technologies. High-risk research does not have a broad comfort zone in which there is a high likelihood of achieving predictable results that often result in little real advancement.

Incremental improvements to existing technology, however, and the pushing of well-known envelopes of knowledge only slightly further, will not be sufficient to develop the procedures, models, and technology mixes that will result in the commercial extraction of hydrate natural gas. Instead, bold new thinking and experimentation is called for. We submit that a possible poor correlation between some predictions and performance in the Mallik hydrate gas conversion and recovery tests does not equate with failure. Rather, we regard the rumored results as confirming our view that more attention should be given to geological modeling of the hydrate reservoir. Special attention should be given to concentration of converted gas (6.6.5) or innovative means for recovery and production of the gas where forming a gas trap is impractical (6.7).

5.10. CONCLUSIONS

Although the rewards of success for commercial hydrate development may be extremely high, a relatively small proportion of high-risk projects are perceived to be able to succeed to the level of breakthrough. Commercial development of hydrate natural gas deposits has the potential to be paradigm shifting because of the amount of natural gas sequestered in hydrate. If only a small part of the gas can be produced, especially adjacent to countries that are presently energy importers, there will be major economic and geopolitical consequences that would dramatically change the present energy supply and consumption system.

The drift from a low cost combustible energy resource base to a higher cost level has now begun because of increasing world demand combined with political insecurity. It is unlikely that low cost energy will ever return, especially if demand outside the United States continues to increase. Therefore, although commercialization of hydrate natural gas would almost certainly have been profitable at somewhat lower energy cost levels than about $40-$50 per bbl, energy costs in excess of that level will inevitably spur development of unconventional energy resources such as hydrate natural gas.

The topic of hydrate as a natural resource is beginning to become considered seriously by industry, although not necessarily by major energy companies. At the 2005 Offshore Technology Conference, for instance, as many people attended the technical session “Hydrate as a Resource” as attended “Hydrate as a Drilling Hazard”, for the first time. Smaller, innovative companies, in which other energy innovations have commonly taken place during the last 40 years, are beginning to consider issues of hydrate recovery.

Chapter 6

Oceanic Gas Hydrate Localization, Exploration, and Extraction

6.1. INTRODUCTION

Although natural gas hydrate is a diagenetic, solid, crystalline economic mineral, it is not a virtually insoluble economic material that must be excavated from the earth in bulk for processing. Hydrate will not be mined in the classical sense of driving a mineshaft into high-grade ore deposits or establishing large open pit mines in low grade deposits. Some non-invasive industrial extraction techniques and practices for the *in-situ* conversion of solid, crystalline minerals residing in weaker geological strata, however, may well provide models for hydrate conversion and gas extraction.

Characterization of economic deposits of gas hydrate is in its very early stages. To date, of those areas from which significant information about oceanic gas hydrate exists, only the Nankai area on the SE Japanese continental shelf has been selected for evaluation of its hydrate energy resources in a determined manner. In the process of developing evaluation criteria and procedures, along with early gas generation testing for the Japanese program, an international consortium that was largely funded by the Japanese program established a test well in a permafrost hydrate deposit in the Mackenzie delta of the Canadian Arctic. Both the Nankai and the Mackenzie delta deposits are examples of high-grade hydrate deposits. These are the only two deposits where commercial interests thus far have driven research.

A further hydrate deposit on Hydrate Ridge on the Cascadia margin of North America is potentially another high-grade deposit associated with focused gas flow. Only one dispersed gas flux, low-grade hydrate deposit is known in any detail and that deposit is at Blake Ridge off the U.S. SE coast. This has been the subject of detailed scientific examination. Exploration of both these areas has been driven by scientific interests in conjunction with the Ocean Drilling Program (under its various guises).

Although only a limited dataset is available from these oceanic locations, and the character of the data is somewhat variable because of the nature of the funding sources and the interests of the researchers, the data inventory has grown enormously in the last ten years. It is thus possible to (1) outline the basics of the modes of localization of oceanic hydrate, (2) speculate about the formation of the hydrate economic sweetspots (of highest grade and good recovery potential) according to hydrate growth models, (3) discuss the

Michael D. Max et al. (eds.), Economic Geology of Natural Gas Hydrate, 207–248.

ways in which deep gas escapes being sequestered in the GHSZ and (4) summarize how these models may lead to identification of economic hydrate deposits, and discuss briefly the main exploration methods.

6.2. GAS HYDRATE PROVINCING

A mineral province is a well-defined area containing resources of a particular type of mineral occurrence in which economic deposits are likely to occur. On a worldwide scale, there are two types of general gas hydrate provinces: Permafrost and Oceanic. Permafrost hydrate has a restricted aerial potential for development. Its presence is dependent on particular geological structures and reservoir conditions, existing gas concentrations, and a supply of water. For permafrost deposits, an outline of the general conditions required is less useful for identifying the best areas for development than is a standard conventional geological exploration approach to individual areas using methodology for identification of conventional hydrocarbon reservoirs.

The oceanic hydrate economic province is mainly in continental slopes below about 800 m water depth (hydrate is stable from about 500 m water depth) where the seawater is cold enough to thoroughly chill the seafloor and its subjacent sediments conditions provide appropriately for a GHSZ (Chapter 3). Some deep or very cold water continental shelves having strata suitable for hosting hydrate and a high gas flux also host hydrate deposits. Economic deposits of hydrate will only occur where the GHSZ is thick enough to host hydrate far enough away from the seafloor so that it does not have the potential to dissolve rapidly by diffusional exchange with the sea water. The base of the hydrate province occurs at some deep water depths on the slope where gas flux is too limited or the sediment is too thin to support economic deposits of gas hydrate. From a practical standpoint, there does not seem to be much gas hydrate in seafloor sediments in more than 4 km water depth. The oceanic hydrate province would likely encompass areas underlain by marine sediment of suitable thickness (Max and Lowrie, 1996) that lie in water depths from about 800 m - 4,000 m on continental margins wherever there is suitable gas flux (Chapter 3).

Gas hydrate itself is confined to GHSZs, which can be predicted by reference to pressure-depth of the seafloor, seafloor temperature, and the composition of the gases that migrate into the GHSZ. The gas hydrate province is thus depth-limited in a manner that no conventional hydrocarbon deposit is. In addition to actual hydrate, there may be large volumes of associated, subjacent gas, which has a genetic relationship with the overlying hydrate. It can be estimated that this broader hydrate - gas zone, where it exists, is no more than about 35% greater than the thickness of the GHSZ alone (as determined from reflection seismic sections across the Blake Ridge (Max and Lowrie, 1996). The maximum gas section below the Blake Ridge, for instance, appears to be no more than about 200 m but significant gas accumulations in sediments below the hydrate cap are more commonly on the order of 100 m depth

(Flemmings et al., 2003). In contrast, the Nankai deposits have a very thin layer of gas and a discontinuous BSR (Tsuji et al., 1998). Below this gas zone, sediment compaction and geotechnical properties are not affected by the presence of either hydrate or gas. Compaction has probably tightened the sediment close to normal density, although sedimentary structures and chemical precipitates associated with the hydrate and gas cycles may be developed. For purposes of designing and carrying out geophysical surveys for economic hydrate exploration, neither geophysical surveys nor drilling likely need be undertaken from below the hydrate - gas zones.

More detailed tectonic, sediment, gas flux, hydrological, and other factors require examination of seismic, drilling, and relevant basinal data to refine the hydrate province and identify specific locations in which to focus the acquisition of further information. This data acquisition can be scaled according to the availability of information. First order evaluations use existing data. Second order evaluations use hydrate-specific data collection focused on the most promising first order evaluations, and third order evaluations involve drilling and quantitative evaluations to generate a 3-D model of the hydrate deposit, establish grade/valuation, and define an extraction strategy.

6.3. SEMI-QUANTITATIVE EVALUATION OF HYDRATE LIKELIHOOD

In order to produce a gas hydrate probability map that reflects a sense of hydrate potential, a semi-quantitative approach has been developed for enumerating the likelihood of the presence of significant accumulations of hydrate. The approach follows energy company practice because these evaluation practices have been developed over a long period of time and have thus been tested against actual performance worldwide. The practices are also based firmly on extrapolation of the best geological information for any area, and it is the geology, along with gas flux and the groundwater system (both of which may be revealed by geological features) that control the development of gas hydrate. This emerging technique for the quantification of gas hydrate in marine sediments is most accurate in locations where high-quality data are available, particularly 3-D seismic data and subsurface data from drilling, coring, and logging. In areas where such data are not available, semi-quantitative analysis is useful as a means of locating likely high-grade areas of interest for additional data collection.

Determining whether a given sediment package is within the hydrate stability zone is simplified if a BSR is present, but may be a challenge where it is absent, for instance, in seafloor parallel sediments where BSR may be masked by seismic reflections and the geothermal gradient and composition of hydrate forming gases are poorly known (or variable). BSRs are often used as the sole first order indicator of hydrate, but the BSR is actually a strong negative impedance contrast that marks the interface between sediment in which hydrate has sealed pore throats and trapped gas beneath. In the absence of nearby well logs, the stratigraphy in an area of interest may be difficult to determine, although a sequence stratigraphic analysis of seismic character may allow the assessment of higher versus lower energy depositional environments from which higher versus lower probability of sand deposition can be inferred. The presence of faulting extending from the near surface to depth along with gas seeps at the seafloor, are good indicators of sufficient gas migration.

There are three main factors governing a first-order evaluation of economic hydrate deposits. 1. Likelihood of being within a reasonably thick GHSZ, which is a function of water depth, seafloor temperature, and geothermal gradient, 2. Presence of reservoir-quality sands or gravels to host the hydrate according to a best-growth model (based on seismic stratigraphy, drilling data, basin or slope stratigraphy), and 3. Likelihood of sufficient gas flux into the GHSZ (based on presence of potential source beds, suitably thick sediments, potential gas and fluid migration pathways. For second and third order evaluations, where more geological and related information is available, for instance, 3-D seismic surveys and established gas and oilfields in the area, these three factors can be split to provide a more refined assessment.

Each factor is assigned a number (0.0 for no chance through 1.0 for definite). These are then multiplied together to yield a probability of hydrate occurrence. For example, if a section is in an area where we see hydrate indicators, a zone of interest just above it might get a 0.9 for GHSZ. If the seismic data show some indication of sands or more coarse-grained beds, especially in a generally fine-grained section, stratigraphy might get a 0.7. If there is sporadic faulting in the area and some of the faults extended from deep in the section up to near the base of the GHSZ, there would be increased likelihood that HFG pathways would exist. Where seafloor venting through the GHSZ is to be identified, hydrocarbons might be assigned a 0.5 (a gas chimney blowing through the GHSZ would be assigned a 1.0). Accordingly, that particular section would get a value of 0.315.

The net result is a number for each section of a seismic line from 0, which is the lowest probability to 1, which is the highest (Table 6.1). These numerical evaluations are valid for defined regions (which are more or less accurate depending on data quality) and are amenable to further statistical and GIS analysis. The magnitude of the scale can be altered as necessary, with 10 or 100 representing unity instead of 1, as the user wishes. As part of the first order analysis, the numerical regions can be contoured, which results in a map that indicates the areas where the greatest chance of an economic high-grade hydrate deposit is likely to occur. The map also allows comparison of the geometry and position of the various zones of likely occurrence with respect to large volumes of different types of geological, geophysical, and other relevant data, and identifies areas for additional data acquisition. Ideally, the data can be contained in a GIS or related data-fusion software (Askari & Max, 1998; Max et al., 1998; Spina et al., 1998) so that further multi-functional analyses involving hydrate probability can be carried out in a numerical environment.

	Area 1	Area 2	Area 3	Area 4	Area 5
HSZ	Geothermal gradient may be high: **0.3**	Well within HSZ: **0.9**	Well within HSZ: **0.9**	Well within HSZ: **0.9**	Thin HSZ due to water depth: **0.3**
Stratigraphy	No data: **0.5**	High energy deposition inferred from seismic: **0.7**	High energy deposition inferred from seismic: **0.7**	Predominantly shale deposition inferred from seismic data: **0.2**	Predominantly shale deposition inferred from seismic data: **0.2**
Source and Migration	No data: **0.5**	Highly faulted: **0.7**	Highly faulted, gas vents observed: **0.9**	Faults from deep section do not extend into HSZ: **0.3**	Section appears unfaulted: **0.2**
Resulting Assessment of Hydrate Potential	**0.075**	**0.441**	**0.567**	**0.054**	**0.012**

Table 6.1. Example of a seismic line analysis study in which areas of hydrate likelihood have been identified. Method and table used with permission of Hydrate Energy International, Corp.

Where petroleum evaluations are made, additional factors are commonly used. A full petroleum play approach using both deterministic spatial and reservoir parameter probabilistic models (Osadetz and Chen, 2005) requires detailed subsurface information that can only be acquired through drilling. Assigning numerical values to various factors is often used in the oil industry to assess relative risk for basins or individual prospects. As many as 30 categories may be employed (such as trap integrity, thermal maturity of source rocks, amongst others). Contouring the numerical potential within a basin is particularly well suited for high grading locations for gas hydrate potential. As more is understood about the factors governing the formation of economic deposits of oceanic gas hydrate, more numerical indicator categories will be used to identify the sweet spots, which are the high grade hydrate deposits that are amenable for economic exploitation.

6.4. REMOTE SENSING FOR THE PRESENCE OF OCEANIC HYDRATE

There are a number of indicators of the likelihood of the presence of oceanic gas hydrate. The principal method of reflection seismics in fact identified the characteristic BSR and other affects prior to an understanding of what were causing the effect (Stoll et al., 1971). Reflection seismics will probably continue to the primary exploration tool for hydrate but more complex seismic or acoustic analysis techniques that will be specific to identification and possibly quantification of hydrate will likely be required. Some seafloor inspection techniques will also be used, especially where there is gas venting from the seafloor.

6.4.1. Seismic Effects of Hydrate Formation

The primary reflection seismic indicator, which can be identified most easily, is the BSR. The increase in pressure wave velocity caused by hydrate formation directly affects the reflection seismic response of sediments, as does any gas that is trapped and concentrated (Fig. 6.1). The distinctive negative impedance contrast that results in the widespread occurrence of BSR, which is a strong negative impedance contrast at the boundary between sediment having hydrate in pores and sediment with gas in pores below the GHSZ (3.3.2). In fact, the BSR indicates the top of a gas horizon, which is presumed to be held in place by hydrate filling sediment pores and decreasing permeability.

The presence of an apparent BSR is not an unequivocal criterion for the identification of hydrate and is less significant in predicting concentrations of hydrate that have potential as economic deposits. BSRs have been associated with opaline, calcareous, and other diagenetic mineralization. Although their position may have been controlled by some factor of hydrate formation or fluid/gas restriction cause by the presence of hydrate, these 'false' BSRs can often occur well below the base of the HSZ. The seismic signature of a low velocity zone beneath a higher one, which would be marked by some diagenetic mineralization could well have the characteristic negative kick on the seismic trace, especially if any gas were trapped beneath the zone of mineralization.

Hydrate formation incorporates water and replaces the liquid with solid hydrate. This changes the proportion of pore to solid volume of the sediment and alters the bulk modulus where:

K is the sediment bulk modulus - the force required to change volume,
μ is the shear modulus – the force required to change shape, and
ρ is the density, then:

$$Vp = \sqrt{((K+4\mu/3)/\rho)} \qquad \text{(Eq. 6.1)}$$

The presence of water-ice or gas hydrate in sediment pores increases pressure wave velocity (Vp) considerably. Winters et al. (2002), for instance, report that water saturated sands had a Vp in the range 1.7 to 1.9 km/s. Freezing the water-saturated samples produced ice that fully enclosed the sediment particles and yielded Vp of 4.23 to 4.33 km/s. The formation of hydrate, however, is rarely as complete as the forming of water-ice. This is because the hydrate must grow in the water matrix and when pore throats close, introduction of further HFG ceases as neither HFG enriched percolating groundwater or diffusion of HFG through water is possible. In contrast, ice can occupy all pore water space when it freezes. The maximum increase in Vp due to hydrate growth, either in laboratory or natural sediments was on the order of 1.85 km/s, in contrast to the increase of about 2.43 km/s caused by freezing water (Winters et al, 2002; Helgerud et al., 2002).

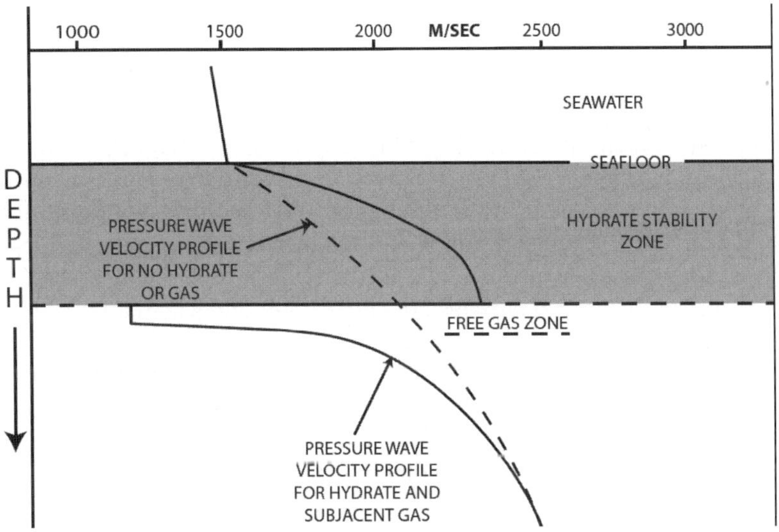

Figure 6.1. Generalized schematic of geoacoustic profile (solid line) showing generalized Vp as a function of depth. After Max (1990). The potential effect of hydrate and gas in sediment is shown by its deviation from the 'normal' geoacoustic profile based on an estimation of compaction-induced Vp alteration (Hamilton, 1980). BSR is at the top of the free gas zone at the Vp step-back, where this occurs. Compare with Figure 3 of Gorman et al. (2002).

Formation of hydrate in pore space increases bulk modulus and Vp while cementation of grain contacts in sediment increases shear wave velocity (Vs). Shear wave (Vs) velocity is:

$$Vs = \sqrt{(\mu/\rho)} \text{ (from equation 1)} \qquad \text{(Eq. 6.2)}$$

Both modes of hydrate formation affect seismic attenuation (Holbrook et al., 2001; Pecher et al., 2001a, 2001b, 2002, 2003; Pecher & Holbrook, 2003. Attenuation and a good identification of the BSR may be achieved from relatively inexpensive single-channel seismic data while more expensive multichannel data can be processed in complex manners to reveal much more information. Analysis involving both single-channel seismic data that may be acquired in special ways (e.g., vertical seismic profile (VSP) data) provides other information. This information shows that Vp > Vs and that the existence of μ not only determines Vp but is necessary for Vs to exist. Vs-Vp ratios give estimates of the degree of cementation, and many other specialist types of processing, all based on mathematical algorithms, are being developed for hydrate analyses. Guerin et al. (2004) use interval velocities calculated from the VSP to identify high velocity zones inferred to contain hydrate. Various elastic models are used to estimate gas hydrate and free gas saturations from sonic velocity and bulk modulus. Although (Vs) yield significant additional information (Waite et al., 2000), combined Vp-Vs analysis and a full spectrum attenuation analysis, will probably also be carried out in the future.

Deep-towed apparatus (Wood et al, 2003; Miles, 2003) resolves finer scale structure (Fig. 4.6) that permits higher resolution analyses, but the higher frequency range used does not resolve broader features well. For instance, BSRs are often indistinct or not as sharply marked as on single channel or conventional multichannel seismic surveys. Seismic survey apparatus is still either configured mainly for acquisition of conventional geology, with a deep-looking capability, or has a frequency range and manner of deployment that usually do not target identification of economic deposits of hydrate. For instance, where conventional multichannel surveys are carried out, the near-seafloor information of most use in identification of hydrate-related features is often reduced in detail by automatic gain control. This has the effect of enhancing deeper responses at the expense of shallow responses. Fortunately, many modern 3-D seismic surveys collect a full range of frequencies for shallow hazard assessment, and the data can be reprocessed later, if funding can be found.

Geophysics has proved thus far to be a crude tool for characterizing hydrate in sediments. Although considerable progress has been made with geophysical interpretation of hydrate deposits, there are as yet no coherent fine-scale seismic methods for determining hydrate concentration to the accuracies required for resource identification and meeting commercial standards. Interval velocities (Guerin et al., 2004) calculated from the VSP identify high velocity zones, for instance, but these can only be inferred to host hydrate. Elastic modeling may be used to estimate gas hydrate and free gas saturations from sonic velocity and bulk modulus, but there is, as yet, no widely accepted technique for estimating gas hydrate or gas saturations from seismic data alone. In addition to determining modulii, acoustic velocities, shear strength, and permeability properties of natural sediments, which contain gas hydrate, acoustic analysis techniques such as frequency-dependent attenuation, and identification of specific signal-modifying affects that may be specific to hydrate may need to be developed. In addition, because acoustic velocities and attenuations often vary with direction both within and across the bedding planes, the detailed physical acoustics of the host sediments may need more careful attention to properly characterize hydrate deposits.

6.4.1.1. Blanking

Blanking (also see 3.3.2) is a condition where hydrate obscures the original impedance structure of a sequence of strata. Blanking can take place where the bulk modulus of more porous sediment is made similar to the bulk modulus of bounding strata that are less porous so that their pressure wave velocities converge and the impedance contrast at their boundary diminishes. Blanking can also result from the general or dispersed formation of hydrate that renders some part of a stratigraphic succession more similar in its seismoacoustic response. Lee & Dillon (2001) modeled dispersed hydrate displacing water preferentially in more porous units, which had the effect of rendering the bulk modulus of initially more and less porous beds more similar. In contrast, Holbrook et al. (1996) did not find the high velocities that would be anticipated for sediment having high hydrate pore filling they identified minimal hydrate development and blanking which they regarded as simply resulting from a sedimentary succession having locally low impedance structure. Hydrate in free water space, however, may not increase Vp significantly unless a large percentage of pore filling exists (Winters et al., 2004).

Blanking may be used to identify sediments where hydrate has formed (Lee & Dillon, 2001) but blanking is not as good an indicator as a BSR because blanking may happen for a number of reasons associated with the original or diagenetic character of the strata or as an artifact of processing of the reflection seismics. There are also possible acoustic interference patterns and frequency- or angle-dependent cases where reflection seismic data are compromised either because of geometry and acoustic response or data processing artifacts. These same factors could also be responsible for blanking. Wood and Ruppel (2000), however, point out that relative Vp amplitude reduction may involve factors other than sediment homogeneity. They note that hydrate only has to occupy a relatively small portion of pore space to affect Vp so that blanking will occur at certain scales on the order of a meter. Dillon et al. (1994) describe blanking that cuts across strata (Fig. 6.2) at the margin of a thermal anomaly that

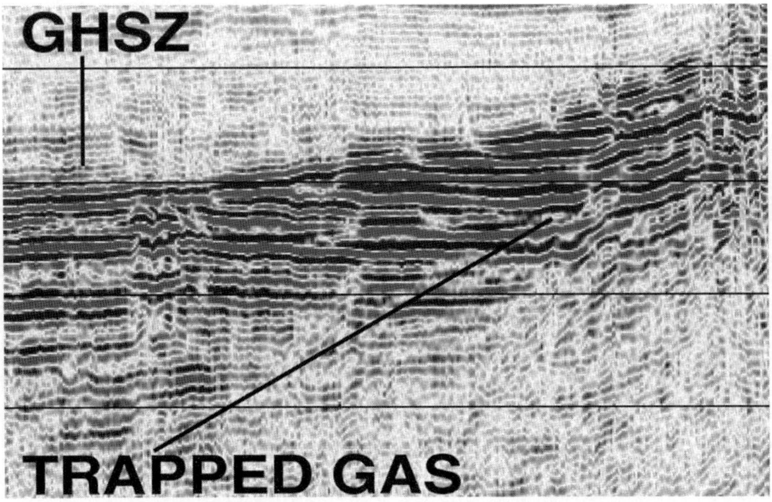

Figure 6.2. True amplitude plot of reflection seismic data from U.S. Geological Survey line 95-20. Supplied by W.P. Dillon. Strong impedance structure where gas is trapped, weak impedance structure in the GHSZ. Note individual reflections can be traced from gas-rich to probably hydrate-rich strata.

pushes up the BGHSZ (4.8). The transition from conditions where hydrate is unstable below the BGHSZ to where it is stable above clearly cuts across bedding and is a younger diagenetic feature. Where individual strata can be mapped across a BGHSZ, however, and the diminution of the impedance structure into the GHSZ is unarguable, the concept of blanking being a function of the development of hydrate can be demonstrated (Fig. 6.2).

6.4.1.2. Accentuation

Accentuation of acoustic structure may occur within the HSZ where enough gas or water replaces pore water in a porous bed. Within the GHSZ, hydrate will introduce relatively higher acoustic velocities. The porous bed, which may well have had a lower bulk Vp than its bounding beds prior to hydrate formation, can develop an overall velocity that will be proportionally much higher than its bounding beds that do not have hydrate developed within them. A group of stacked, discontinuous reflective GHSZ horizons (Fig 4.16, A) in a generally blanked region are an abnormal impedance structure and may represent the response of strata containing dense hydrate separated by beds without significant hydrate. In addition, within a GHSZ that includes a number of tight, upright anticlines, (Fig 6.3, left-most anticline) abnormal acoustic impedance structure in the culmination of the left anticline may indicate the presence of accumulations of hydrate. However, much of the apparent blanking of the reflection structure in the anticlines (Fig. 6.3) may be acoustic artifacts caused by interference at steeper dipping seafloor over the anticlines. Note that the apparent blanking extends across the otherwise very strong BSR and weakens the strength of its reflections.

Accentuation of impedance contrasts below the BSR in both anticlines and synclines indicate gas trapped in porous beds by the impermeable hydrate cap (Fig. 6.3). Gas traps on the flanks of anticlines closer to the syncline axial trace than to the anticline axial trace would not normally be a conventional drilling target, unless there were suitable porous beds beneath an unconformity (4.5.1). Some of the blanked areas above the BSR may indicate hydrate while some of the blanked areas below could be, at least in part, gas wipeout.

In general, hydrate-rich strata that have relatively sharp boundaries with adjacent sediments having little hydrate will have a strong impedance contrast. Such conditions should result in strong reflections, even when located well away from the BGHS (Fig. 4.16). Low-grade deposits, however, may have very gradational margins in that the percentage of hydrate may decrease gradually away from areas of highest concentration. These deposits may not be particularly amenable to identification using reflection seismics methods (seismic refraction and velocity analysis using multi-channel data may prove a more reliable method) because there is no reflector developed by the presence of hydrate. Some yet to be proven seismic attribute or combination of attributes, such as Vs, attenuation, or pressure wave refraction, as well as velocity analysis, may prove to be critical for the identification of economic deposits of gas hydrate.

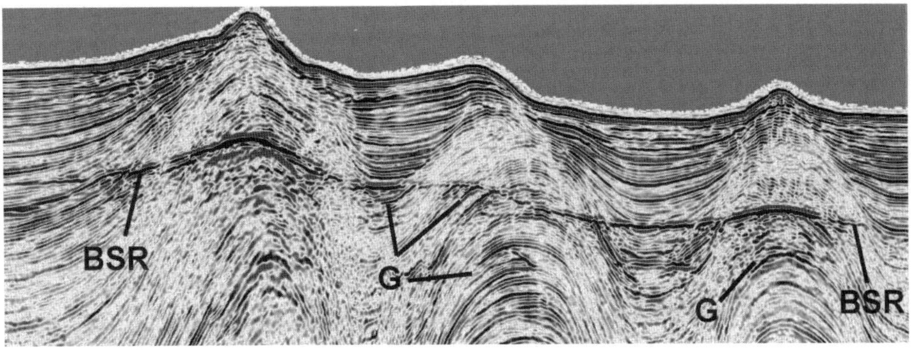

Figure 6.3. Unscaled diagram of reflection seismic lines showing BSR (strong reflector cutting across upright active folds that deform strata and the seafloor. Seismic survey from the Makassar Straits, Indonesia. Seismic data provided courtesy of TGS-NOPEC Geophysical Company. G, gas in sediments below BSR terminated at BSR permeability barrier. Note that all three anticlines appear to have gas concentrated within folded strata that are not associated with BSR. The apparent blanking in the anticlines may be an artifact of interference caused the seafloor bathymetric highs.

6.4.1.3. Seafloor Acoustic Imagery

Side-scan sonar systems use variations in the intensity and travel time of back-scattered acoustic energy to produce acoustic images of the seafloor. These images are the main tool for identifying seafloor features formed by gas and fluid venting. These features include pockmarks, mounds, and rough ground (Fig. 4.18) Bathymetric side-scan systems are currently capable of measuring the topography of the seafloor to a precision of approximately 3% of the water depth (Miles, 2003) and can provide data for seafloor and immediate sub-seafloor classification. These systems are used primarily for identifying areas of seafloor venting involving natural gas.

6.4.2. Sulfate Reduction Identification

Hydrate is often absent from sediments near the seafloor, even in a dispersed form, because of the effects of anaerobic oxidation of methane and sulfate reaction where sulfate reducing bacteria and archaea that feed on methane effectively remove methane from the sediments (Borwski et al., 1997; 1999; Dickens, 2001; Treude et al., 2003) (3.3.1). Steep sulfate gradients are typically interpreted to be the result of greater methane flux and have been related to the abundance of subsurface hydrate in some regions of gas hydrate exploration (Borowski et al., 1999; Pohlman et al. 2004). Measurements in the upper few meters of sediment may not fully constrain the processes occurring deeper in the section, but do constrain processes in the boundary region between the deeper sedimentary section and the overlying water column.

6.4.3. Natural Gas Analysis and Application

Knowledge of gas composition in a hydrate province can be important, especially where thermogenic gas is present. Analysis of gas venting from the seafloor can be important in predicting likely BGHS for different mixtures of gas, which will help in reflection seismic analysis. Anomalously thin zones of hydrate enrichment may develop where higher density hydrocarbons occur alone with methane. Examples of anomalous seafloor behavior may be related to the development of compound hydrate, which could trap gas and form a low shear strength zone at shallower depths than would be expected from methane hydrate alone.

Although they do not infer hydrate, for instance, Huvenne et al. (2002) identify a seafloor parallel zone of very low shear strength in the western Porcupine Basin, offshore SW Ireland. The combined analysis of high-resolution 2D seismics and an industrial 3D seismic survey from that area revealed an unusual picture of a buried sediment collapse and slope failure. A proportionally thin (85 m) but vast (> 750 km^2) slab of consolidated sediments began to slide down slope and break into hundreds of vertically undisturbed blocks, up to 500 m in across. Some of these have upturned edges similar to dried mud flakes that are often seen in an intertidal environment. The underlying horizon was most probably overpressured and seems to have liquefied and acted as a slide plane until the excess pore pressure had dissipated. Then, still early in the slide development, the process stopped, freezing the failure at its initial stage and not allowing a mass flow deposit to develop.

Although the depths to the slide surface are too shallow for methane hydrate in the area of the sediment redistribution noted by Huvenne et al. (2002), mixed hydrocarbon gases are a possibility in the Porcupine Basin due to known oil and gas shows. Cherkis et al. (1999) proposed that persistent gas venting along major faults was both responsible for repeated landslips and the apparent solidity of surface sediments north of Svalbard. If, for instance, the mass flows in the Porcupine basin formed for reasons similar to those proposed by Cherkis et al. (1999), a compound hydrate would have formed, for instance, from 90% methane, 7% ethane, and 3% propane would form a BHSZ that would cross the seafloor at about 450 meters. The depth beneath the seafloor of the BHSZ for such a natural gas hydrate is conjectural due to the lack of good geothermal data. A thermogenic gas hydrate could be stable to about 200-250 ms beneath the seafloor in the areas depicted. These depths into the seafloor are where the slumping effects are focused.

Capture of gases venting from the seafloor, coring and analysis of groundwater for its hydrate forming gas components are an important part of slope analysis for the likelihood of hydrate deposits. It cannot be assumed that the HFG is composed solely of methane. Knowledge of the mixture of hydrocarbon gases allows the calculation of the depths to BGHS. This information is important for analysis of reflection seismics, especially where BSRs are weak or where sediment bedding is parallel to the seafloor in areas where heat flow shows little lateral variability.

Sniffers are sensors that are towed near the seafloor. They measure seimiquantitative the dissolved natural gas components in the water. Sniffer profiling is an established technology that has been used for early identification of gas provinces by the hydrocarbon exploration industry for many years (Dunlop & Hutchinson, 1961; Anonymous, 1964; Jones & Drozd, 1983). Sniffers are relatively small and are ideal for implementation on autonomous underwater vehicles (AUV) for remote seafloor mapping. Fully quantitative *in-situ* analysis for hydrocarbon gas using Raman spectrography is presently in the final stages of development (Peltzer et al., 2004) and offers considerable potential for determining natural gas compositions as a hydrate exploration technique.

6.4.4. Heat Flow / Vent-related Seafloor Features

Hydrate deposits may induce small variations in heat flow because of their higher thermal conductivity than the sediments in which they occur, especially when the effect of water displacement is taken into account. Detailed heat flow measurements are not likely to reveal directly the locations of subjacent hydrate-enriched sediments because of (1) their typical distance below the seafloor, (2) adjustment of isograds that occurs between hydrate-enriched-sediments and the seafloor, and (3) the insensitivity of the method to distinguish hydrate from heat flow anomaly caused by other materials, such as salt plugs. Nonetheless, where high heat flow anomalies can be mapped, they can indicate venting of gas and fluids derived from beneath the GHSZ (4.8). Larger scale salt diapirs can warp the BGHSZ upward over a large enough area to cause a potential gas trap to form (Fig. 3.7).

Vanneste et al. (2002), Wood et al. (2004), and Hagen (2004) identify abnormal heat flow near seeps but the physical features on the seafloor are the first-order indicators of subjacent natural gas. Methane venting is almost certainly associated with bioherms as it provides nutrition for the base of the food chain. Henriet et al. (2001) found carbonate mounds that are associated with seafloor venting along the eastern slope of the Porcupine Seabight. These mounds occur primarily between the depths of 600 and 900m in two surveyed areas in the northern and eastern Porcupine Seabight (Huvenne et al., 2002) and support identification of this region as a gas province with an open groundwater system. Vent-associated associated gas hydrate may also have an association with deep carbonate reefs (Henriet et al., 1998). Authigenic high-magnesium calcite (14 to 19 mol% $MgCO_3$) and aragonite that are intercalated with massive gas hydrates in sediments of the Cascadia margin occur at least in the uppermost 50 cm of sediments on the southern summit of the Hydrate Ridge (Bohrmann et al., 1999).

6.4.5. Electromagnetic Methods

Electromagnetic methods that may be used in identifying concentrations of hydrate include frequency domain, induction, controlled source audio-frequency magnetotellurics, and time domain (EM). EM methods map electrical resistivity, conductivity and conductivity fields, and are useful for detecting brine layers beneath oil in sediment pore space (Duckworth and O'Neill, 1989). Transient electromagnetic (TEM) surveys have been used in conjunction with seismics to characterize resistivities in the vicinity of potential conventional petroleum traps (Tarso et al., 2003), where the resistivities of the fluids in the trap distinguish water from oil. Related studies have identified conductive anomalies associated with pyrite halos over hydrocarbon reservoirs that have been produced by seepage of hydrocarbons. The electrical character of subsurface sediments is largely controlled by the porosity and resistivity/conductivity of pore fluids and the sediment constituents. Because temperature affects response, modeling is difficult in sediments in which stratigraphy, sediment composition, and geothermal gradient is not well known.

EM methods have only recently been applied to field characterization of hydrate deposits (Weitmeyer et al. 2005; Yuan & Edwards, 2000). These methods appear to be of limited value in hydrate exploration and characterization when used as a sole technique, but when combined with seismics, considerable perspective may be gained. Willoghby et al. (2005) substantiated seismic blanking (3.3.2) as coinciding with anomalously high resistivity values exceeding 5 Ωm, where background resistivities ranged from 1.1 to 1.5 Ωm. They concluded that a higher gas hydrate concentration within the blanked zones caused the increase in resistivity and indicated the presence of hydrate on order of 25% of the available pore space (rising locally to 50% or more), although they noted that free gas might also contribute to the increase in resistivity. Assumptions about the host sediments may entirely control results, however, particularly where uniform response is required from sediment in order that anomalies may correlate with the presence of hydrate. EM of natural gas hydrate deposit localities appear to contain much less detail than seismic data, and survey is slower, owing to the requirement for seafloor EM receivers to provide station data for correcting time-series EM data.

In addition, for imaging deeper, lower frequencies are required and resolution decreases. Hydrate, except in dispersed low-grade (4.6.2) Class 4 (Chapter 7) deposits may not have a significant effect that can be determined systematically, especially where GHSZs are crosscutting to bedding. Of the EM methods, induction appears to be the technique that would be most useful for shallow investigations between the seafloor and depths in the 5 to 40 meter range. In this depth range deep-tow shallow seismics may resolve small amounts of hydrate and conventional seismics will not resolve the interval less well. EM methods may be most useful for assessment drilling safety, especially where the consistency of the host sediment is the same throughout any study area.

6.5. EXPLORATION FOR NATURAL GAS HYDRATE DEPOSITS

In many ways, exploration for economic deposits of gas hydrate should be fundamentally similar to searching for conventional gas deposits. Evaluating a commercial gas hydrate prospect begins with assessment of all of the basic components of any conventional natural gas prospect. These components include regional geology (tectonic framework, structural history, sedimentology, depositional environments, and stratigraphy) identification of hydrocarbon source and reservoir lithology porosities and permeability and overall geometry. The basic principals of petroleum geology apply to the formation of hydrate deposits; gas is produced in source beds and migrates to a trap. The basic principles of extractive methods and technology will apply to hydrate as to any other mineral deposit with the proviso that the character of the valuable commodity, natural gas, must be changed from that of a solid, crystalline mineral deposit to something more closely resembling a conventional gas deposit prior to extraction. Core evaluation methods for hydrate characterization and valuation are still in the formative stage, although conventional down-hole techniques appear to have adequate sensitivity (Chapter 5).

In many areas knowledge about oceanic hydrate is uneven because no area has been specifically studied for its constituent hydrate except for the Nankai area of SE Japan and the Storegga area off the central Norwegian coastline. Hydrate indicators on most reflection seismic records only accidentally image hydrate-related features because the surveys usually were undertaken to reveal other, and deeper, geological information. In the Blake Ridge area off the U.S. SE coast, for instance, the location of most of the scientific research and knowledge about the deposit come from marine sediments in water depths between 2.5 and 3 km. This location was originally chosen for scientific and historical reasons. The main early research interest was the sedimentation and modern depositional environment of the nose of the Blake Ridge. Subsequent hydrate work, including the ODP Leg 164 drilling, was carried out there because that is where the quantity and quality of existing geophysical data was best. It might not, however, be the location that should be the target of exploration for economic deposits of hydrate.

6.6. ISSUES CONCERNING RECOVERY OF GAS FROM HYDRATE DEPOSITS

Oceanic hydrate generally occurs in unlithified, geologically weak marine sediments between about 150 meters and 750 meters below the seafloor, rather than in geologically strong traps already containing conventional gas and petroleum deposits at elevated pressures and temperatures. Because the host is structurally weak, special consideration must be given to the changing nature and strength of the trap and to reservoir relationships as recovery of gas proceeds and parts of the trap itself may be produced. In deposits where hydrate forms a diagenetic cement that may strengthen the sediment matrix, containment of the free gas produced through dissociation will lead to significant weakening of the trap and sediment matrix. In addition, production of gas from hydrate results in a decrease in overall matrix volume because water constitutes only 80% of the volume of hydrate, which may lead to sediment collapse, while overly high gas overpressures may lead to blowout. Therefore, a balance between maintenance of gas pressures and back flooding may be necessary to prevent breaching of the dissociation areas. Understanding growth and dissociation mechanisms of solid hydrate masses, which occur in high-grade deposits having high pore filling of hydrate is key to commercial extraction of natural gas from hydrate in a safe manner.

Permafrost hydrate is usually contained within strong geological structures and is not liable to be any more of a hazard or have potential for pollution than a conventional gas deposit, which may or may not be associated with subjacent oil. Oceanic hydrate has little or no condensate or liquid hydrocarbon fraction. It is generally not associated with more than rare traces of oil except in locations such as the Gulf of Mexico where there is considerable thermogenic gas and other petroleum products. Thus, it has little potential for pollution, except to add a greenhouse gas to the atmosphere if the natural gas is not dissolved in the sea. Only very large and dramatic natural gas venting or venting in shallow water, reaches the atmosphere directly.

6.6.1. Reservoir Characterization

The oceanic hydrate reservoir is unlike common reservoirs known to the hydrocarbon industry, although gas deposits trapped below hydrate-sealed strata can be dealt with as a conventional gas deposit. Conventional petroleum traps are either gas or liquid, or a combination. Hydrate deposits, in contrast, are solid crystalline materials, indistinguishable in many ways from the solid clastic material of the sedimentary host. Hydrate comprises both the trap and reservoir of a hydrate deposit. Unlike conventional hydrocarbon deposits, hydrate deposits have many commonalities with both metallic and non-metallic mineral deposits hosted in permeable strata. Because the hydrate-enriched strata will almost certainly have variable concentrations of hydrate, its mechanical strength may vary, especially in low-grade hydrate deposits where the strata are mechanically inherently weaker than they are in high-grade deposits. Thus, the strength of trap strata may be very difficult to characterize. This situation is very different from a classical reservoir where cap strength is usually not an important consideration because of its overwhelming geological strength. The reservoir character of permafrost hydrate, however, will vary, depending mainly on the nature of the geological reservoir and trap configuration.

In a high-grade deposit, the hydrate may not contribute much to the strength of the reservoir, even where substantial cementation exists because the clastic framework is strong enough to resist considerable compressional stress. High-grade host beds are unlikely to suffer significant collapse or compaction, and under normal conditions, significant reduction in permeability by compressional stress is unlikely. In general, extensional stresses that might be induced by gas overpressures will more likely equilibrate along porosity by gas and water flow, especially after some dissociation has opened or cleared permeability. Thus, blowout of gas from a high-grade deposit is not likely, unless the rate of dissociation becomes excessive. High-grade deposits have the strong potential for safe extraction of the natural gas sequestered in hydrate without the imposition of any 'heroic' or special circumstances. Following conversion of the hydrate and extraction of the natural gas, and the abandonment of the prospect, it is likely that the thermodynamic equilibrium responsible for the establishment of the GHSZ will reestablish. If there is an adequate flux of HFG-enriched groundwater, there is every likelihood that another high-grade hydrate deposit will form in the same location. Although the kinetics of the reestablishment of a hydrate deposit are uncertain, this attribute gives high-grade hydrate deposits the character of a new type of renewable energy.

In a low-grade deposit hydrate may contribute substantially to the strength of the reservoir, especially where there is substantial cementation of the sediment grains. In general, especially after some dissociation has taken place, the strength of the reservoir has the potential to weaken considerably, particularly where high pore pressures may fluidize very fine-grained sediments. Compressional stress is more likely to cause sediment compaction and collapse while extensional stress induced by overpressured gas has the potential to blow out along crack-propagation secondary porosity. The permeability of the host sediments is likely to be too low to allow pressure to equilibrate fast enough to keep the strain field below the brittle fracture limit of the sediments. Because of the inherent weakness of low-grade hydrate deposits, blowout of gas is much more likely than from a high-grade deposit. Because of the inherently lower porosity of a low-grade deposit, combined with the likelihood that sediment compaction and further reduction of permeability will take place during hydrate conversion and extraction of natural gas, it is unlikely that reestablishment of the deposit could take place in a short enough time to give the deposit a renewable energy character.

In both high- and low-grade deposits, reservoir character must first identify the likely host beds in a stratigraphic succession and determine the nature of focused or dispersed HFG character. These parameters provide the basic or first-order 'likelihood' characterization, as well as an indication as to whether a high-grade hydrate deposit might exist. In order to evaluate a deposit and characterize its three-dimensional form on a mineral grade basis, some remote technology that allows estimation of hydrate distribution and form must be applied, or close drilling similar to that commonly carried out in metallic mineral deposits will be required. During this phase of 'economic' characterization, the form and value of a deposit is determined using concepts such as 'cut-off values' that will identify the margins of the deposit. Cut-off values for hydrate mineralization will vary in the same manner that they do for conventional economic mineral deposits, based on other economic projections and assumptions, not on an a particular percentage of hydrate filling.

6.6.1.1. Contrasts between Hydrate and Conventional Gas reservoirs

An important difference between a conventional gas deposit and a hydrate deposit is the effect of the form of the gas on the trap. A conventional gas deposit is entirely contained within a geologically entrapped reservoir. In other words, a geological framework exists where a gas deposit may form, depending only upon migration into the existing physical situation. A hydrate deposit, in contrast, may commonly form where there is no preexisting geological trap. A hydrate deposit develops when HFG-mineralizing solutions pass upward from natural gas source beds into the GHSZ where hydrate will spontaneously nucleate and grow. A hydrate deposit is likely to form best where there is an open groundwater circulation system that is capable of bringing in a supply of both gaseous and dissolved HFG. The groundwater would likely migrate to the seafloor if the GHSZ were not there to cause sequestration of substantial HFG into hydrate. A conventional gas deposit, in contrast, is a groundwater flow dead-end in which gas may concentrate and from which escape is only possible if the deposit is breached by some geological feature, such as a fault or by drilling.

One of the main differences between conventional and hydrate deposits is that conventional deposits may be found in continental shelves and slopes in suitable geological situations, with no regard for confining pressure or seafloor temperature. Conventional deposits may be found beneath considerable thicknesses of very hard rock and rocks and sediments of very different character, which require heavy-duty drilling capability. In contrast, hydrate deposits must be found within the upper km of sedimentologically and mechanically similar marine clastic sediment worldwide, within the local GHSZ, whose thickness is determined by water depth, seafloor temperatures, and geothermal gradient. From the standpoint of exploration, designing and carrying out geophysical surveys or drilling for economic hydrate deposits, is not necessary below the GHSZ and into its subjacent gas deposit (if any). Conventional gas deposits, in contrast, may be found at depths considerably in excess of where hydrate-related deposits may be found.

A conventional hydrocarbon trap will be pressurized and often strongly artesian in character (when a drilled connection is made with the reservoir it will flow to the surface spontaneously and often at very high pressures). An oceanic hydrate trap, except for some lightly overpressured gas that may be held subjacently below the hydrate-saturated strata, requires artificial stimulation to generate gas from the hydrate before recovery of the gas is possible. Gas from a subjacent deposit, however, commonly may be transferred through the GHSZ in chimneys (4.7.1) and vents (4.7.2). Colossal amounts of natural gas appear to be vented from marine sediments (Judd et al., 2002; Etiope and Milkov, 2004), which include mud volcanoes even in restricted seas (Bohrmann et al, 2003).

Any gas reservoir must have appropriate traps and seals. Probably the main contrast with a conventional gas deposit is that hydrate deposits, especially where they are of a high-grade character, may provide an excellent natural gas trap but they may not provide a good seal that will force the gas to concentrate once it has been converted from the hydrate. These deposits may be set in an open groundwater system through which gas may flow away from the location where it has been converted from hydrate.

There are a number of important differences between conventional and hydrate deposits that require particular attention.

1. Hydrate can provide an element of a trap that cannot be predicted by conventional geological analysis. A conventional deposit is completely dependent on the development of a geological trap and reservoir. Hydrate forms according to temperature and pressure controls, in the presence of sufficient gas flux, and may develop in locations that conventional hydrocarbon exploration practices do not identify.

2. Pressure analysis may not apply to predicting presence of hydrate.

3. Whether or not groundwater flows have entrained gas cannot directly indicate whether the water is driven by gas.

4. An initial pressure drop in a hydrate-gas reservoir does not necessarily indicate poor production prospects.

5. The presence of hydrate may be indicated by pressure events and other responses that cannot be attributed to, or caused by, a conventional reservoir

6. Hydrate dissociation-driven water flow has the potential to flow against to hydrostatic pressure or natural head, which will not happen in conventional reservoir.

6.6.2. Producing Gas from Oceanic Hydrate In-Situ

Recovering the methane from oceanic hydrates has two distinct aspects. First, the solid hydrate must be converted to gas (+water and sediment) under controlled conditions, and then the gas must be recovered from reservoirs that will have geological and engineering character that may be very different from conventional gas deposits. The gas must then be brought to the seabed in deep water, but within water depths in which conventional marine gas reservoirs are found today. The engineering, and in places such as the northern Gulf of Mexico, the infrastructure for extracting the gas and transporting it to shore already exist.

Although oceanic gas hydrate deposits are likely to become economic targets for the production of methane, they are very different from conventional gas deposits, and also from those other unconventional gas deposits where methane is present as gas. Although oceanic hydrate is a diagenetic mineral deposit, it is very different from unconventional methane sources such as tight gas and coalbed methane in which the gas phase is already fully formed as methane. Kerogen shale deposits can be gasified, but they must be first mined and then processed, often using considerable heat. For hydrate, no intermediate process, such as combustion, is necessary to produce methane from other materials, although much of the engineering experience with producing gas from unconventional gas sources might be applied to methane extraction from hydrates.

Moridis (Chapter 7) demonstrates that economic recovery of natural gas from high-grade hydrate deposits which are open to groundwater circulation is a possibility. Converting the hydrate to its component natural gas and water, however, is only part of the challenge. Conversion of solid hydrate to its component gas and water must be carried out *in-situ* in order to recover the natural gas by conventional means. Collett (Chapter 5) discusses the secondary or unconventional methods for accomplishing this conversion and Moridis (Chapter 7) demonstrates that conversion of high-grade Class 2 oceanic hydrate deposit is a commercial possibility. A relatively small lowering of pressure or rising in temperature near the phase boundary at the base of the GHSZ will invoke an almost immediate dissociation response. Where the hydrate deposit is found higher in the GHSZ, it is more stable, and proportionally more energy must be applied to achieve conversion.

Gas produced from hydrate may tend to migrate along geological structures, for instance up dip in porous strata, rather than collecting in the vicinity of its dissociation. This is especially true if the groundwater regime in which the hydrate deposit is set is open and there is no up dip seal to prevent gas from migrating away from its parent hydrate. For instance, it appears that the 2002 Mallik dissociation test that resulted in a prominent flare (Cover) converted much more gas than was extracted through the perforations in the drill pipe. It appears that much of the gas moved farther into the formation and away from the drill hole. Given the high permeability of the Mallik reservoir, this observation is not surprising, but without a pressure differential towards the well bore, it is somewhat surprising that gas was produced at all. Cased hole logging conducted after the Mallik test showed that the dissociation was laterally quite uneven. These results are still not well understood. That is, much of the gas that was converted from hydrate was not collected because it migrated away from the collector in the open groundwater / porous strata system. Contrarily, the conditions for best hydrate formation near the BGHSZ, which depends on an open groundwater system, are not the best conditions for recovery of the gas from the deposit because the gas will not tend to concentrate spontaneously during and following dissociation.

In addition, hydrate may dissolve into surrounding water if the gas saturation of the water is sufficiently low, even though the hydrate may still reside within its field of stability. If hydrate is converted to gas slowly and groundwater movement is relatively rapid, the groundwater has the potential to become nearly saturated. The dissolved HFG-rich fluid has the potential to migrate up-dip into colder regions where hydrate could either redeposit or the dissolved HFG could exchange with seawater.

It is important that dissociation driven by depressurization takes place above one atmosphere. At this pressure, the stable configuration of the water molecules is in the form of water ice (Sloan, 1998). Because dissociation of hydrate in the natural environment will usually take place at pressures just below the *in-situ*, ambient pressure that allows for a suitable rate of dissociation, it is likely that pressures will be high enough to insure that hydrate dissociation will produce liquid water. Therefore, where depressurization is used to cause dissociation and produce gas for recovery, it is important to avoid letting the pressure at the dissociation surfaces fall below five atmospheres to insure that an ice phase does not form at the dissociating hydrate front.

6.6.2.1. Extraction Methodology

The primary factor in extracting gas from a hydrate reservoir, once gas can be produced, is that it must flow, and, ideally, concentrate is a location amenable to extraction. Where geological conditions do not naturally favor good gas flow, these conditions must be created (6.6.4). In both cases in oceanic hydrate, an open groundwater situation should apply (Chapter 7).

There are two general types of hydrate that may form potential methane reservoirs: (1) those where hydrate alone is present, or at least where no significant naturally occurring body of gas exists in contact with the hydrate, and (2) those where bodies of gas-charged sediments exist below the BSR. Where a gas reservoir exists, controlling hydrate conversion to gas extraction can at least in part be managed by depressurization. In addition, where a gas deposit exists, the conditions for concentrating additional gas converted from hydrate would largely exist naturally. Addition of warm water, especially near the BGHSZ, however, can be expected to result in a downward flow into the gas zone and away from the hydrate.

Where gas must first be produced from the solid hydrate in order to form a gas zone, measures may have to be taken to insure that the gas will concentrate. No matter how high-grade or rich the hydrate deposit is, if the gas converted from it cannot be concentrated, then it cannot be recovered in economic quantities. It may be necessary to provide an up dip seal in the porous horizons in order to force gas to concentrate. These seals or flow dams would require a very detailed geological analysis in order to locate the seals in the optimum situation. Where a number of hydrate-rich horizons are closely associated, provision might have to be induced so that gas would flow to the highest point of multiple porous beds and concentrate at a single sealed horizon. Up dip seals of porous horizons could be induced through cementing or hydrate formation, for instance. It may also be necessary to install an intensively strata-conforming collector system with one or more wells and lateral courses (Chapters 1, 7). The best commercial extraction model will allow the most rapid extraction of gas consistent with maintaining reservoir integrity and safety.

The extractive process may utilize analog mining techniques that are designed to preserve mine integrity and prevent collapse during hydrate gas extraction. In extracting gas and oil from a conventional trap there is very little danger of collapse of the gas reservoir, even if all the gas were to be removed. Extraction strategies that were derived in part from adapting mining methodologies, which are concerned with preventing collapse of the workings, should have application for the hydrate conversion process. Newer, more closely monitored and controlled exploration and extraction practices than are used with conventional petroleum deposits may be used. For instance, strain gauges may be used to indicate the on-going state of structural integrity of any openings that are created.

In low-grade hydrate deposits, sediment failure, creep, and liquefaction are the three main mechanisms likely to affect the stability of hydrate reservoirs during hydrate conversion and gas extraction. These mechanisms are usually associated with sediment mass flows (Potter et al., 2005) but are also important for designing extraction strategies that maintain the physical integrity of the sediment reservoir, because its physical strength may alter appreciably as the sold hydrate is replaced by water and gas. Much more care will have to be taken with hydrate conversion and gas extraction from a low-grade deposit than from a high-grade deposit.

Pressures within the gas reservoir will almost certainly need to be managed by varying the rate of gas extraction to balance maximum extraction with respect to maintaining reservoir and 'trap' stability. This will also call for new approaches, but technology currently exists that can be adapted for the oceanic hydrate deposits. The method may require lowering reservoir pressures more for oceanic hydrates than for the much shallower permafrost hydrates, because at the greater depths the hydrate is stable at higher temperatures and the phase boundary stability curve is steeper than it is for Polar hydrate (Fig. 2.1). Also, the pressure gradient as a function of total pressure is lower at greater water depths.

Effective porosity that can hydraulically transmit pressure changes for long distances along porous strata is more important in converting hydrate than is total porosity. Where isolated pore space is filled with gas or water, or a combination of gas, water, and hydrate, heat alone may initiate dissociation of hydrate. Because the thermal properties of sediment-hydrate-water-gas mixtures will vary considerably between and within potential hydrate reservoirs, the response to dissociation provided by application of heat alone is more difficult to predict than the effect of pressure changes within a communicating pore system.

6.6.3. Drilling

Deep-water drill ships and other platforms that are capable of operations in deep water (Fig. 1.16) where the GHSZ and its potential subjacent free gas deposits occur are highly complex, heavy duty, high cost systems. The most recent Integrated Ocean Drilling Program (IODP) drill ship is the massive, $550 million, 210 meter long, 57,500 ton displacement *Chikyu* that is 45% longer and 2.4 times the weight of the previous drill ship, the JOIDES Resolution Normile, 2005). The *Chikyu* is presently being fitted out to be the most comfortable, best equipped floating laboratory in the world and is expected to be engaged in systematic ocean drilling by the summer of 2007. These capabilities are necessary because of the two main ocean drilling requirements: (1) to drill in a wide variety of geological materials, and (2) to drill to a considerable depth below the seafloor. Currently available drill strings, which are composed of segments of steel pipe, can be up to 8-10 km in length.

Hydrate deposits will occur only within the GHSZs in the upper km of marine sediment worldwide. The sediments will be largely unconsolidated and only partly compacted. Because drilling depth and drilled materials will vary only within narrow ranges, a common, and not particularly heavy drilling capability may suffice for both exploration and extraction of hydrate gas deposits. High-cost conventional drilling practices commonly applied to ocean drilling are almost certainly not required. For instance, even where a conventional drilling platform or ship is used, it may be possible to use lightweight aluminum drill string instead of conventional steel pipe.

A hydrate mission-specific platform for studying, sampling, and drilling hydrate localities at lower cost could be designed not only for research drilling of hydrate but possibly also to carry out commercial drilling. Mission-specific drilling platforms envisaged by IODP (Austin, 2004), that will compliment the new full-riser capability provided by the 'Chikyu', are oriented primarily in study of climate issues through analysis of marine sediment cores. Although hydrate-specific drilling capability is being discussed by IODP (Rack, pers comm.), in combination with studying deep biosphere, there are no plans currently to provide this capability to the international community. These vessels are available at costs that are at orders of magnitude lower than conventional deep water drilling systems. In addition to being used for drilling directly from the sea surface, a mission-specific vessel of this type could also be used to control special seafloor-mounted drilling capability. A low cost drill rig deployed from such a vessel would allow the drilling of a large number of shallow holes into the GHSZ. By gathering production from a large number of wells, the gas production rates required to justify the investment of drilling could be achieved.

Low-power, seafloor-mounted drilling rigs could be operated from a relatively small, mission-specific vessel, which could be as simple as a mud boat with a large deck workspace. Obviously, ships of this size have no substantial sea-keeping ability in foul weather so drilling operations would have to be suspended when weather deteriorated. A seafloor mounted drilling rig, however, has the potential to be operated independent from weather conditions on the surface. It also could be left and reacquired following return of the surface vessel to the drilling site following improvement of weather.

Seafloor drilling can be carried out in two ways, in addition to rotary drilling. First, a high-pressure hose could power a low-thrust, low-torque hydraulic drill on the seafloor. Low thrust, low torque drilling is possible in shallow sediments, especially where they are typical unconsolidated to semi-consolidated hydrate host sediments wherein hydrate may provide most of the mechanical rigidity. A second method would involve compressed water pulse generators that would essentially drill by pulses of ultra-high pressure water blasts (Kolle, 2000). This method would be particularly applicable for drilling hydrate-strengthened sediments because the hydrate in the direct path would dissociate and produce gas that would act to drive the water and sediment materials out of the drill stem. Both light to medium duty methods would be suitable for drilling the semi-consolidated sedimentary host for hydrate deposits. The hose drilling system would be equipped with an integrated, telescopic surface conductor capable of protecting the first 100 m of the well. Drilling, which would almost certainly involve lateral and multibranching directional drilling, would commence with the surface conductor and wellhead in place on the drill string (Kolle & Max, 2000).

Existing drilling technology may well be adapted to hydrate exploration and natural gas recovery with only small modifications in the equipment or only practices (Hannegan, 2005). In the first instance, managed pressure drilling (MPD), which is currently practiced, will be required (Hennegan, 2005). Rotating control devices, manifolds, and choke manifolds are available to carry out MPD as are the insulated pipes, special mud handling, and operational procedures. Small risers and BOPs adapted for the lower pressures in hydrate drilling and gas recovery would also involve aeasily achieveable downsizing of existing equipment.

Because of the shallow depths below the seafloor and the relative homogeneity of hydrate deposits worldwide, a hydrate-specific drilling strategy for the recovery of natural gas from hydrate would be widely applicable. The same equipment, if not almost the same layout of stimulation and collector wells (Chapter 7), would have wide application. Depending on the local geological situations, a multiplicity of different well strategies may be used (for instance, Figs. 1.17; 1.18; 1.19; 1.20). Detailed understanding of the hydrate field geology will be required to optimize natural gas extraction.

Multiple use of the same equipment and extraction strategy would have the effect of reducing overall cost of gas recovery from oceanic hydrates. Arrays for both stimulation and collection at a deposit, with relatively short (no more than 2-4 km) lateral branches would be practicable because of the relatively short distance from the surface to the hydrate pay zone (Fig. 6.4). Although vertical wells are shown in Figure 6.4 penetrating the hydrate deposits from the surface, collector arrays will lie along strike of beds. The collector arrays may be associated with up dip permeability dams to prevent escape of gas. Capability to swap collector and stimulation functions (not shown) may exist so that 'secondary' recovery techniques may take advantage of the multiplicity of wells and underground piping. Where flux of the natural groundwater containing dissolved gas is high enough so that successor hydrate will form in about the same hydrate physical position in the porous strata from which original hydrate has been removed, the well array may be left in position for re-use at a later time if corrosion and biofoul issues can be overcome.

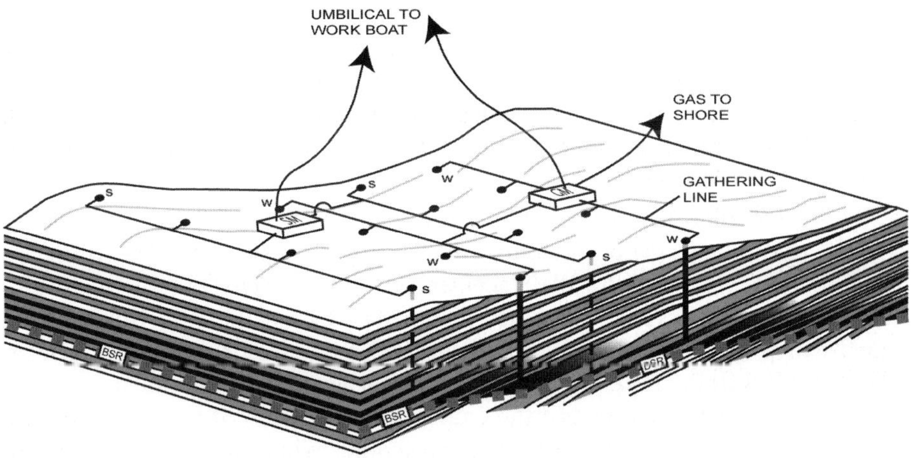

Figure 6.4. Perspective sketch of proposed seafloor array of stimulation and collector wells for recovery of natural gas from seafloor hydrate. GM, gas manifold for water-gas separation and conditioning before being transported via pipeline to the surface. SM, stimulation manifold that controls the pressure and/or thermal stimulation for dissociation of the hydrate.

Coiled tube drilling (CTD) may prove to be particularly adaptable for drilling into the GHSZ. CTD replaces a conventional rotary drill rig that uses jointed pipe, greatly simplifying the drilling process and allowing drilling into a pressurized well bore. The CTD technique uses a drill motor or a hydraulic drill assembly and bit (along with other components) near the head of the drill 'stem'. The drill stem is a continuous length of steel tubing with a diameter of 25 mm to 100 mm or more that is coiled onto a reel and unreeled to follow the drill head. The tubing is inserted into a producing oil or gas well through a production wellhead. CTD is commonly used to place production tubing or to carry out workover operations. This technology has been used to drill small diameter holes from an existing well bore since 1990 (Fultz and Pittard 1990), although CTD is generally considered to have begun in 1991 with a horizontal sidetrack re-entry drilled by Oryx Energy in the Austin Chalk of Texas (Williams et al., 2001). Following a rapidly growing trend, approximately 600 wells were drilled with CTD rigs in 1997 (Leising and Walton, 1998) and by 2001 about 900-1,000 wells were completed yearly, including 120 directional re-entries and 800 new shallow vertical wells. In 2004, BP set a CTD record for lateral sidetrack drilling to 4,000 feet and also achieved a 17,000 foot vertical depth record in another well at the Niakuk field in Alaska (Nelson, 2004). Lateral sidetrack horizontal wells are probably the best configuration for hydrate recovery.

Drilling might also be carried out using heavy ROV systems, which can provide in excess of 75 kW hydraulic power in water depths of up to 2500 m (Kolle, pers. comm.). Such a system could power a coiled tube drilling unit designed to operate from the seafloor. The ROV umbilical provides electrical power and real-time high-bandwidth data communication to the surface via fiber-optic links. All controls for drilling could be provided by the hydraulic systems on existing vehicles. The primary issues in powering a system such as this are the limited torque and thrust capacity of small diameter tubing. The addition of high-pressure jets would significantly add to the penetration capabilities of these systems. Such a system would need to incorporate automated casing installation tools.

In addition, new down-hole development of hydrate reservoir-specific MEMS technology (Micro-Electro Mechanical Systems) will have to be developed that will allow for the placement of sensor arrays for reservoir strain, fluid chemistry, pressure, and other parameters within the reservoir. These arrays will probably lie mainly along the collector pipes but possibly also at other locations in the HSZ and below it. Sensor arrays will provide a means to monitor the reservoir and permit the controlled conversion of gas from hydrate. This real-time, down hole information should allow the reservoir to be adequately managed and its size maintained through planned and controlled hydrate gasification. Artificially forming hydrates or backflooding certain volumes to optimize the size of the gas reservoir while maintaining the integrity of the HSZ will be the main technique that will be employed to minimize blow out effects, and help prevent run-away reservoir volume growth.

6.6.4. Artificially Induced Permeability

Permeable material contains interconnected cracks or spaces that are both numerous enough and large enough to allow fluid to move freely. In some permeable materials groundwater may move several meters in a day; in other places, it moves only a few centimeters in a century. In contrast, groundwater moves very slowly through relatively impermeable materials such as clay and shale. Formation of hydrate in porous material may strongly alter this original sedimentary permeability.

Where little or no natural porosity occurs, such as is the case in the highest grade hydrate deposits where hydrate fills most of the sediment porosity; depressurization may not propagate far from the location where it is induced near the drill/collector assembly in the deposit. In this situation, the surface area of hydrate exposed to the depressurization, which causes dissociation and produces water and gas from the hydrate, may not be great enough to provide for an economic rate of gas production. For instance, during the 2002 Mallik gas production test, there was less dissociation in the originally highly porous conglomerates where the more massive hydrates occurred. The surface area of the hydrate exposed to dissociation was small in comparison to the mass of hydrate. The dissociation front that was artificially created during the test preferably followed fractures to a significant distance away from the well bore. It follows, then, that where permeability and porosity are strongly reduced by the formation of hydrate, the process of artificially inducing a higher surface area of hydrate to promote more rapid conversion and gas production can be accomplished by causing fractures to form within the deposit.

Fraccing, which is the drilling industry term for the process of artificially inducing fractures in a reservoir to increase permeability, is a common secondary recovery technique for conventional gas and petroleum deposits. In these conventional deposits, fracturing is usually induced in the reservoir surrounding a borehole casing through the use of high-pressure water injection. A clastic material, such as sand grains is often introduced into the fractures that are opened so that the fractures remain open following pressure reduction. The enhanced permeability allows petroleum or gas to flow more easily and at a higher rate, to the collector and thence to the surface. Crack propagation (Broek, 1984; Kanninen and Popelar, 1985) works through an artificial wedging aside of an existing fracture or connected porosity and is most efficient where a natural softway (tendency for a material to fracture or break more readily in a particular plane) already exists in the reservoir materials (Johnson et al., 2002). Where the reservoir consists of well-bedded sediments that have not undergone significant diagenesis or lithification, fractures tend to propagate along bedding, often for considerable distances (Fig. 6.5, a). Where sediments are more mechanically isotropic, the induced fractures tend to be more complexly interconnected and more concentrated near the drill hole (Fig. 6.5, b). There are many other complexities to inducing artificial fractures, among which are: (1) generation of interconnectiong networks, (2) the character of fracture surfaces, (3) openness of fractures and the persistence of openness, (4) production of suitable flux orientation and volumes, and (4) modeling percolation theory to control flows (Adler and Thovert, 1999).

Conventional fraccing techniques might well be used to artificially introduce fracturing that would not only increase permeability but would also increase the ratio of surface area to mass of hydrate. This technique would expose more hydrate surface to the interconnected permeability along which gas would flow from the hydrate into the collector drill hole. Because dissociation of hydrate is primarily a surface phenomenon (Chapter 2), the volume of gas produced will increase proportional to the surface area of the hydrate that is exposed to original and artificial permeability.

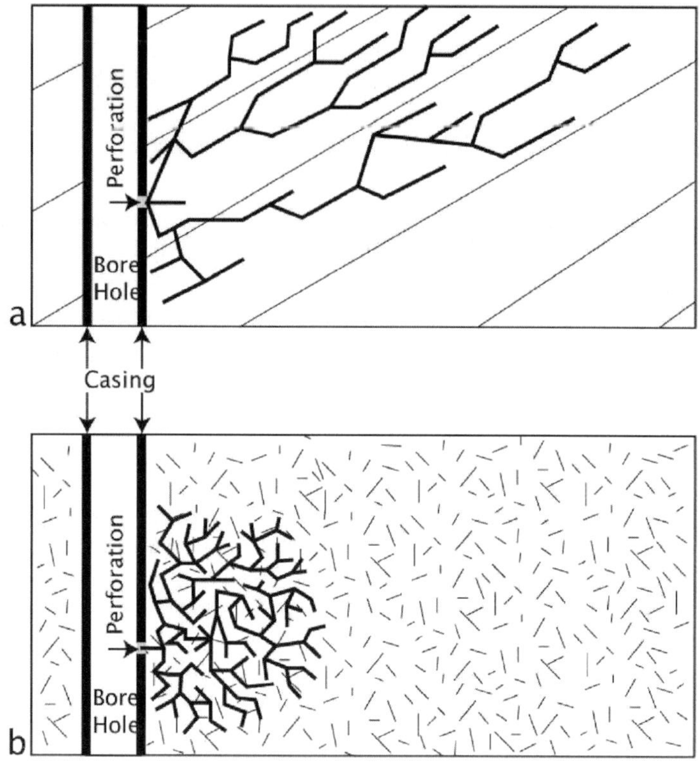

Figure 6.5. Fraccing results using high-pressure fluid injection through perforations in the bore hole casing (dark lines on sides of bore hole). a. Normal secondary fracture pattern in bedded strata with preferential dilation of fractures along bedding (softway). b. Radiating fracture pattern in material with no preferred fracture orientation, such as massive sandstone.

Fraccing, by periodically inducing fracturing by high pressure water injection, could be expected to propagate away from the stimulation points of penetration of the drill casing (Fig. 6.5) better in a hydrate reservoir than in a conventional reservoir. Whereas the actual porosity of a conventional reservoir will only be increased proportionally to the secondary or induced fracture porosity, the porosity of the hydrate reservoir will increase disproportionably in a positive manner as hydrate dissociates and liquid water and gas replaces the hydrate. A conventional reservoir, on the other hand, may actually compact as gas and petroleum are removed, and the reservoir tends to depressurize. Because gas is produced as hydrate dissociates, the actual supply of gas increases in proportion to the solid mass of sediment plus hydrate in any given volume of converted reservoir.

It is also possible to envisage conditions under which artificial permeability could be induced in a high-grade hydrate reservoir. It may not be necessary, for instance, to drive an open secondary porosity fracture through the reservoir. It may only be necessary to stimulate the dissociation of hydrate in either a planer or a linear fashion. Artificial dissociation and collector courses could be produced without artificial fracturing of the reservoir as a whole. Hydrate dissociation weakens sediment by inducing a dilatational force (Winters et al., 2002) normal to the shear plane, which has the affect of promoting crack propagation. By allowing the pressure to vary, it is possible that short periods at higher pressure could provide a type of self-fraccing that would force open cracks in order to present more hydrate surface area.

A number of methods can be speculated upon for both opening gas and water reservoirs within solid hydrate that is not in contact with gas or open to groundwater circulation. One method might involve the use of a small thermal probe that could be inserted into the sediment. The probe would move itself into the reservoir by a variety of mechanical means. An electrical connection, and possibly also an air or oxygen supply where combustion was the objective, could be pulled behind. This would allow the required heating to cause hydrate to dissociate, in a manner similar to that used by a wire-guided missile. As the hydrate dissociated and the porosity increased, an open passage would be established along the connector cable. Flow assurance issues would be important, of course, in the relatively small, unlined passages. Industry, however, already has considerable experience in flow assurance in the presence of unwanted hydrate. Because of the history of the energy industry in solving this type of problem, yet another innovative solution not currently being used would have to be developed. A second method could involve the stimulation through the use of microwave transmitters that would focus energy along particular paths or planes through constructive reinforcement of wave fronts. This technique would cause the formation of local hotspots at some distance from the actual microwave projectors. A further method could involve the use of lasers. There are a number of models for dealing with these possibilities, but they will not be considered here.

Wherever a gas pocket could be stimulated, the gas in the pocket would be pressurized to the local ambient pressure of the hydrate phase boundary. Where local heating could be increased, the pressure in the gas space would increase. The effect of raising temperature would force the local P-T condition down the phase boundary curve (Fig. 2.1). Relatively small increases in temperature at the pressures within a normal oceanic hydrate reservoir would raise pressure disproportionably as the negative slope of the phase boundary increases with increasing depth. Where increasing pressure could stimulate a train of overpressurized gas pockets, an auto-fraccing event would be possible. The general mechanics of causing a fracture to form in this manner would be similar to the technique of line drilling and wedging practiced in stone quarries for the recovery of stone gallets. The individual overpressured gas pockets would form a line or zone of weakness along which a fracture could propagate as triaxial strain developed, and the brittle fracture limit was reached.

In addition to mechanically induced permeability enhancement, the dissociation of hydrate may naturally ease the movement of gas through sediment pores. McGuire et al. (2005) note that lowering the salinity of waterflood in petroleum reservoirs enhanced recovery. They attribute this LoSal enhancement alteration in sediment wetability similar to those induced by alkaline or surfactant injection, but without the addition of chemicals, while noting that Tang and Morrow (1999) regard the enhancement as possibly due to detachment of mixed-wet clay particles from the pore walls. When hydrate dissociates, it releases very low saline water (3.3.1: Fig. 3.4), which may coincidentally contribute to a beneficial LoSal effect on gas migration in a hydrate reservoir.

6.6.5. Hydrate and Natural Fracturing

Sub-vertical faults interpreted on reflection seismic profiles are often commonly clustered in the lower part of the GHSZ and in the gas zone immediately below it (3.4.1; Fig 3.11; Fig. 4.6). Although some of these faults may penetrate to near the seafloor, most occur where the sediments are most enriched with hydrate in the lower third of the GHSZ. Because of their prevalence, these faults are probably associated with sets of smaller fractures that are not seen on the reflection seismic profiles. These faults appear to be more associated with the hydrate and the BGHSZ than to regional structures.

McLellan et al. (2003) identified a number of sets of sub-vertical to steep-dipping fractures in their downhole analysis of permafrost hydrate in the Mallik 5L-38 drill hole. The fractures are much more common in hydrate-enriched strata between 890 and 1110 meters than they are in intervening or immediately adjacent strata within a vertical distance of 60 m from the base and top of the hydrate. McLellan (2005) further points out that the presence of these partially open, permeable fractures, combined with the overpressures associated with hydrate dissociation, could have important implications for natural gas production from horizontal wells in the hydrate.

The apparent clustering of faults and fractures in both permafrost and oceanic hydrate-enriched sediment may well be related to the alteration of the mechanical properties induced in the sediment by formation of hydrate acting as a cementation agent. Formation of hydrate that displaces pore water or acts as part of the sediment framework increases the bulk modulus and resistance to compressibility, whereas cementation strongly affect the shear modulus (Dvorkin et al., 2003). The presence of hydrate increases the mechanical strength of the sediments and, in effect, makes it more brittle. Whereas triaxial stresses acting on sediment without any hydrate may be resolved by intergranular adjustments of sediment grains, which approximates an elastic response in a coherent material, the hydrate-enriched sediment is more likely to fracture than bend or adjust.

Thus, although there are too little data on the specific relationship between the presence of hydrate in sediment and the distribution of fracturing and faulting to be confident at this time, it is possible that there may be a natural fracture porosity associated with hydrate-enriched sediments. If a natural secondary permeability actually proves to be generally associated with both permafrost and oceanic hydrate, then McLellan's (2005) observation regarding their fracture concentration could be important for the production of natural gas from hydrate deposits. A natural fracture porosity within hydrate-enriched strata would mean that artificial fraccing might be unnecessary. Natural gas hydrate deposits could be intrinsically associated with a secondary fracture permeability that would allow substantial and continuous flow in the deposit upon conversion of gas from hydrate. Because considerable hydrate would be exposed to the fracture porosity, high dissociation rates could be anticipated during forced dissociation.

In contrast to faulting being better developed in hydrate-enriched strata, faulting also appears to be common in the gas zone below a GHSZ, especially one that has probably been subjected to a considerable number of hydrate formation and dissociation cycles owing to environmental change (3.4.1.1), such as the Blake Ridge. These faults have considerable effect on the reflection events and probably indicate that they have considerable throw. In terms of their scale, they are almost certainly larger features that the fracturing seen in the permafrost hydrate enriched strata at Mallik. This clustering of faults is well seen on broad profiles across the Blake Ridge (Fig. 6.6). These faults may die out downwards through the gas-enriched and become indistinct in the sediments beneath, although the seismic section data becomes indistinct downward on this section. A number of the interpreted fault could be extended upward into the hydrate-enriched zone through gaps in the more horizontal events almost certainly related to the orientation of strata, which would suggest that the faults are not restricted to the region beneath the GHSZ. On the Mohican Channel site on the Atlantic margin of Canada (Mosher et al., 2005, Fig. 2) faults clustered below the BGHS die out about half way to the surface within the GHSZ. Where the strata are alternately and repeatedly subjected to formation of hydrate, which enhances rigidity, and gasification, which may significantly weaken the strata and cause block movement along faults because of gas pumping and venting, a dynamic mechanism exists for the initiation and maintenance of faults related to the surface-parallel zone of transition between hydrate- and gas-enriched sediment. As the transition zone moves up and down the strata are subjected to a far more active stress regime that would be associated with sediments on a passive continental margin, such as the eastern continental flank of the United States in which little hydrate concentration or recycling has taken place. Fractures in soft sediments may also host significant quantities of gas hydrate (McGee et al., 2005).

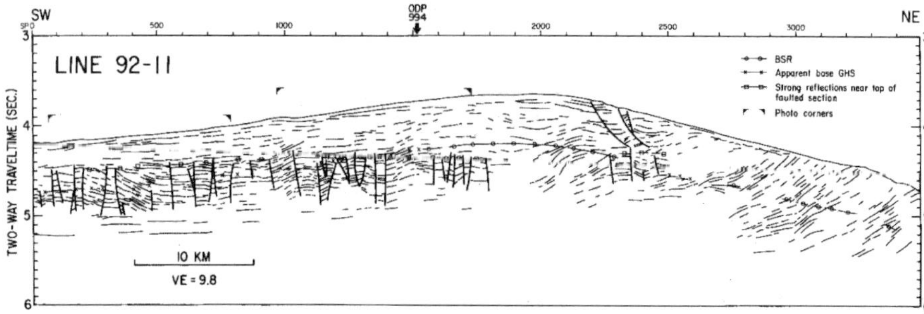

Figure 6.6. Interpreted reflection seismic section across the Blake Ridge, SE U.S. continental margin. From Dillon et al. (1996), Figure 4. Note that breaks in reflection events within the GHSZ immediately above faults suggest that they may continue across the BGHSZ.

High permeability associated with high-grade hydrate deposits also introduce recovery issues that must be taken into consideration when converting hydrate to form gas concentrations. If no geological trap exists for the converted gas to concentrate within, a trap must be created or a gas collection method must be devised that will stop the converted gas from escaping from the immediate area where its precursor hydrate was located. Escaped gas would travel along the secondary porosity or the primary porosity of the more porous strata in which the hydrate occurred.

6.6.6. Volume-Pressure Relationships for Hydrate Dissociation at Depth

Conventional gas deposits have reservoir porosity mainly filled with gas that is pressurized to at least their ambient pressure depth. Max and Dillon (1998; 1999) pointed out that the volume or percentage of free gas produced from any particular amount or percentage of hydrate is related to the ambient pressure of the hydrate reservoir. This relationship is a principal factor in valuing hydrate deposits. There are important implications for economic hydrate deposit exploration because the ambient pressure of the hydrate deposit must produce gas that flows spontaneously. Where gas must first be converted from hydrate, it will not flow if the produced volume is small and does not displace enough water to allow the gas to coalesce and fill a discrete volume of the porosity, preferably in the immediate vicinity of a collector pipe. The aim of converting gas from hydrate is to produce in the hydrate reservoir a gas deposit that resembles a conventional gas deposit. If the resultant converted gas is below some pressure and volume, for any particular permeability, gas flow will not take place.

Simple calculations can be made that show how the pressure-depth of the converted hydrate generally affects various volume percentages of hydrate in a reservoir. The assumptions are: 1. All the gas converted from the hydrate is recoverable, and any that goes into solution with the water will be recovered: thus the volumes represent total gas held in hydrate. 2. Only Boyle's law is used. No corrections for temperature, partial pressures of gas in fluids or any other physical-chemical factors that would actually operate in a real reservoir are considered. 3. Calculations are made entirely for methane (sI) hydrate. Some heavier density thermogenic hydrocarbon gases that may be present in some hydrate deposits may liquefy. All dissociated HFG is treated as gas phase.

A volume of 164 m^3 of HFG (at STP) is held within the crystalline structure of a cubic meter of hydrate. For every m^3 of hydrate in pore space, 0.8 m^3 is water, leaving 0.2 m^3 for gas. Where 1% of hydrate occurs in m^3 of pore space, which is otherwise occupied by groundwater, 99% of the pore space is occupied by water. Furthermore, for every 1% per m^3 of hydrate, the porosity is filled with 99.8% water (0.2% water = 0.02m^3 water). Thus, when the gas dissociates, it has only a relatively small space into which to fit. The main factor that will control the actual pressure and volume of the gas produced from each given volume of hydrate is the degree to which the gas can expand against ambient pressure and displace water in the reservoir. Where groundwater displacement is difficult, gas volumes as a percentage of bulk or porosity will be low, and pressures will be high. Where groundwater displacement is easy, relative gas percentages will be higher, and pressures will be relatively lower.

Where pressure equilibration is perfect, that is, where the pressure of the gas displaces water and achieves the ambient pressure (Table 6.2), the increasing ambient pressure with depth compresses the gas produced into smaller volumes. For different types of reservoir and its internal gas and water flow characteristics, the lower cut-off for hydrate percentage will depend not only on the percent of hydrate present, but also on the pressure-depth of the hydrate reservoir. For instance, at 1 km depth, 5% hydrate produces 0.082 m^3 of pressure-equilibrated gas while at 3 km depth (a little shallower than the Blake Ridge hydrates studied); only 0.027 m^3 gas is produced. If a cut-off volume of 0.5 m^3 is taken as being the cut-off, then the area that has been studied in deeper water would not be economic whereas hydrate deposits of the same percentage in shallower water closer to the U.S. East coast (Dillon et al., 1994) would be. The hydrate deposits nearer shore also likely include more course grained sediments having higher porosity and permeability. Very high pore fill of hydrate, however, will produce significant volumes of gas to substantial depths.

The greatest relative compression of gas being removed from its environment and sequestered in hydrate takes place in shallower water depths (Fig. 6.7). When the hydrate is dissociated, greater volumes of gas are produced at shallower depths because of the compression factor of ambient pressure.

D	P	Vol	OverP	2%	5%	7%	10%	20%	40%	80%
km	atm	m^3	%	Volume of Free Gas in Pore Space (m^3)						
0.5	50	3.2	1640	.065	.164	.230	.328	.656	1.312	2.624
0.8	80	2	1025	.041	.103	.143	.205	.41	.82	1.64
1	100	1.64	820	.033	.082	.115	.164	.328	.656	1.312
2	200	0.82	410	.016	.041	.057	.082	.164	.328	.656
3	300	0.54	273	.011	.027	.038	.054	.109	.219	.437
4	400	0.41	205	.008	.021	.029	.041	.082	.164	.328
5	500	0.32	103	.007	.016	.023	.033	.066	.131	.262
6	600	0.27	137	.005	.014	.019	.027	.055	.109	.219

Table 6.2. Pressure and volume relationships for selected percentages of hydrate in pore space. D, depth; P, pressure, OverP, overpressure (where there is no displacement of water) as a percentage of ambient pressure. The table is taken to 6 km depth to demonstrate the pressure relationships; the authors believe that recovery of oceanic hydrate from depths greater than about 4 km water depth is unlikely or very far in the future.

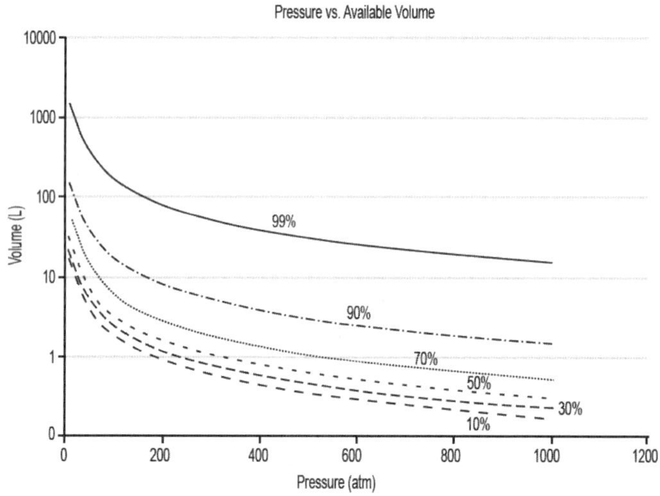

Figure 6.7. Relative volumes of gas produced from 1 liter of (methane) hydrate at different ambient pressures.

In order for gas to coalesce, separate from water, and occupy discrete volumes of pore space, water must be displaced. Where displacement of water proceeds very slowly in relation to the dissociation rate of the hydrate, the high pressure of the gas will have the tendency to force the gas rapidly along the pore space. Recovery of the gas may prove difficult as it may migrate rapidly away from the zone of dissociation in small, overpressured packets rather than forming contiguous volumes in the pore space. When locating stimulation and recovery wells (Fig. 6.3), it also may be necessary to remove groundwater from down-dip wells in strata in which hydrate is being dissociated. This would allow the dissociated gas to be able to expand and coalesce and remain at pressures at which it could be managed in the reservoir so that its recovery might be optimized. Run-away dissociation pressurization will almost certainly have a detrimental effect on overall gas recovery.

For every particular volume of hydrate, those deposits in shallower water will tend to yield higher percentages of gas in porosity than the same volume of hydrate in deeper water so long as displacement of water can be achieved. Therefore, if the primary objective is to recover gas from hydrate through its *in-situ* dissociation, shallower water exploration targets will undoubtedly prove to be better. The shallow sites will have inherently stronger gas flows than the same hydrate dissociated in deeper water. Alternatively, if the primary target for gas production is from an existing free gas reservoir trapped below hydrate, then deeper deposits may be better. The gas will be at higher pressures and more strongly artesian with respect to surface pressures than free gas in shallow deposits.

6.6.7. Safety

The trapping of gas beneath the hydrate-bearing layer may result in near-seafloor structural failure. Catastrophic gas release could result in mass sediment flows (Dillon et al., 1993). Even with no attempted extraction, these hydrate-gas reservoirs may be metastable, at least with respect to the 10,000-to-100,000 year time scale of the glacial-interglacial cycle. Hydrate engineering must focus on developing different, and probably more closely monitored and controlled exploration and extraction practices, than are used in dealing with conventional hydrocarbon deposits. Where dissociation is forced, especially by thermal means, overpressured pore water is likely, and management of both water and gas will be necessary. Keeping the volume of free gas in the reservoir of the hydrate system to a minimum can mitigate the danger of blowout. Forced hydrate dissociation (Max and Cruickshank, 1999) and gas collection and extraction should be as temporally and spatially associated as possible. Beyond these specifics, the practices commonly used by the offshore industry for good safety should continue to be practiced.

6.7. UNCONVENTIONAL GAS RECOVERY FROM HYDRATE

Ideally, once hydrate has been converted into its constituent gas and water, gas pressures and volumes will be sufficient to allow the gas to flow and concentrate in such a way that it can be recovered into collector apparatus and removed from the seafloor. Where the conversion of a gas phase results in its unwanted dispersion, another means of recovering the gas from the hydrate other than gasification may be required. One means may be to provide for an artificial 'trap', another may be to remove water so as to pull gas out with it, yet another possible means of unconventional recovery is promote and control the dissolution of hydrate.

6.7.1. Dissolution

Where the water around hydrate has sufficiently low methane saturation, hydrate will dissolve even though the pressure and temperature conditions of hydrate stability are maintained. This effect has been observed in seafloor hydrate where hydrate forms in the presence of active seafloor seeps and redissolves when the seeps cease and the local water is no longer saturated in the hydrate forming gas. The dissolution effect (Rehder et al., 2004) has also been observed experimentally by Brewer et al. (1999; 2002a; 2002b), for both methane and CO_2 hydrate in natural ocean waters. We propose that direct dissolution of hydrate into pressurized water and recovery of the gas in a gravity driven circulating water system is possible, and may provide an alternate recovery process to the formation of a gas phase.

Pore water at depth can contain significant dissolved gas under pressure (Fig. 6.8) that can be removed by the simple process of lowering pressure. Paull and Ussler, for instance, note that 1 liter of methane-saturated water at 3 km water depth will evolve 4.24 liters of methane at surface conditions. At the surface, the same water contains only 0.02 liters of methane at full saturation (at 20 °C). Introduction of a tendency toward dissociation of hydrate, that is, where the conditions surrounding the hydrate are altered so that the hydrate approaches instability, may also provide for enhanced direct dissolution. The process of dissolving hydrate also self-limiting has a self-governing aspect, as do the thermal and pressure buffering effects of dissociation. The rate of dissolution will decrease as relative saturation equilibrium of the hydrate and the surrounding water is approached and conditions favoring formation of a gas phase are reached. The water in the vicinity of the hydrate would have to be replenished at a high enough rate to preclude the formation of a gas phase (assuming conditions of hydrate stability) while yielding rapid enough rates of transfer of methane from hydrate to water to establish dissolution as an extraction method.

Water in the region of the dissolving hydrate can be maintained at sufficiently low levels of saturation to provide a diffusion gradient between the hydrate and the surrounding groundwater that will allow the hydrate to dissolve unimpeded. No significant gas phase will form until the water is depressurized. Direct dissolving can be carried out using less energy for the conversion of hydrate to its constituents where heat would otherwise have been added to the system (Fig 6.9).

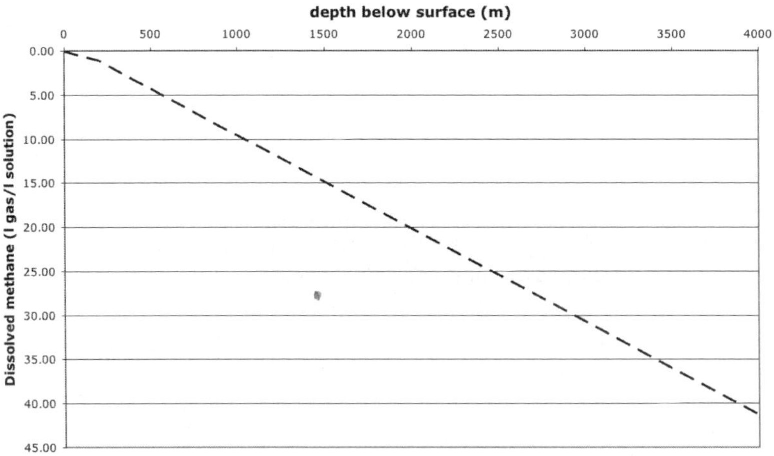

Figure 6.8. Graph of gas volume that can be contained in water at different pressure depths. Dashed line, water at 5 °C.

Causing the breakdown of hydrate by promoting dissolution may provide an alternative means of converting hydrate and transporting the gas in a largely dissolved form. An advantage to dissolution over dissociation is that when water is circulated in the flow loop to collect the dissolved gas under relatively high pressure, free gas will not collect and migrate away from the collection system. Pumping water through a large flow loop could be carried out at a very low cost.

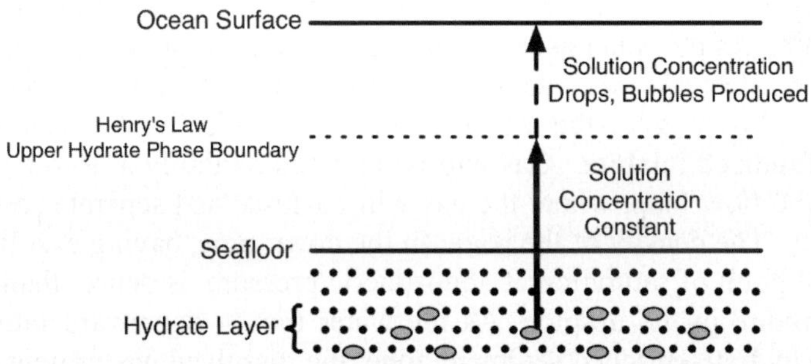

Figure 6.9. Diagram of solution-exsolution with change in depth in ocean.

The exposure of hydrate to pure water results in endothermic dissolution. The heat of dissolution (hydrate → aqueous methane) is not as positive as the heat of dissociation (hydrate → gaseous methane + water). Dissolution is mass limited: only a relatively small amount of hydrate dissolves before the water is saturated with methane. Because of this, temperature depression is low. At 50 MPa and 280 K 1.0 l of water will dissolve 66.1 mM of methane and be cooled 0.7 K by the dissolution action, not including the heat of mixing. The initial conditions are close to the phase boundary so a small drop in pressure will begin the process of bubble nucleation within a relatively short distance from the seafloor during ascent.

Where the surrounding gas or water media is undersaturated, the HFG dissolves on a molecular basis; where there is saturation or supersaturation, a gas phase may be produced. This is because coalescing of dissolved (molecular) gas consumes energy, in addition to that required for dissociation. The energy required for dissolution is about 20% less than for dissociation (Fig. 6.10).

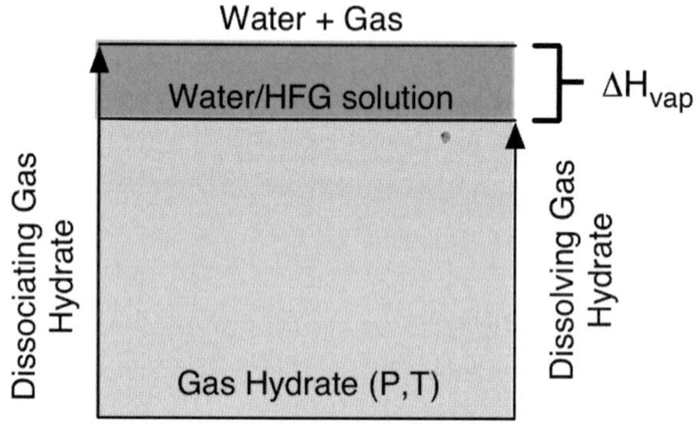

Figure 6.10. Diagrammatic relationship of energy required for dissolving hydrate and vaporizing it from solution.

In this dissolution model, a closed flow loop is established that passes through a region of hydrate-enriched sediments, and then passes to a ship or other platform from the seafloor. The flow loop then returns to the region of hydrate-enriched sediments (Fig. 6.11). At the pressure-depths of the hydrate-enriched sediment, considerable gas will dissolve into the water that is under high pressure. As the water ascends, the gas exsolves, first as a fine fizz of small bubbles that decrease the density of the gas-water mixture and provide considerable lift to give the water momentum. Ideally, the gas and water become a fluidized mixture. Gas will be released naturally at lower pressure at the top of the flow loop, where the gas will coalesce and separate so that it can be extracted. The density of the water in the down-pipe, having exsolved its gas to about the point of saturation at atmospheric pressure, is denser than the water and gas mixture in the up-pipe. As the water moves downward into the flow loop in the hydrate-enriched sediment zone, the dissolved gas present at the top of the flow loop causes water to be strongly undersaturated and capable of absorbing more dissolved gas. Dissolution-gravity lift could provide a potentially inexpensive method for the recovery of natural gas from hydrate deposits, especially where energy is either difficult or very expensive to deliver

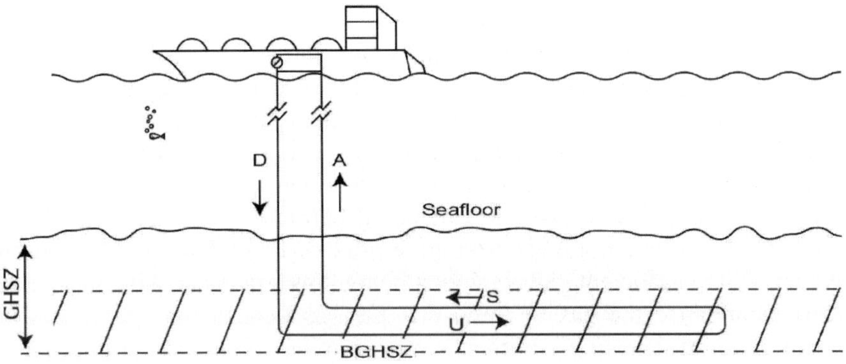

Figure 6.11. Closed loop, gravity-driven deepwater hydrate gas recovery system. S, saturated, A, ascending; D, descending, U, undersaturated. Because it is gravity-driven, and may produce hydroelectric power, the energy balance over time has the potential to be positive.

to the hydrate enriched sediments. Because the overall density of the fluid in the upper region of the upstream circulation system is less than the overall density of the water in the water down-course, a natural, gravity-driven circulating system is established.

The circulatory flow pattern functions in effect as a "gas elevator." The circulation-generating pressure in the system as a whole is equivalent to, or a function of the difference in pressure head between the manifold water down-course and the up-course. A pressure head will exist in the loop as a whole owing to the greater density of the fluid in the down-course. Thus, the differential in densities within the two portions of the apparatus automatically creates a circulatory flow pattern that brings the dissolved gas from hydrate up into the upper reaches of the installation at a significant rate.

Circulatory flow pattern in the flow loop may be initiated and regulated using appropriate flow-rate control devices, e.g., controllable blade-pitch pumps or thrusters that are located between uppermost part of the system where gas is removed and the down-course pipe. Under conditions of large flow rates of water, once the natural circulation system has been established and stabilized at optimal operating conditions for gas production, the difference in densities will continue to drive circulation in the system. A tremendous amount of energy will be entrained within the gravity-driven flow loop and it will be important to control water flow and prevent run-away and blowout. At a point where water flow must be slowed, a mechanical load can be placed on the moving water of the circulating water in the system. A thruster/pump assembly restricts the flow of water by removing mechanical energy and may be used to generate hydroelectric power. Power production would be directly related to the mass of water and its velocity.

6.7.2. Low-Grade Deposit Special Issues

Low-grade deposits (4.6.2) have a far higher potential to collapse than high-grade deposits (4.6.1, Fig. 4.12) as the hydrate is produced, because of the mechanical weakening that occurs within them as hydrate is dissociated. Because of the likelihood for sediment collapse and difficulty of controlling gas pressures and flows in mechanically weak sediments, it is not likely that low-grade hydrate exploitation will be attempted before considerable experience is gained in recovering natural gas from high-grade hydrate deposits. It is possible, however, where sufficient gas is trapped below low-grade deposits, in sediments porous enough for the gas to flow, that the gas deposit may be treated as a conventional gas deposit for purposes of extraction. The hydrate in the trap to these gas deposits, however, may dissociate when the gas deposit is depressurized. Because the interface between the gas-charged and hydrate-charged sediments would be very large and diffuse, dissociation of hydrate would be slow locally. Thus, the natural geothermal heat supply would probably be sufficient to allow dissociated gas to coalesce with the existing subjacent gas as the BGHS prograded upward into the hydrate-enriched sediments. This could be characterized as a secondary recovery technique for conventional gas that would have no energy input and a positive recovery rate in that more gas would be added to the gas reservoir than originally was present in the reservoir. This simple-appearing model for recovery of natural gas has many difficulties that will not be discussed in detail here because without real production data and experience from high-grade deposits, which are the most likely for early exploitation, this discussion would be too speculative. Controlling sediment collapse of both the gas and hydrate reservoir at their interface would involve controlling mechanical adjustments by backflooded.

Where dissociation of hydrate within a low-grade, mechanically weak host takes place without gas already in contact, the possibility exists to apply analogs of mining methods. The primary mining method for low-grade mineral deposits is not available for oceanic hydrate deposits. Open pit mining, where economic methods for the removal of huge volumes of rock having low mineral grades, is unlikely to be practiced on the seafloor. The hydrate mineralization will almost certainly be too deep with too much overburden to be removed (most open pit mines have little or no unmineralized cover over the payzone), and the methane has too low a value to justify wholesale mechanical removal.

One of the main issues in underground mining is the removal the mineral deposit without opening up so much volume underground that collapse of the workings takes place. In areas where the host rocks are mechanically strong, surprisingly large open volumes can be left open. When the mines are abandoned, they are usually flooded. The hydrostatic pressure contributes to the integrity of the openings. Where the host rock in conventional mineral deposits is relatively weak, mechanical strengthening is necessary. Back-bolts, metal reinforcement, rock backfill, and the leaving of pillars of unmined ore are among the common techniques for mining in these circumstances.

The best analog for the application of mining techniques to low-grade hydrate deposits, where *in-situ* mechanical strengthening is difficult because the workings are remote, is the establishment of a room and pillar structural framework. Room and pillar mining is commonly carried out in flat or gently dipping mineralized or pay zones. Room and pillar methods are commonly used to mine coal, potash, phosphate, and bedded uranium ores, amongst others. Rooms are cut into the coal or other bed leaving a series of pillars, or columns, to support the mine roof (back). Generally, in a coalmine, rooms are 20-30 feet wide and the pillars are up to 100 feet wide. A grid-like pattern of rooms and pillars is formed as the mining faces advance into the deposit (Fig. 6.12).

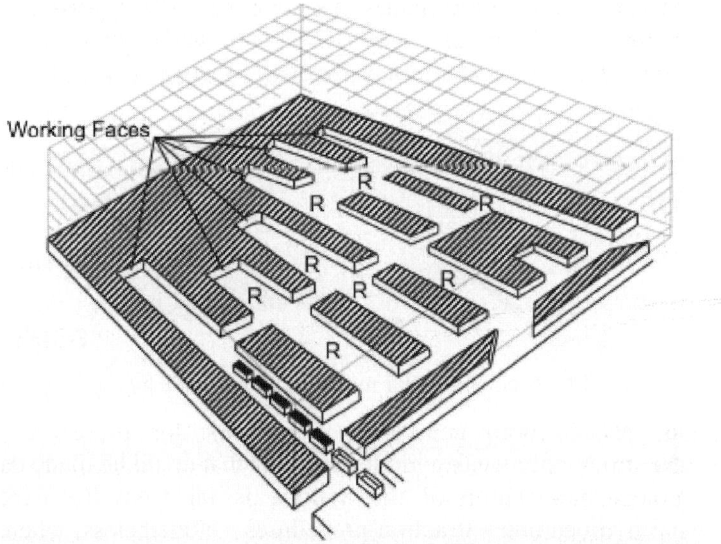

Figure 6.12. Example of a room and pillar mine in a coal deposit. After a diagram provided by the AFL-CIO United Mineworkers of America. R, rooms, pillars shown as unremoved masses with pattern at the elevation of room tops. Rooms enlarged by removal of mined material. Mined material being removed in mine rail cars, lower right. No mining apparatus shown.

Conventional mining, cutting, blasting and mechanical removal in cars of some sort is generally being replaced by continuous mining methods wherein a continuous mining machine, which is similar in many ways to modern tunneling machines, cuts the pay zone face. When mining reaches the end of the working, a phase of retreat mining may be initiated. In retreat mining, the rooms may be supported by rock backfill or artificial supports while as much of the economic mineral as possible is removed from the remaining pillars. In certain circumstances where general collapse of the overlying ground is not a hazard, such as in deeper workings, the retreating workface is allowed to collapse under controlled conditions. Abandoned mines are allowed to flood as the hydrostatic pressure helps to support the remaining parts of the mine that have not collapsed.

A pseudo-room and pillar system can be established within a low-grade hydrate deposit by the carefully controlled zone dissociation of hydrate. Reservoir engineering would be merged with mining engineering practices to provide a composite framework that would allow structurally weak zones to be created which would not destroy the integrity of the hydrate reservoir as a whole. Pillars of hydrate-enriched sediment would be allowed to remain between pseudo-rooms. These would not collapse (Fig. 6.13). Indeed, slight compaction of sediment within the pseudo-rooms may be desired as it may assist in the separation of gas from water in the pore space and in the management of the gas and water to facilitate coalescing of the gas. The back (roof) of undissociated hydrate-enriched sediment is supported by the pillars. Stimulation of dissociation would likely follow the richest parts of the deposit.

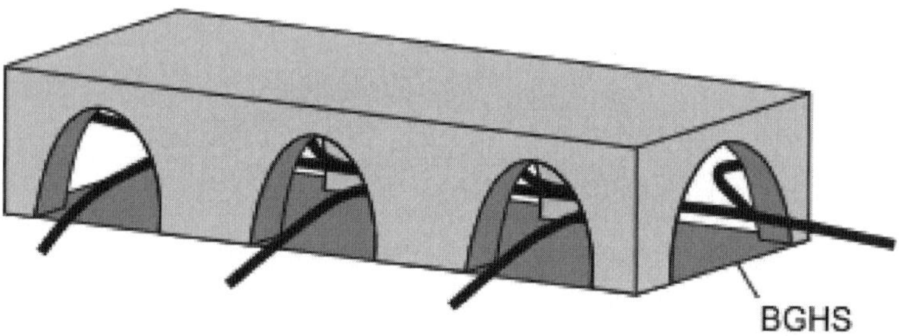

Figure 6.13. Room and pillar model in a low-grade hydrate deposit.

This pseudo-room and pillar framework for preserving structural integrity in a structurally weak hydrate reservoir that could be made dangerously weaker through dissociation of the hydrate is far from the experience of conventional hydrocarbon extraction procedures. Nonetheless, where there are no strong geological structures in the hydrate reservoirs, new approaches will be required to safely and rapidly extract the natural gas from hydrate.

6.8. CONCLUSIONS

Means for identifying the most likely locations for economic deposits of gas hydrate will depend, as with conventional hydrocarbon deposits, on an understanding of the processes that control the generation, the migration, and the concentration of the natural gas. Methodology for identifying target areas exists. Current exploration technology, such as the processing of multi-channel and 3D seismic surveys developed for conventional hydrocarbon exploration and now being used in water depths where hydrate deposits are liable to be found, is directly applicable, as are some EM analysis techniques. Drilling capabilities, especially CTD technologies, are directly applicable to hydrate exploration and exploitation, as are the seafloor completion and processing technologies. Little technique or technology needs be invented for hydrate exploration; considerable innovation, new modeling, and focus, however, must be applied to make the economic exploitation of ocean hydrate a reality.

Chapter 7

GAS PRODUCTION FROM UNCONFINED CLASS 2 OCEANIC HYDRATE ACCUMULATIONS

George J. Moridis and Michael Kowalsky

7.1. INTRODUCTION

Unconfined Class 2 hydrate accumulations in the oceanic subsurface are characterized by mobile saline water zones enveloping the hydrate-bearing formation and by the absence of impermeable layers to vertical flow. We evaluate the gas production potential of such deposits using both single-well and five-spot well configurations. Single-well production is based on depressurization-induced dissociation of the hydrates, while the five-spot configuration involves both depressurization at the production wells and thermal stimulation at the injection wells. The results of the study indicate that unconfined Class 2 hydrate accumulations, which are the common type of oceanic hydrate deposit, are among the most challenging targets for gas production because (1) the absence of confining boundaries limits the effectiveness of depressurization, (2) gas production is accompanied by the production of very large volumes of water, and (3) thermal stimulation, when employed, requires substantial energy inputs. The amount of produced gas is limited in both the single-well and the five-spot configurations, and is significantly smaller than the total volume of gas released in the formation. For the five-spot configuration, hydrate dissociation releases relatively large amounts of gas into the reservoir, but this gas is not readily recoverable. Gas production is significantly affected by the initial hydrate saturation.

7.2. BACKGROUND

Gas hydrates are solid crystalline compounds in which gas molecules are encaged inside the lattices of ice crystals. Vast amounts of hydrocarbons are trapped in hydrate deposits (Sloan, 1998). Such deposits occur in two distinctly different geologic settings where the necessary low temperatures and high pressures exist: in permafrost regions and in deep ocean sediments.

Current estimates of the worldwide quantity of hydrate natural gas vary widely, and a range between 10^{15} to 10^{18} m^3 has been reported (Sloan, 1998). Note that these estimates are not the result of a systematic attempt to evaluate hydrate reserves, but are based on extrapolated from a limited number of hydrate deposits. Even by the most conservative estimates, the total quantity of gas in hydrates may surpass, by a factor of two, the energy content of the total fuel fossil reserves recoverable by conventional methods (V). The magnitude of this resource commands attention because it could make hydrate reservoirs a

Michael D. Max et al. (eds.), Economic Geology of Natural Gas Hydrate, 249–266.

substantial future energy resource. The potential importance of hydrates is further augmented by the environmental attractiveness of gas (as opposed to solid and liquid) fuels. Although the current energy economics cannot support gas production from hydrate accumulations, their potential clearly demands further evaluation.

The three main methods of hydrate dissociation for gas production (Chapter 5) are: (1) depressurization, in which the pressure is reduced to a level lower than the hydration pressure P_H at the prevailing temperature, (2) thermal stimulation, in which the temperature is raised above the hydration temperature T_H at the prevailing pressure, and (3) the use of inhibitors (such as salts and alcohols), which causes a shift in the P_H-T_H equilibrium through competition with the hydrate for guest and host molecules (Sloan, 1998).

The numerical studies of gas production in this paper were conducted using the TOUGH-Fx/HYDRATE model (Moridis et al., 2005a), the successor to the earlier EOSHYDR2 code (Moridis, 2003) for the simulation of the system behavior in hydrate-bearing geologic media. TOUGH-Fx/HYDRATE can model the non-isothermal hydration reaction, phase behavior and flow of fluids, and heat under conditions typical of common natural CH_4-hydrate deposits (i.e., in permafrost regions and in deep ocean sediments) in complex formations. TOUGH-Fx/HYDRATE includes both equilibrium and a kinetic model (Kim et al., 1987; Clarke and Bishnoi, 2001) of hydrate formation and dissociation. The model accounts for heat and up to four mass components, i.e., water, CH_4, hydrate, and water-soluble inhibitors such as salts or alcohols. These are partitioned among four possible phases: gas phase, liquid phase, ice phase, and hydrate phase. By solving simultaneously the coupled equations of mass and heat balance, hydrate dissociation or formation, phase changes and the corresponding thermal effects are fully described, as are the effects of inhibitors. The model can describe all possible hydrate dissociation mechanisms, i.e., depressurization, thermal stimulation, salting-out effects, and inhibitor-induced effects.

7.3. DESCRIPTION OF THE GEOLOGIC SYSTEM

Moridis and Collett (2004) have developed a simple classification system for naturally occurring gas hydrate deposits, describing three classes on the basis of the phase distributions in the immediate vicinity of the hydrate-bearing layer. Class 1 and Class 2 hydrate deposits are characterized, respectively, by a hydrate-bearing layer (hereafter referred to as the HBL) underlain by (1) a two-phase zone involving mobile gas and (2) a single-phase zone of mobile water. Class 3 accumulations are composed of a single zone, the HBL and are characterized by the absence of an underlying zone of mobile fluids.

In terms of gas production, Class 1 is the most desirable exploitation target because of the favorable relative permeability regime and the thermodynamic proximity to the hydration equilibrium at the highest possible T_H (necessitating only small changes in pressure and temperature to induce dissociation). The desirability of Class 2 and 3 accumulations as gas production targets is less well defined than for Class 1 deposits. Class 2 and 3 hydrate deposit can be a complex function of several factors, including thermodynamic proximity to hydration equilibrium, initial conditions, environmental concerns, and economic considerations (Moridis et al., 2004; Moridis, 2004).

Production from Class 1, Class 2 and Class 3 hydrates from confined permafrost accumulations has been discussed by Moridis and Collett (2004), Moridis et al. (2004), and Moridis (2004). In this paper, we focus on gas production from a Class 2 hydrate deposit in the oceanic subsurface. A particular feature of this gas hydrate accumulation is the absence of confining geologic formations. Thus, the HBL is enveloped by permeable sediments that are fully saturated with ocean water. Such a gas hydrate accumulation (hereafter referred to as a Class 2-OU deposit, with the 'O' denoting the oceanic environment and the 'U' the unconfined type of the deposit) can be formed from supersaturation in dissolved CH_4 (emanating from deeper in the oceanic subsurface). Its evolution can be facilitated by the presence of a lower permeability layer, which causes gas to accumulate and to begin forming hydrates. Generally, the bottom of such deposits coincides with the bottom of the hydrate stability zone at the prevailing pressure and temperature.

Class 2-OU hydrate deposits appear to be challenging targets for gas production because the absence of barriers to vertical flow can severely limit the effectiveness of depressurization, the fastest and most efficient method of hydrate dissociation. An additional complication in such deposits is the difficulty in focusing and directing flow through the hydrate (thus allowing an appropriate pressure drop to develop). The low-permeability HBL can be bypassed if faster flow pathways (through the more permeable bounding layers) are available. If permeability is not a limiting factor (otherwise, cavitation may occur), the production efficiency appears to be adversely affected by the ready availability of practically limitless amounts of water. This condition requires very large (and potentially uneconomical) water production rates to affect a pressure drawdown sufficient to induce hydrate dissociation at an acceptable rate. At the same time the flow of the evolving gas is hampered by an unfavorable relative permeability regime. Note that in Class 2-OU deposits, the permeability and extent of the water-saturated formation underlying the HBL are generally more important than those factors in the overlying layer because of the low permeability in the HBL. This limits the amount of water that flows to the well from the upper boundary and through the HBL (at least during the initial stages of production), especially when the hydrate saturation and the thickness of the HBL are large.

Although clearly there are disadvantages, there is a potential advantage in the availability of large amounts of water during gas production from Class 2-OU hydrate deposits. Because permeability within the hydrate-bearing layer is generally limited, the large proportion of water in the production stream flows mainly from farther and deeper in the underlying formation toward the producing well. By appropriate placement of the wells, this water (which has a high heat capacity and relatively high temperature by virtue of its large flow rate) can provide some of the heat needed to fuel the strongly endothermic hydrate dissociation reaction. An additional potential advantage is that disposal of the produced water in an oceanic environment is expected to pose far less of an environmental challenge than disposal in the sensitive arctic environment. The produced water need not be brought to the ocean surface, but can be safely released above the ocean floor, thus reducing the cost of production.

7.4. OBJECTIVES

The aim of this study is not to develop a design for efficient gas production from Class 2-OU hydrate deposits. Instead, the main objectives are: (1) to provide a first-level estimate of the production potential of such deposits using conventional technologies, (2) to identify the major issues and limitations affecting production, and (3) to obtain a measure of their relative appeal against other hydrate deposit classes.

7.4.1. Case 1: Gas Production from a Single-Well System

In a single-well system there is only one drill hole through which to operate the secondary recovery methods required for gasification of hydrate deposits. In addition, all recovered gas must be brought up the same well. This places certain constraints upon the timing of stimulation and recovery operations, unless a double-cased hole that effectively provides two separated flow paths within the same well, is established.

7.4.1.1. Geometry and Conditions of the System

The geometry of a single-well gas production from a Class 2-OU hydrate accumulation is relatively simple. The thickness of the HBL is a uniform $H = 50$ m over the footprint of the reservoir. The absolute (intrinsic) permeability of the HBL and of the underlying water-saturated stratum is $k = 10^{-12}$ m^2 (1 Darcy), whereas the absolute permeability of the overlying layer is $k = 10^{-14}$ m^2. The porosity of all three strata is $f = 0.38$. The pressure follows a hydrostatic distribution and is $P = 10.24$ MPa at the bottom of the HBL. The temperature follows the geothermal gradient and is $T = 11.25$ °C at the bottom of the HBL. The top boundary (i.e., at $z < 0$, immediately above the HBL) and the bottom boundary (at $z \leq -350$ m) were maintained at constant temperature, pressure, and salinity conditions. The HBL has an aerial extent of 1 km^2. In the HBL, the hydrate is pure CH_4-hydrate, and the initial hydrate and water saturations are $S_H = 0.75$ and $S_w = 0.25$, respectively.

Fluids are produced from a single well at a rate of $Q = 36.8$ kg/s (corresponding to an initial rate of 20,000 bbl/day of water) and are distributed according to their motilities at the well. The well is completed in the -45 m to -55 m interval. The initial conditions in the HBL and its enveloping layers, as well as the basic pertinent hydraulic and operational parameters, are listed in Figure 7.1.

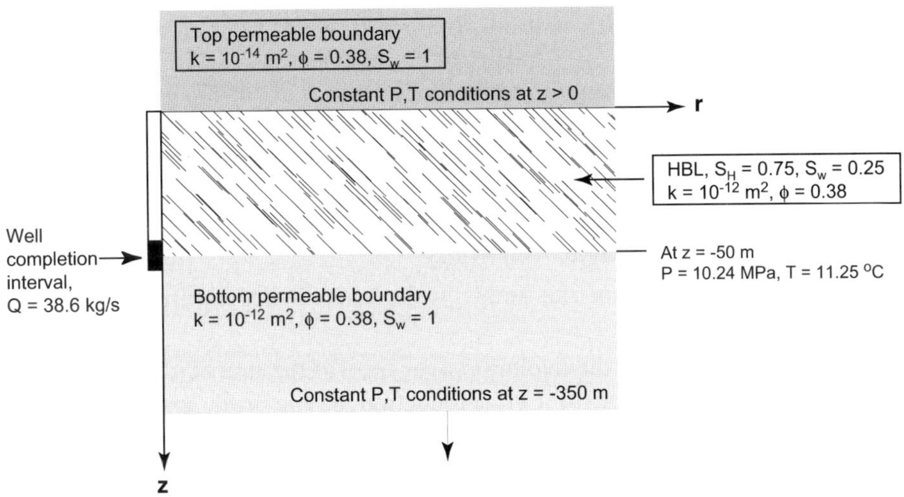

Figure 7.1. The characteristics and properties of the Class 2-OU formation studied in Case 1.

The water released during dissociation continuously dilutes the salinity of the native reservoir water. Because this is a localized phenomenon that is concentrated where flow (and, under the conditions of the proposed scheme, where there is maximum dissociation) occurs, salinity could not be ignored in this analysis. Note that the salinity level in the native pore water of an oceanic hydrate accumulation can be significant, causing a 1.4 °C decrease in the dissociation temperature (Dallimore et al., 1999) at the prevailing pressure.

7.4.1.2. Domain Discrimination and Simulation Specifics

To describe this single-well problem, a cylindrical system was used, extending to an outer impermeable boundary at R_b = 564.19 m, which corresponds to an area of 1 km^2. The domain was gridded in 38 x 36 subdomains in (r, z), resulting in a total of 1214 grid blocks. Because the hydrate dissociation reaction was assumed to occur at equilibrium, the four equations per cell (mass balance equations of H_2O, CH_4 and salt, plus the heat balance equation) resulted in a system of 4800 equations.

The well is represented as a domain of porosity f = 1, very high vertical permeability ($k_z = 10^{-8}$ m^2), and of horizontal permeability equal to that of the formation in the completed section of the well and was assigned zero elsewhere. The distribution of fluid production along the completed section of the well is determined by the phase permeability regimes in the vicinity of the wellbore when fluids are withdrawn at a mass flow rate Q applied at the well cell immediately above the completed section of the well. The production period is 4 years.

The phase relative permeabilities follow a modification [Moridis et al., 2005a] of the model of Stone [1970], with an irreducible water saturation of S_{wr} = 0.2, irreducible gas saturation S_{gr} = 0.02, and an exponent n = 3.572. The capillary pressure was computed using the method of Parker et al. (1987) with S_{wr} = 0.2, a = 3 m^{-1}, and n = 1.65. These values were based on data obtained during the Mallik field test of gas production from a permafrost hydrate deposit (Moridis et al., 2005b).

7.4.1.3. Results of the Single Well Study

The fluid production from the single well leads to depressurization-induced hydrate dissociation.

Figure 7.2 shows the evolution over time of the rate of CH_4 release into the reservoir and the rate of CH_4 production at the well, expressed as ST (standard) m^3/day. The gas production rate at the well exhibits oscillations, but remains roughly constant during the entire production period. The gas release rate into the reservoir is marked by an initial steep increase, followed by a decline and eventual stabilization. The oscillations in both curves are caused by gridding effects and heat transfer limitations affecting the dissociation reaction. Dissociation is followed by a pressure increase (caused by the gas release) and a drop in temperature (due to the strongly endothermic nature of the dissociation reaction). This process results in the steep drop in the release rate and the production rate because of a shift in dissociation P_H-T_H relationship (see Fig. 7.3). As more heat becomes available (through conduction and fluid advection), the dissociation rate begins increasing. The natural oscillations are exacerbated by the finite size of the elements into which the simulation domain is subdivided.

Figure 4 shows the cumulative volumes (expressed as ST m^3) of hydrate-originating CH_4 released during the depressurization-induced dissociation and the cumulative gas volume converted from hydrate in the reservoir over the four-year duration of the study. A comparison of these curves to the ones of the corresponding rates in Figure 7.4 indicates that the rate fluctuations are attenuated in the cumulative volume curves, which appear remarkably smooth.

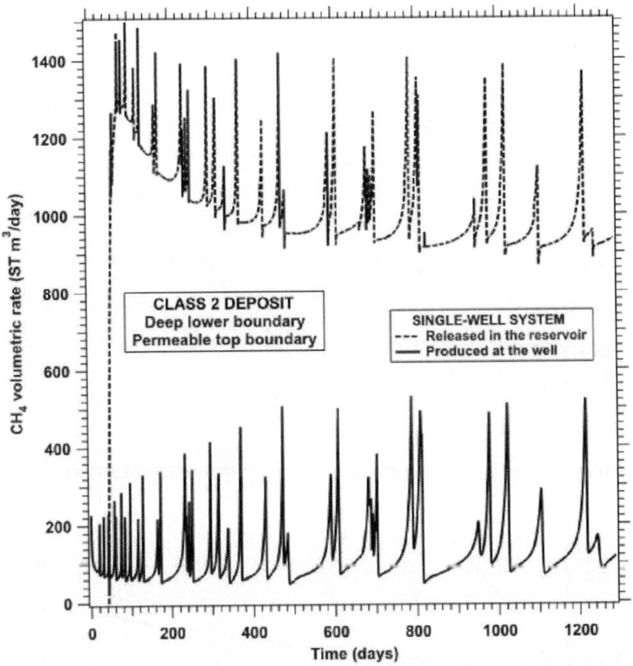

Figure 7.2. Evolution of the CH_4 production rate and of the rate of CH_4 release (from depressurization-induced hydrate dissociation) into the reservoir during production from a single well in Case 1 ($S_H = 0.75$).

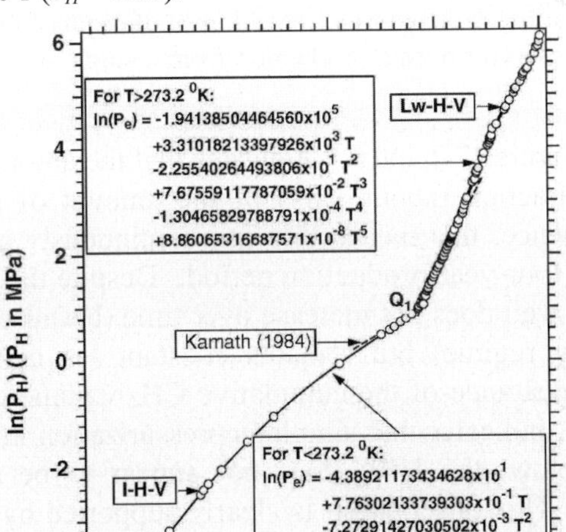

Figure 7.3. Pressure-temperature equilibrium of methane hydrate (Moridis, 2003).

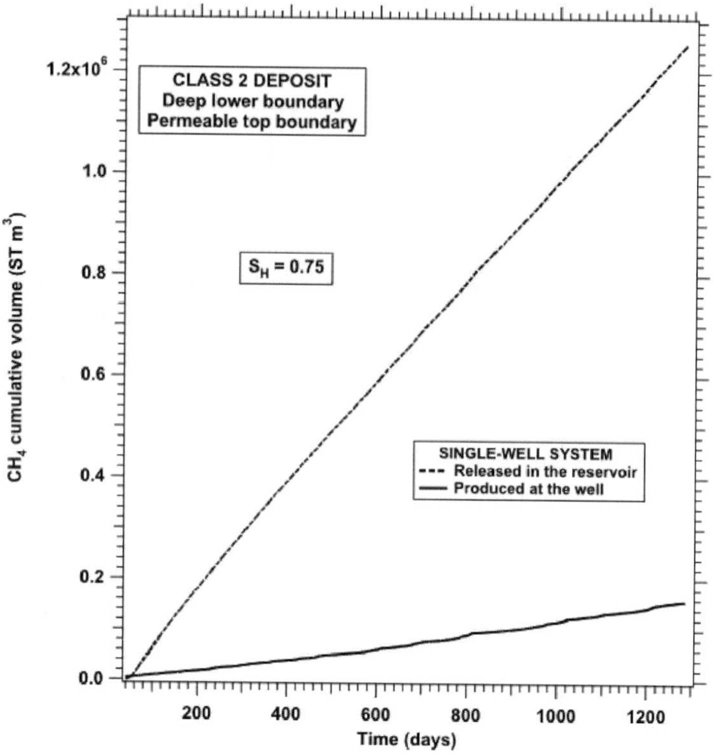

Figure 7.4. Evolution of the cumulative volumes of produced CH_4 and of CH_4 released into the reservoir during production from a single well in Case 1 (S_H = 0.75).

Review of Figures 7.2 and 7.4 indicates that the amount of produced gas represents a small fraction (about 1/8th) of the amount of gas released from dissociation. In essence, this means that gas continuously accumulates in the reservoir during the four-year production period. Despite the gas accumulation, the gas reaching the well does not increase over time (because of an adverse gas relative permeability regime) but remains constant (as demonstrated by the practically linear appearance of the cumulative CH_4 volume curve). This is an undesirable scenario, indicates that simple depressurization in the lower part of, and immediately below, the HBL does not appear to be a very promising production method. This observation is clearly supported by the magnitude of the CH_4 production rate (about 100 ST m^3/day), which is very low compared to the roughly 3,150 m^3/day of water produced in the process. The lack of appeal of gas production from the single-well depressurization of a Class 2-OU deposit would persist even if the production rate equaled the entire CH_4 release rate in the reservoir.

7.4.1.4. Effect of the Initial S_H in the HBL

Figure 7.5 shows the evolution of the cumulative volumes (expressed as ST m^3) of produced CH_4 and of the CH_4 released in the reservoir over the four-year production period when the initial S_H = 0.375, i.e., half of that in the base simulations of Figures 7.2 and 7.4. Comparison of Figures 7.4 and 7.5 indicates that production from a Class 2-OU deposit with a lower initial S_H is even less appealing. The cumulative gas production decreases with S_H. In this case, the lower initial S_H leads to higher S_w and consequently higher water permeability. The result is greater flow through the HBL (most of which is bypassed in the higher S_H case) and more effective depressurization of the hydrate. The increase in gas production (as a larger HBL volume becomes available for gas release) is countered by the smaller amount of hydrate (i.e., the CH_4 source) and

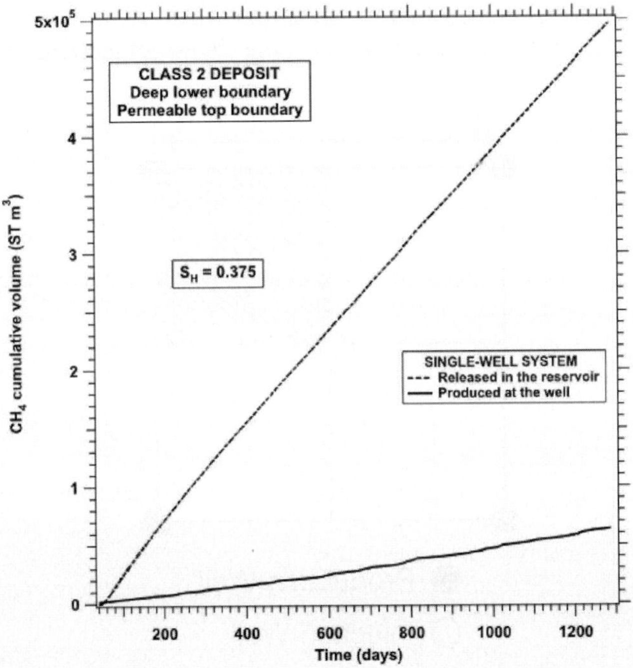

Figure 7.5. Evolution of the cumulative volumes of produced CH_4 and of CH_4 released into the reservoir during production from a single well in Case 1 (S_H = 0.375).

the increasingly limited flow through the HBL as the hydrate near the well dissociates. As in the case of higher initial S_H, the produced CH_4 volume represents a rather small fraction of the released volume, and increases roughly linearly with time.

Note that, because of differences in the formation process (and the fact that the presence of hydrates in oceanic porous media limits mass transfer and further hydrate formation), low hydrate saturations are far more common (and probably the norm) in ocean deposits. This situation further limits the attractiveness of the Class 2-OU deposits.

7.4.2. Case 2: Gas Production from a Five-Spot Well System

7.4.2.1. Geometry and Conditions of the System

In Case 2, the geology, geometry, properties, and initial distribution of saturations of the Class 2-OU hydrate accumulation and its boundary formations remain as in Case 1. The stencil in Figure 7.6 represents the five-spot well configuration, involving production and injection wells. The injected fluid was hot water at a temperature of 41 °C. Hot water was chosen over steam because the parametric study of McGuire [1981] indicated that the amount of produced gas is less than the estimated fuel consumption when steam is employed. Hot water was injected into the center-well of the five-spot pattern, and reservoir fluids were produced from the four production wells. The obvious advantage of this scheme is that it combines the two most important mechanisms of hydrate dissociation, i.e., depressurization at the production well and thermal stimulation at the injection well.

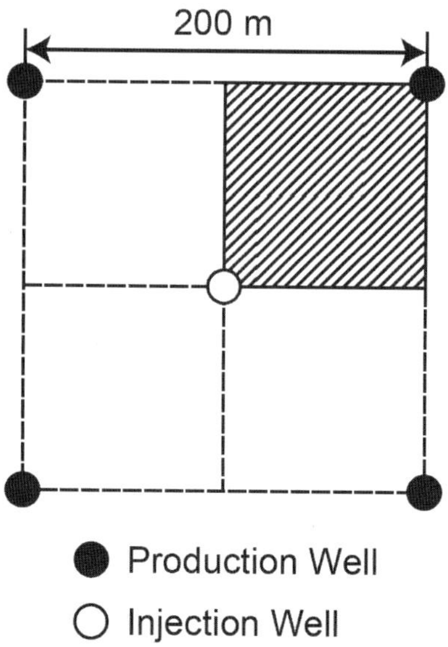

Figure 7.6. Five-spot well stencil (pattern) for modeling a 1/4 symmetry subdomains (shaded) in the simulations of Case 2.

In an effort to focus the thermal stimulation and depressurization effects in the vicinity of the hydrate interface, the production well was completed in the 0 to -55 m interval, and the injection well in the 0 to -50 m interval. This configuration offered the advantage of limiting mixing of the injected hot water with the colder native reservoir water, while maximizing the thermal advantages of buoyancy that tends to concentrate the warmer water immediately below (and in contact with) the hydrate interface. Additionally, the completion of the wells in the entire hydrate interval took advantage of the gas buoyancy and the maximized contact with the receding hydrate interface nearby. However, the upper parts of the well intervals did not contribute significant amounts of any fluids in the early stages of gas production because of adverse relative permeability conditions.

7.4.2.2. Domain Discrimination and Simulation Specifics

Because of symmetry, only a quarter of the domain was simulated using a 3-D Cartesian system. The side of the simulated quadrant was 100 m. The domain was gridded in 15x15x25 unequally spaced subdivisions in (*x,y,z*), resulting in a total of 5,625 elements. Assuming equilibrium dissociation, four equations (i.e., components) were considered (CH_4, H_2O, salt, and heat) in each element, leading to a system of 22,500 simultaneous equations. The large size and the complexity of the simulated system made the solution of this problem very computationally demanding.

The well representation, as well as the relative permeability and capillary pressure models and parameters remained as in Case 1. Because only a quadrant of the five-spot pattern in Figure 7.6 (corresponding to 1/4th of the rates in the full system) is simulated, and each five-spot configuration occupies 1/25th of the 1 km^2 area of the footprint of the hydrate deposit, the initial mass production rate was $Q = 36.8/(4x25) = 0.368$ kg/s. However, the water injection rate was equal to the rate of water withdrawal from the production well. Thus, the total production rate from the entire hydrate deposit was equal to that in the single-well system of Case 1. The gas and aqueous phase production rates were determined by the phase relative permeabilities in the production well elements. The simulation was allowed to continue for 4,000 time steps, when the results were to be evaluated and a decision about further continuation was to be made.

7.4.2.3. Results of the Five-Spot Study

After 4000 time steps, the simulation had lasted over three days of continuous execution and had covered a production period of 212 days. Because the results and their pattern were rather well defined, and because the execution time requirements of this problem were very substantial, we decided to terminate the simulation at that point.

Figure 7 shows the evolution over time of the rate of CH_4 release into the reservoir and the rate of CH_4 production at the well, expressed as ST m^3/day. These results correspond to the entire 1 km^2 hydrate accumulation. Comparison with the rates from the single-well production in Case 1 leads to the obvious conclusion that, while the rate of gas release in the reservoir is substantially larger than that in Case 1, the rate of CH_4 production at the well is much lower. Additionally, although the rate of gas release continues increasing (albeit slowly) over the 212-day period of the study, the rate of CH_4 production at the well is either constant or declines slightly.

The lower production rate and the difference in the long-term trend are attributed to two reasons. The first reason is that the injection and production rates are roughly equal during this period (as evidenced by the very low gas production). This rate parity does not allow significant depressurization at the production well because of the relatively short distance between the injection and production wells, and the speed at which the pressure front advances. Thus, the warm water re-injection prevents a significant pressure drop at the production well, leading to lower dissociation and a lower production rate at the well. Moreover, after a maximum pressure imbalance early in the production period, the pressure tends to a steady-state distribution (though continuously disrupted by the gas release from hydrate dissociation), leading to the constant (or slight decline of) the production rate of CH_4 that originated almost exclusively from depressurization-induced hydrate dissociation near the production well.

The second reason for the low gas production rate (and for the increasing trend of release rate) is because gas from the thermally induced dissociation in the vicinity of the injection well has not yet reached the production well because of adverse relative permeability conditions. Thus, the origin of most of the gas released in the reservoir is expected to be thermal dissociation. The slowly and continuously expanding thermal front leads to the slowly increasing (over time) release rate observed in Figure 7. The disparity between CH_4 release (in the reservoir) and production rates, brought about by low depressurization and the limited mobility of the gas from thermal dissociation, are expected to lead to a significant gas accumulation in the reservoir.

The differences in the fluctuations of the rate curves (Fig. 7.7) stem from the origin of gas they represent and from the different impact of gridding effects. Thus, the release rate curve is expected to affect a relatively large number of elements as the warm water front advances toward the production well, as indicated by the large number of oscillations (denoting finite element size). Conversely, because of the limited impact of depressurization, dissociation is expected to occur in a limited number of elements, as implied by the few and distinct oscillations in the production rate curve.

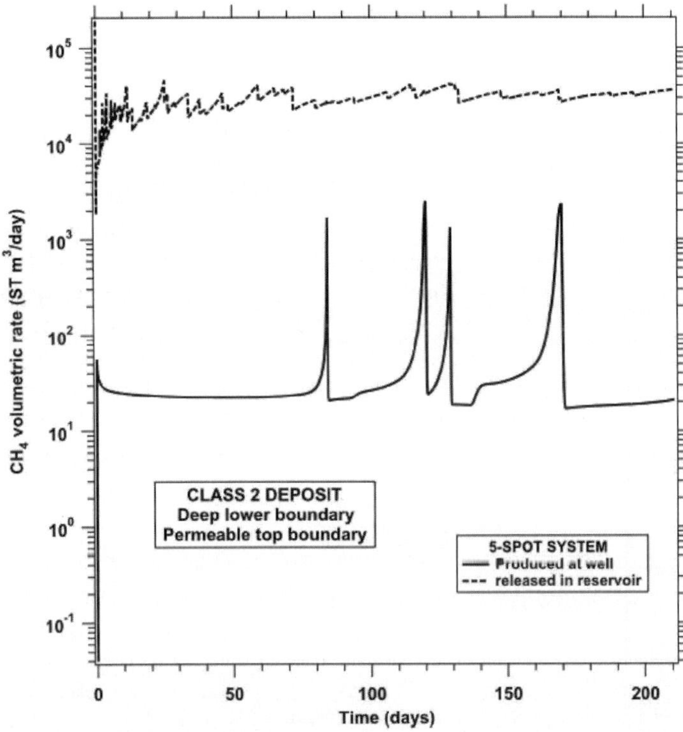

Figure 7.7. Evolution of the CH_4 production rate and of the rate of CH_4 release into the reservoir during production from a five-spot well system in Case 2 (S_H = 0.75).

The expectations from Figure 7.7 are confirmed in Figure 8, which shows the evolution over time of the cumulative volume (in ST m^3) of produced CH_4 and of CH_4 released in the entire 1 km^2 deposit. The curves in Figure 7.8 point to a very substantial gas accumulation in the reservoir, which is not accompanied by a commensurate increase in gas production during the 212 days of simulation. If the reason for this disparity is that the gas from thermal dissociation has not reached the production well during that period, then it is almost inevitable that this gas will reach the well at a later time, and lead to a surge in production. However, the question that has to be asked before further pursuing this approach is whether this model holds any appeal, as it tends to indicate over 0.6 of a year of continuous water production, heating and circulation with practically no gas production. Note that, as in Case 1, the cumulative volume curves attenuate the oscillations in the corresponding rate curves and have a remarkably smooth appearance.

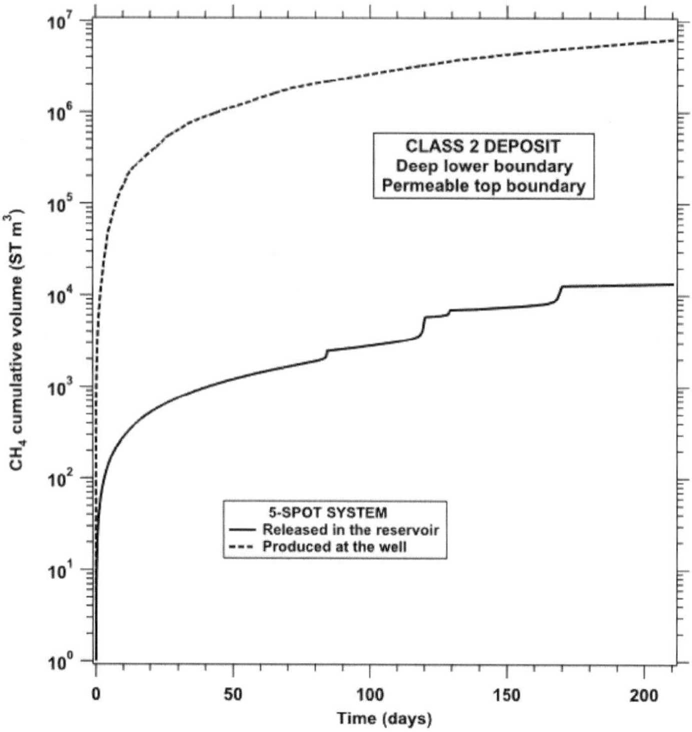

Figure 7.8. Evolution of the cumulative volumes of produced CH_4 and of CH_4 released into the reservoir during production from a five-spot well system in Case 2 ($S_H = 0.75$).

Further consideration of Figures 7 and 8 indicates that the rate of CH_4 release in the reservoir and the corresponding cumulative volume are large and within the range of commercial viability. The obvious issue is whether an appropriate production system can be designed that enables the early access to, and production of, the majority of the released gas. Such a system may involve asymmetric production and injection rates (with injection rate being a fraction of the production rate) or horizontal wells. Such a system is likely to require higher levels of management. Although these studies can hold substantial scientific interest, their appeal is eclipsed by far more promising hydrate targets for gas production, e.g., Class 1 deposits.

Figure 7.9 shows the pressure distribution in the (1) 3-D five-spot quadrant and (2) along the $x = y$ plane, i.e., the plane passing through the injection and production wells. As expected, the highest pressures are observed in the vicinity of the injection well, and the lowest and in the immediate neighborhood of the production well. The temperature distribution in Figure 7.10 exhibits high temperatures at the injection well and low temperatures (below the initial T = 11.25 °C) near the production well, where some depressurization-induced dissociation occurs (see Figure 7.7).

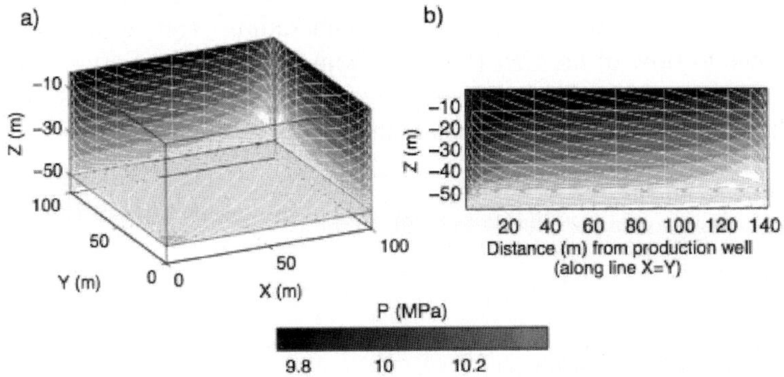

Figure 7.9. Pressure distributions (1) in the simulated 3-D domain and (2) in the plane defined by the injection and production wells in Case 2. The lighter shading in the right and right rear corner of the diagram indicates pressures in excess of 10.2 MPa.

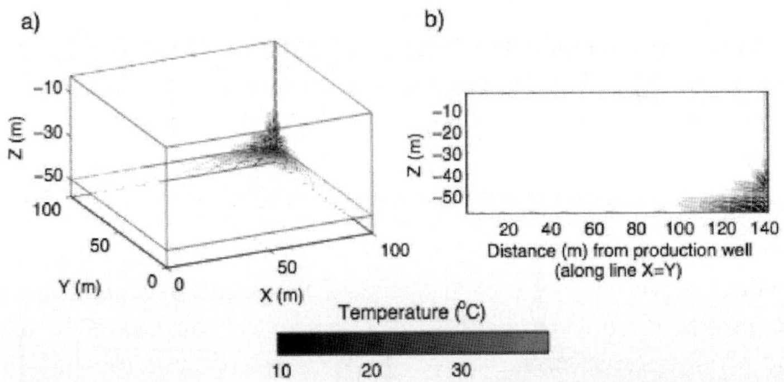

Figure 7.10. Temperature distributions (a) in the simulated 3-D domain and (b) in the plane defined by the injection and production wells in Case 2. The dark shading in the right and right rear corner of the diagram are temperatures in excess of 25 °C.

The gas and hydrate saturations in Figures 7.11 and 7.12, respectively, are consistent with each other, with the pressure and temperature distributions of Figures 7.9 and 7.10, and with the rates and cumulative volumes of Figures 7.7 and 7.8. Minor hydrate dissociation is observed along the bottom of the HBL, consistent with the limited depressurization to which this region is exposed. The highest gas saturation corresponds to the location of maximum hydrate dissociation and is located near the well of warm water injection. This observation is consistent with the expectations and deductions from Figures 7.7 and 7.8. The main body of the gas saturation front in Figure 7.11 has not reached the production well after 212 days of production (actually, it has covered less than half the distance between the wells), thus explaining the reason for the low production rate and confirming the source of the majority of the released gas. The maximum pressure is observed at a location above the hydrate interface and at the location of the highest temperature. This result is caused by the resistance to flow of the CH_4 (released from thermal dissociation of hydrate) into the low permeability HBL. The increased pressure shifts the equilibrium T_H and results in hydrate persistence despite the higher temperatures.

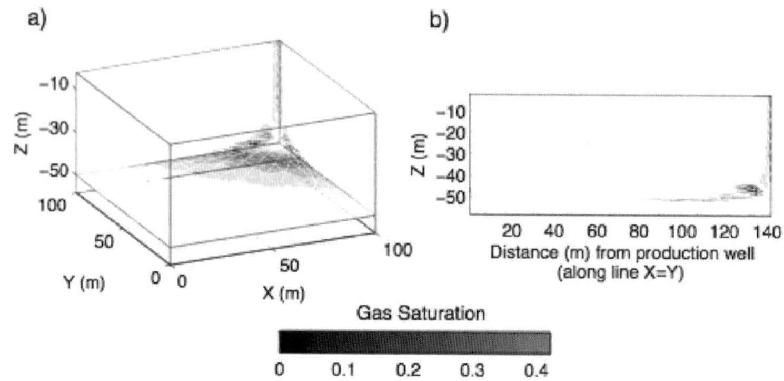

Figure 7.11. Gas saturation distributions (a) in the simulated 3-D domain and (b) in the plane defined by the injection and production wells in Case 2. The dark shading shows gas saturation levels in excess of 0.2, rising to 0.4.

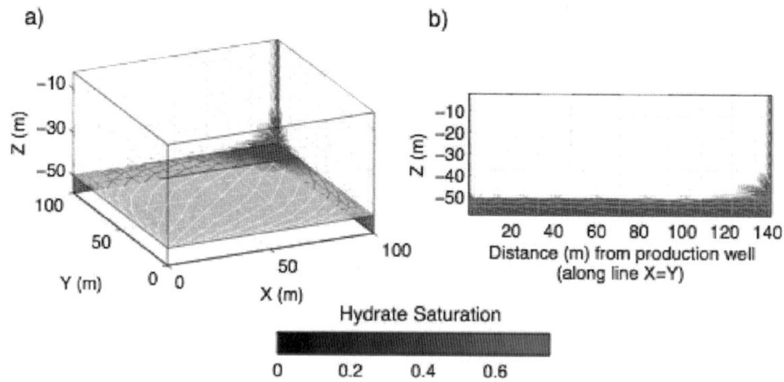

Figure 7.12. Saturation distributions of the gas hydrate (a) in the simulated 3-D domain and (b) in the plane defined by the injection and production wells in Case 2. Dark shading in right and right rear shows hydrate saturation in excess of about 0.4, rising to about 0.5.

7.5. SUMMARY AND CONCLUSIONS

This chapter focuses on the study of gas production from Class 2-OU hydrate accumulation, i.e., oceanic deposits characterized by a HBL enveloped by permeable geologic media fully saturated with mobile water. The objectives of this study were twofold: (1) to provide a measure of the production potential of such deposits using conventional technologies and (2) to identify the factors limiting production. Thus, a knowledge base is developed for meaningful comparisons of the relative value of such deposits as production targets against the values of other hydrate accumulations as gas production targets.

Using a Class 2-OU deposit with an aerial footprint of 1 km^2 and a HBL of pure CH_4-hydrate 50 m thick, two production strategies were investigated. In the first, fluid production from a single well at the center of the reservoir effected depressurization, which led to gas production by inducing hydrate dissociation. The second strategy employed combinations of injection and production wells in a five-spot pattern, and involved both depressurization (at the production wells) and thermal stimulation near the injection wells, through which the produced water (heated to 41 °C) was re-injected.

The results of these studies lead to a number of conclusions:

(1) In both cases, the CH_4 production rates were practically constant over time, and significantly lower than the rate of gas release into the reservoir. At no time during the simulation period do the release and production curves show any tendency toward convergence or even constant deviation. This indicates that gas accumulates in the reservoir at a rate that increases with time, but the adverse relative permeability regime does not allow ready gas recovery.

(2) As indicated by the trend in the rates, in both cases, the cumulative volumes of produced CH_4 were significantly lower than cumulative volumes of CH_4 released into the reservoir.

(3) Gas production decreases with a decreasing hydrate saturation of the HBL because of reduced availability of hydrate.

(4) The production rate was higher in the case of single-well production because, in the five-spot pattern, the (1) gas produced from the thermal dissociation of hydrate caused by the warm water re-injection is very slow to reach the production well, and (2) the water re-injection does not allow a significant pressure drop, thus reducing the driving force of the depressurization-induced dissociation near the production well.

(5) Conversely, the CH_4 release rate into the reservoir was significantly higher in the five-spot well pattern because of the large volumes of re-injected water as an agent of thermal dissociation. The spatially uniform injection of the warm water (a result of the small footprint

of the five-spot stencil) increases the effectiveness of thermal dissociation.

(6) Although the volume of CH_4 released from dissociation is large (and within the realm of economic viability) in the case of the five-spot well system, this volume is not readily recoverable using the vertical wells of this conventional configuration. A higher-level of management and/or different well systems (e.g., horizontal wells) may be needed to achieve a more efficient (and economically attractive) recovery from such hydrate deposits.

(7) The production rates in both well configurations are very low, and cannot justify considering Class 2-OU hydrate accumulations as economically viable targets for gas production. Further study will be needed to develop appropriate production strategies if such deposits are to be targeted for gas recovery.

(8) For all the aforementioned reasons, Class 2-OU hydrate accumulations do not appear to be appealing targets for gas production using conventional technologies, especially when considered against far more promising candidates such as Class 1 deposits.

(9) The main reasons for the limited potential of Class 2-OU hydrate deposits as a gas source are (1) the ineffectiveness of depressurization as the driving force of dissociation in the absence of confining layers, (2) the ready availability of practically limitless amounts of water, which requires large water production rates for an effective pressure drop, (3) the challenge of focusing and directing water flow through the HBL (easily bypassed if higher permeability pathways through the enveloping boundary layers are available), and (4) the adverse relative permeability to gas flow, as gas attempts to emerge as a mobile free phase in a flowing or fast percolating water-saturated geologic medium.

7.6. ACKNOWLEDGMENTS

This work was supported by the Assistant Secretary for Fossil Energy, Office of Natural Gas and Petroleum Technology, through the National Energy Technology Laboratory, under the U.S. Department of Energy, Contract No. DE-AC03-76SF00098. The authors are indebted to Timothy Kneafsey and John Apps of LBNL for their insightful review comments.

Chapter 8

Regulatory and Permitting Environment for Gas Hydrate

George Taft[1]

8.1. INTRODUCTION

Oceanic natural gas hydrate (mainly methane) occurs predominantly in underconsolidated marine sediments around the edges of continents that form the outer part of deep continental shelves and the upper two-thirds of continental slopes. Hydrate is likely to continue to be found in continental margins where the depth of water exceeds 500 m. It now appears that deposits of economic significance are unlikely to be found in water depths much shallower than the 900+ m water depths such as those of the Japanese outer SE continental margin, which hosts the Nankai Trough deposit. That deposit, which is currently being evaluated as a source of indigenous natural gas, is unusual in that it is on a deep continental shelf rather than a slope where most hydrate deposits are likely to be found (Chapter 5). The abnormal depth of the shelf here, deeper than the usual ~ 200 m shelf break, and the cold currents from the Bering Sea area create the special conditions for the Nankai and other deposits adjacent to Japan.

The tendency for oceanic hydrate to accumulate at continental margins means that much of it will fall within the exclusive economic zones of coastal nations, many of which are not presently energy producers. In addition to new energy supplies based on methane hydrate in coastal zones of countries that are already producing deep-water conventional hydrocarbons, it is possible that countries, such as Chile, which has not been a significant producer of petroleum, may hold significant energy resources in the form of oceanic hydrate. Therefore, this possibility of developing oceanic hydrate as an energy supply has the potential to change the distribution of global energy resources and their economics.

The search for conventional hydrocarbons has now taken the energy industry into waters deep enough to support hydrate in the marine sediments. Because of this, infrastructure for recovering deep water gas deposits including hydrate-derived natural gas is now in place and moving into increasingly deeper water. Potentially economic oceanic hydrate appears mainly to occur in water no shallower than about 1 km (Chapter 3) whereas considerable conventional hydrocarbon prospects remain in shallower water. Thus, the entire focus of exploration for oceanic hydrate as an energy source is in relatively deep water, which is often far from land.

---

[1] The views expressed are those of the author and do not necessarily reflect the views of the United States Department of State.

Michael D. Max et al. (eds.), Economic Geology of Natural Gas Hydrate, 267–288.

In general, oceanic hydrate may occur in both narrow and broad continental slopes, whose geomorphology is a function of plate tectonics, sediment supply and depositional environment through time. The structural framework may deform, compress, and strongly alter the marine sediments in which hydrate may occur. Some of the best known deposits, however, occur in sediments that have partly been subjected to compaction, such as the deposits on the eastern continental margin of the United States (Chapter 3). Whatever the geological and diagenetic history of sediments that could host hydrate concentrations, a combination of the distance from shore and disposition of geological components of the sea floor are the determining factors for control of hydrate occurrences. These factors also apply to other potential natural resources of economic value in the marine area.

8.2. REGULATORY AND PERMITTING FRAMEWORK

The domestic and international legal framework for the exploration for and exploitation of gas hydrates depend, for the most part, on whether the minerals are found within; (1) the territorial sea, (2) the exclusive economic zone, (3) the continental shelf beyond 200 nautical miles (nmi) from the baselines from which the territorial sea is measured or (4) the deep seabed beyond 200 nmi or the outer legal limit of the continental shelf where it extends beyond 200 nmi, whichever is further seaward. The coastal State's rights over non-living resources generally are greatest in the territorial sea and are minimally constrained in areas of national jurisdiction seaward thereof. Rights of coastal States are very limited in the deep seabed, in which area such rights exist only where there is a pool of resources straddling national and international jurisdiction. The rights and obligations of States, including coastal States, are contained in the 1982 United Nations Convention on the Law of the Sea (the Convention) and the 1994 Agreement on the Implementation of Part XI of the Convention, as well as in customary international law. The main element for defining coastal zones is distance from shore (Fig. 8.1.), although geophysical data is also important for establishing the outer limit of the continental shelf beyond 200 nmi (Singh et al., 2004).

8.2.1. Territorial Sea

A coastal State may establish a territorial sea to a maximum distance of 12 nautical miles from the baselines off its coasts. In the territorial sea the coastal State has full sovereignty, as it has within its land territory, subject to the right of innocent passage for foreign flag vessels. Rights of States regarding the resources within the territorial sea are all encompassing. A State may exercise these rights in accordance with its domestic laws and policies, including, *inter alia*, licensing/leasing, laying pipelines, constructing and operating artificial islands and installations for extraction purposes, as well as designating safety zones around such installations. At the same time, a State Party to the relevant international conventions relating to the establishment of sea lanes and traffic separation schemes must observe the provisions thereof. In this regard, such international obligations may impact on the decisions of where to locate installations for mineral extraction. In addition, coastal States may undertake to develop and adhere to international environmental standards applicable to activities in their territorial seas. Nevertheless, the rights of the coastal State are broad, with limited obligatory derogations under international law.

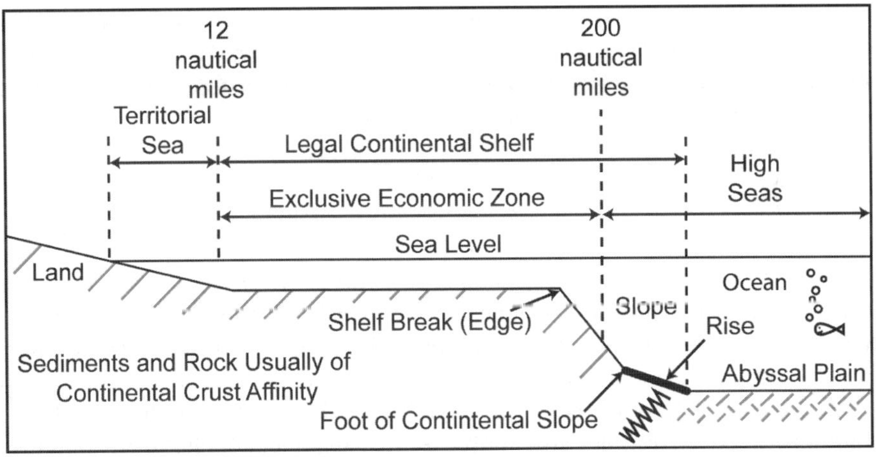

Figure 8.1. Geomorphic regions and legal regimes.

8.2.2. The 200 Nautical Mile Exclusive Economic Zone

A coastal State may establish an exclusive economic zone to a maximum of 200 nautical miles from the baselines from which the territorial sea is measured. In the exclusive economic zone, the coastal State does not have sovereignty. Rather, it has sovereign rights and jurisdiction for economic purposes and for the conduct of marine scientific research.

In particular, a coastal State has, in accordance with Article 56 of the Convention, extensive sovereign rights for resource purposes. These sovereign rights are limited by other provisions of the Convention, other international obligations the State may undertake and rules of customary international law. It has broad discretion regarding licensing and leasing for both exploration for, and exploitation of, the non-living resources. It has broad discretion regarding establishing royalties or other payments from the licensee or lessee. It may establish its own environmental standards, subject to any international legal obligations undertaken in other international agreements. The coastal State is also subject to the provisions of the Convention relating to the environment, including the general duty to protect and preserve the marine environment and to customary international law. It must undertake its exercise of sovereign rights with due regard to the rights and duties of other States in the exclusive economic zone. Other States have the right, for example, to exercise the high seas freedom of navigation (and overflight) in the exclusive economic zone. That freedom includes not only transit, but maneuvers and exercises. Thus, the coastal State exercise of its rights for the purpose of exploring for, and exploiting, of resources must be tempered by the necessity of co-existing with the broad rights of other States. Nevertheless, a coastal State may pursue a vigorous program regarding resources, as, for example, the US has in the Gulf of Mexico.

A fuller explanation of the regime applicable to non-living resources of the seabed and subsoil of the exclusive economic zone is contained in the following section regarding the continental shelf. Any differences will be noted between the regime applicable to the seabed and subsoil within the 200 nmi exclusive economic zone and the regime applicable to the seabed and subsoil of the continental shelf beyond 200 nautical miles.

8.2.3. The Continental Shelf

The continental shelf geologically extends from the shore to the shelf break, which is a prominent change of slope gradient from the slightly sloping shelf to the more steeply sloping continental slope. Many shelf breaks occur between 200 - 300 m water depth but locally the shelf break may occur in water depths between 900 and 1,200 m. Article 76 of the Convention addresses in detail the legal definition and limits of the continental shelf. It contains a complex, but workable, formula for establishing the outer limits where the continental shelf extends beyond 200 nautical miles.

The Convention uses the term 'continental shelf', in a legal sense, to include the geological shelf, slope and rise. It generally defines the legal continental shelf in Article 76 as, "*1. The continental shelf of a coastal State comprises the sea-bed and subsoil of the submarine areas that extend beyond its territorial sea throughout the natural prolongation of its land territory to the outer edge of the continental margin, or to a distance of 200 nautical miles from the baselines from which the breadth of the territorial sea is measured where the outer edge of the continental margin does not extend up to that distance. 2. The continental shelf of a coastal State shall not extend beyond the limits provided for in paragraphs 4 to 6. 3. The continental margin comprises the submerged prolongation of the land mass of the coastal State and consists of the seabed and subsoil of the shelf, the slope, and the rise. It does not include the deep ocean floor with its oceanic ridges or the subsoil thereof.*" However, the continental shelf of a coastal State shall not extend beyond certain limits (Fig. 2).

In the 1969 North Sea Continental Shelf cases, the International Court of Justice clearly recognized the continental shelf and its relationship to the natural prolongation of the land territory. This concept was subsequently included in paragraph 3 of Article 76. The continental shelf thus defined in Article 76 captures the essence of the geological shelf, slope and rise. It explicitly excludes the deep seabed beyond 200 nmi, with its oceanic ridges, which fall within the non-living resource jurisdiction of the International Seabed Authority. Thus, basaltic oceanic crusts and ridges are not subject to coastal State sovereign rights beyond 200 nmi.

The sweeping definition of continental shelf is significantly different from that of the 1958 Geneva Convention on the Continental Shelf, which provided in its Article 1 that the continental shelf is "…the seabed and subsoil of the submarine areas adjacent to the coast but outside the area of the territorial sea, to a depth of 200 meters or beyond that limit, to where the depth of the superjacent waters admits of the exploitation of the natural resources of said areas."

This definition was imprecise. Adjacency had no objective or agreed geographical extent. Exploitability was subject to technological progress. Furthermore, no provisions addressed the scientific aspects of the continental shelf, including the natural prolongation of land territory. The formula was adequate in recognizing coastal State rights, in light of the technological realities of the time. Moreover, the formulation did not prejudice the negotiation and development of Article 76.

The provisions of Article 76 of the Convention addressed the uncertainties of the 1958 Convention by an expansive definition of continental shelf subject to complex, but necessary, limitations.

In paragraph 4 of Article 76, criteria are provided for qualifying the sweep of paragraphs 1 and 3. Paragraph 4 provides:

(a) For the purposes of this Convention, the coastal State shall establish the outer edge of the continental margin wherever the margin extends beyond 200 nautical miles from the baselines from which the breadth of the territorial sea is measured, by either:

(i) a line delineated in accordance with paragraph 7 by reference to the outermost fixed points at each of which the thickness of sedimentary rocks is at least 1 per cent of the shortest distance from such point to the foot of the continental slope; or

(ii) a line delineated in accordance with paragraph 7 by reference to fixed points not more than 60 nautical miles from the foot of the continental slope.

(b) In the absence of evidence to the contrary, the foot of the continental slope shall be determined as the point of maximum change in the gradient at its base.

The limitations contained in paragraphs 4(a)(i) and 4(a)(ii) may be applied by a coastal State in a manner to encompass the greatest extent of seabed and subsoil subject to its sovereign rights. The coastal State may thus apply paragraph 4(a)(i) to certain parts of its continental shelf and paragraph 4(a)(ii) to other parts. The so-called sediment thickness formula of paragraph 4(a)(i) recognized the relationship of sediments to hydrocarbon resources and the interests of the coastal State. The formula found in paragraph 4(a)(ii) recognized the significance for hydrocarbon resources of the area not more than 60 nautical miles seaward of the foot of the continental slope. Together, the limitations addressed satisfactorily the expectations of most coastal States.

Paragraph 5 further qualifies the extent of the continental shelf for the purpose of the coastal States exercise of sovereign rights over resources. "*The fixed points comprising the line of the outer limits of the continental shelf on the sea-bed, drawn in accordance with paragraph 4 (a)(i) and (ii), either shall not exceed 350 nautical miles from the baselines from which the breadth of the territorial sea is measured or shall not exceed 100 nautical miles from the 2,500 meter isobath, which is a line connecting the depth of 2,500 meters.*"

The coastal State may apply either the 350 nautical mile limit or the 100 nautical miles from the 2500 meter isobath limit in different parts of its continental shelf in order to maximize its entitlement. Moreover, there may be a number of 2500 meter isobaths offshore, the coastal State having the discretion to apply the most seaward, subject to paragraphs 4(a)(i) or (ii), as appropriate.

Article 76, paragraphs 1 to 5, while complex, are scientifically based, legally defensible, and politically acceptable, if not necessary, (1) to address the range of continental shelf configurations globally, (2) to take account of the fact that the provisions were negotiated in the late 1970's, (3) to recognize that our scientific knowledge would increase over time, and (4) to establish a regime the soundness of which would withstand the test of time. It was necessary to accommodate States with broad continental shelves, but also to provide the basis for delineating the outer limit of so-called national jurisdiction, within which the coastal State exercised sovereign rights. Also it was necessary to define the landward boundary of the deep seabed beyond the limits of national jurisdiction, the so-called Area. Within the seabed and subsoil of the Area, the International Seabed Authority, established by the Convention, exercises jurisdiction over the exploration for, and exploitation of, the non-living resources.

The first sentence of Paragraph 6 of Article 76 is obscure in its language, creates tension with the plain meaning of paragraph 3 of Article 76, has no apparent application and certainly no agreed interpretation. The melding of law and science in the first sentence of paragraph 6 is tortured at best. "*6. Notwithstanding the provisions of paragraph 5, on submarine ridges, the outer limit of the continental shelf shall not exceed 350 nautical miles from the baselines from which the breadth of the territorial sea is measured. This paragraph does not apply to submarine elevations that are natural components of the continental margin, such as its plateau, rises, caps, banks and spurs.*"

It is noteworthy that the first sentence of paragraph 6 only seeks to qualify paragraph 5. The submarine ridge must meet the requirements of paragraphs 1, 2, 3, and 4. Article 76 recognizes features of the continental shelf, which may include ridges of continental origin, as subject to the sovereign rights of the coastal State, if the parameters of paragraphs 1 through 4 are met. Article 76 states that ridges of the deep ocean floor, i.e., ridges of oceanic origin, and ridges that are not the natural prolongations of the land territory, are not so included. Paragraph 6 seems to suggest that there is a third category or other categories of ridges to which it applies. However, scientists with the relevant expertise have expressed no consensus. While the origin of certain ridges may not be entirely clear, scientists over time will, no doubt, learn more and be able to determine whether a ridge should be considered continental or oceanic in character or whether a third and separate category of submarine ridges exists.

The first sentence's limitation does not apply to submarine elevations that are natural components of the continental margin, such as its plateaux, rises, caps, banks, and spurs. If the first sentence only pertains to paragraph 5, how can a "submarine ridge" be qualified to fulfill the requirements of paragraphs 1 and 3? There is no meaningful *traveaux preparatoires* that gives any insight as to what and where the first sentence applies, if anywhere.

In this regard, it is widely understood that mid-oceanic ridges do not give rise to coastal State sovereign rights beyond 200 nautical miles. It is widely understood as well that islands do not have different legal entitlements than continental landmasses. The same rules of application apply to both.

Paragraph 7 provides a useful tool for a coastal State undertaking the planning of its detailed scientific efforts to establish the outer limit of its continental shelf beyond 200 nautical miles. Paragraph 7 reads: "*The coastal State shall delineate the outer limits of its continental shelf, where that shelf extends beyond 200 nautical miles from the baselines from which the breadth of the territorial sea is measured, by straight lines not exceeding 60 nautical miles in length, connecting fixed points, defined by co-ordinates of latitude and longitude.*"

The outer limit of the continental shelf might be highly irregular if a coastal State followed the many determinations of the depth of sediment calculations of paragraph 4(a)(i) or the foot of the continental shelf plus a maximum of 60 nautical miles calculations of paragraph 4(a)(ii), subject to paragraph 5. Connecting points of such an outer limit by straight lines not exceeding 60 nautical miles in length creates a significant degree of regularity. That process is important in delineating the boundary between coastal States sovereign rights over the continental shelf and the jurisdiction of the International Seabed Authority. This provision, thereby, also recognizes that the coastal State sovereign rights, in this limited respect, may include areas that would otherwise be included within the Area.

Recognizing that the determination of the outer limits of the continental shelf is technically taxing and that the coastal State should not have carte blanche to impose its determination on the international community, the Convention established a Commission on the Limits of the Continental Shelf. This Commission makes recommendations to a coastal State, which has made a submission of its proposed outer limit to it. The Commission cannot require a coastal State to accept its recommendation. However, if a coastal State establishes the outer limit based upon the recommendation of the Commission, the limit is final and binding. It is not subject to further review. There is no provision in the Convention for the compulsory and binding settlement of disputes regarding the technical aspects of establishing the outer limits.

The certainty of the outer limit not only provides the landward limit of the International Seabed Authority's jurisdiction, but also provides stability of expectations for the coastal State and its licensees and lessees. Stability and certainty are essential preconditions for making significant investments necessary for the expected mineral exploration and exploitation. Thus, over time there is an incentive for the coastal State to come to terms with recommendations of the Commission or to provide additional data to support its position on the outer limit.

Paragraph 8 of Article 76 regarding the Commission reads: "*Information on the limits of the continental shelf beyond 200 nautical miles from the baselines from which the breadth of the territorial sea is measured shall be submitted by the coastal State to the Commission on the Limits of the Continental Shelf set up under Annex II on the basis of equitable geographical representation. The Commission shall make recommendations to coastal States on matters related to the establishment of the outer limits of their continental shelf. The limits of the shelf established by a coastal State on the basis of these recommendations shall be final and binding.*"

The Commission serves the further purpose of promoting consistency in applying the provisions of Article 76 (Fig. 8.2). However, in this regard it is essential for the Commission to exercise restraint in making recommendations regarding unusual occurrences, not limited to the first sentence of paragraph 6 with respect to submarine ridges. The Commission should look not only to the expertise of its members; it should look to the broader relevant scientific community. There should, in those circumstances, be rigorous debate in journals, symposia and other venues, to test assumptions, analyses, and theories. Considering the knowledge of the scientific community in the late 1970's compared to that possessed today, and recognizing so much is unknown or not sufficiently mulled over by the scientific community, such a prudent approach seems warranted. There is nothing in the Convention to preclude such caution. Indeed, caution can be expected to promote stability in the oceans.

Furthermore, the Commission is composed of experts in the fields of geology, geophysics and hydrography. The Commissioners may also possess other skills. However, they are not judges or arbitrators, empowered to make pronouncements of law directly. They are intended to be scientific experts. In the absence of any mechanism for the provision of authoritative and binding legal guidance to the Commission, the Commission should also take a prudent approach to, for example, any recommendations relating to paragraph 6. In the absence of any likely exploitation in such problematic areas in the near or medium term at the very least, such caution will not prejudice coastal States, the International Seabed Authority, or private interests.

Should a coastal State establish its outer limit based upon the recommendation of the Commission, those limits are final and binding. Should another coastal State, sometime after the aforementioned coastal State, make a submission regarding similar features off its coast, but at a time when scientific understanding has advanced substantially, differing results in similar circumstances will contribute to the very instability which the Convention sought to ameliorate. Consistency of results, based upon tested solid science, over time, subject to peer review, is important. Thus, it is essential for the Commission to balance a range of interests and not rush to make recommendations, except when it is confident that such recommendations will stand the test of time from both a legal and scientific perspective.

It should also be recognized that there is a range of accepted scientific opinion. It thus behooves the Commission to recognize in its recommendations the "on the basis of" language of paragraph 8, thereby limiting the coastal State freedom to deviate, while recognizing the need for some unspecified flexibility. This is quite limited compared to a "taking into account" standard, which was proposed and rejected in the negotiations.

Should the coastal State decide not to establish the outer limit based upon a recommendation of the Commission, the coastal State shall make a subsequent submission. There are no legal limits as to how long this process of submission and recommendation, resubmission and recommendation may go on. While the coastal State's rights are not prejudiced by the process, as such, any continuing disparity between its views and those of the Commission create uncertainty in the geographical area of disagreement. Such situations may be expected to discourage investments regarding such areas and perhaps in areas nearby. The coastal State has sovereign rights over the continental shelf, even in the absence of an outer limit that is final and binding. However, setting aside paragraph 6, there is some degree of uncertainty as one approaches the outer limit in exploration and exploitation, in the absence of a recommendation of the Commission.

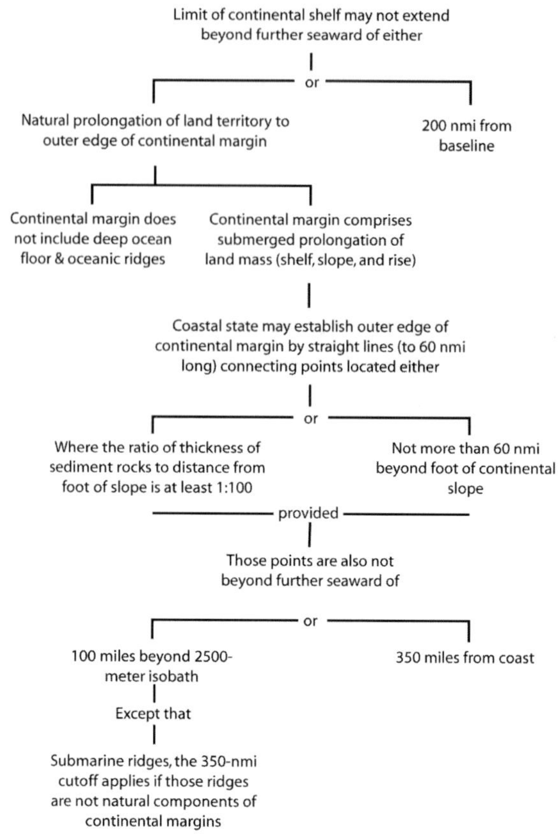

Figure 8.2. Flow chart for identification of geomorphic zones under Article 76.

Paragraph 9 ensures that the world community has specific knowledge of the outer limit of the continental shelf established by the coastal State based upon the recommendation of the Commission. Paragraph 9 reads: "*The coastal State shall deposit with the Secretary-General of the United Nations charts and relevant information, including geodetic data, permanently describing the outer limits of its continental shelf. The Secretary-General shall give due publicity thereto.*"

This paragraph promotes stability by ensuring that all States, the International Seabed Authority, and private interests have access to critical information implicating their rights and interests. The relationship of paragraphs 8 and 9 is noteworthy. While paragraph 8 uses the term "final and binding," paragraph 9 uses the term "*permanently describing the outer limits of its continental shelf.*"

The charts and relevant information deposited with the Secretary-General are all that is necessary to determine the outer limit. The information deposited does not include the underlying information submitted to the Commission for its assessment of the basis for the coastal State's conclusions.

It should be noted that the Commission may not prejudice matters relating to the delimitation of offshore boundaries between adjacent and opposite States. Paragraph 10 reads: "*The provisions of this article are without prejudice to the question of delimitation of the continental shelf between States with opposite or adjacent coasts.*"

8.2.4. The Commission on the Limits of the Continental Shelf

Annex II to the Convention addresses the particulars of the Commission. The Commission shall consist of 21 members who shall be experts in the fields of geology, geophysics or hydrography. There is no further explanation of the requisite qualifications. They shall be elected by States party to the Convention from among their nationals. Nationals of non-Parties to the Convention may not be elected to the Commission. Not less than three members shall be elected from each geographical region, i.e., Western Europe and others, Eastern Europe, Latin America and the Caribbean, Africa and Asia. Thus, the Eastern European Group has three members of the Commission, notwithstanding the fact that there are a relatively small number of States in that region, only one with a continental shelf that extends beyond 200 nmi. This anomaly dates from the immediate post-World War II realities. While the major player in the Eastern European Group is essentially guaranteed a seat, several major broad margin States in other regions must vie for a limited number of seats.

The members are mandated to serve in their personal capacities, although the Annex takes into account the possibility of a different reality. Moreover, the State Party, which submitted the nomination of a member, shall defray the expenses of that member while in performance of Commission duties.

The members are elected for five-year terms and are eligible for reelection. The first Commissioners were elected in March 1997.

The Commission has two primary functions. First and foremost, it is to consider data and other material submitted by coastal States concerning the outer limits of the continental shelf and to make recommendations as noted previously. Second, it may provide scientific and technical advice, if requested by the coastal State concerned during the preparation of data for a submission. In that event, the coastal State shall defray the expenses incurred with respect to the rendering such advice.

The Commission may also cooperate with the Intergovernmental Oceanographic Commission of UNESCO, the International Hydrographic Organization, and other competent international organizations with a view to exchanging scientific and technical information which might be of assistance in discharging its responsibilities.

A coastal State intending to make a submission shall submit the particulars of such limits to the Commission, along with supporting scientific and technical data, within 10 years of entry into force of the Convention for that State. As the Convention entered into force in 1994, the 10 year period for the initial States Parties would have ended in late 2004. In light of the burden that might be placed on certain States, a Meeting of States Parties to the Convention decided that the earliest that any State would have to make a submission would be 2009. Nevertheless, there is no legal penalty for not making a submission within the 10-year or extended period. In any event, a State may make a partial submission regarding its continental shelf beyond 200 nautical miles. A partial submission is implicitly permitted under certain circumstances including: (1) where there are boundary disputes regarding the continental shelf between opposite or adjacent States, where it would be inappropriate to seek a recommendation of the Commission in the absence of agreement of the States concerned, and (2) where the coastal States could not gather the necessary data and other information to make a submission.

The coastal State, when making a submission, shall give the names of any Commission members who have provided scientific and technical advice. If the coastal State has a national on the Commission, the Commissioner whom it nominated would probably be included.

Unless the Commission decides otherwise, it shall function by way of subcommissions, composed of seven members. Nationals of the coastal State making the submission who are members of the Commission and any Commission member who has assisted a coastal State by providing scientific and technical advice shall not be a member of the subcommission dealing with that submission. Such individuals have the right, however, to participate as members of the Commission concerning the said submission. It is essential for Commissioners to avoid conflicts of interests and act with impartiality as they serve in their personal capacities. It may be useful for the Commission to consider elaborating a code of conduct or ethics or at least guidelines to enhance confidence in its work.

In any event, the coastal State making a submission may send its representatives to participate in the relevant proceedings without the right to vote.

The subcommission submits its recommendations to the Commission. The Commission may approve the recommendations of the subcommission by a majority of two-thirds of the Commission members present and voting. The recommendations of the Commission shall be submitted to the coastal State, which made the submission, and to the United Nations Secretary-General.

The Annex further provides that in case of disagreement by the coastal State with the recommendations of the Commission, the coastal State shall, within a reasonable time, make a revised or new submission to the Commission.

Moreover, although the Convention is silent, the Commission may not address questions relating to the underlying sovereignty of the land territory, from which any rights regarding the continental shelf flow. Thus, the existence of a dispute (however denominated, either in fact or in an international agreement or otherwise) over the offshore boundary of the continental shelf between opposite or adjacent States or underlying sovereignty will *per force* require the Commission to refrain from making a recommendation. In the case of an offshore boundary, the limitation on the Commission's competence is explicit. In the case of an issue relating to the underlying sovereignty of the land territory, the limitation flows from the fact that such is beyond the scope of the Convention.

It is unnecessary for all parties to the dispute to agree that a dispute exists. It is sufficient that one State makes a credible comment to the Commission that it is involved in a dispute, or the Commission may take notice of the *defacto* or legal state of affairs. Indeed, the Commission's practice is such as to request comments from all States upon receiving a submission. Such comments may address the existence of a dispute or address, for example, technical aspects of the outer limit. The weight to be given to such comments should vary according to their merits.

8.2.5. Rights of the Coastal State over the Continental Shelf

The coastal State exercises over the continental shelf sovereign rights for the purpose of exploring it and exploiting its natural resources. Those rights are exclusive. No one may undertake those activities without the express consent of the coastal State. Furthermore, the rights of the coastal State do not depend on occupation, effective or notional, or on any express proclamation. The rights flow, as a matter of law, from the coastal States sovereignty over the relevant land territory. The natural resources consist of the mineral and other non-living resources of the seabed and subsoil together with living organisms belonging to sedentary species.

Thus, while a coastal State must declare an exclusive economic zone, by whatever name, to a maximum distance of 200 nautical miles from baselines in order to exercise rights over fisheries, for example, in the water column, no such declaration is necessary as a prelude to exercising resource rights over the continental shelf, whether within or beyond 200 nautical miles.

8.2.6. Legal Status of the Superjacent Waters and Air Space and the Rights and Freedoms of Other States

The rights of the coastal State over the continental shelf do not affect the legal status of the superjacent waters or of the air space above those waters. Moreover, the exercise of those rights of the coastal States must not infringe or result in any unjustifiable interference with navigation and other rights and freedoms of other States as provided for in the Convention and customary international law. Indeed, the waters superjacent to the continental shelf within the exclusive economic zone, and beyond 200 miles, continue to embrace such high seas freedoms and rights. Furthermore, high seas freedoms of all States are recognized on the continental shelf and co-exist with coastal States rights.

8.2.7. Submarine Cables and Pipelines on the Continental Shelf

All States are entitled to lay submarine cables and pipelines on the continental shelf, both within and beyond 200 nautical miles. The coastal State may not impede the laying or the maintenance of such cables or pipelines, subject to its right to take reasonable measures for the exploration and exploitation of its natural resources and the prevention, reduction, and control of pollution from pipelines. The delineation of the course for the laying of such pipelines (but not cables) is subject to the consent of the coastal State. At the same time, when laying submarine pipelines and cables, States shall have due regard to cables or pipelines already in position. In particular, possibilities of repairing existing cables or pipelines shall not be prejudiced. The coastal State retains clear rights to establish conditions for cables or pipelines entering its territory or territorial sea (areas of sovereignty). It has clear rights and jurisdiction over cables and pipelines constructed or used in connection with the exploration of its continental shelf or exploitation of its resources or the operation of artificial islands, installations, and structures under its jurisdiction.

8.2.8. Artificial Islands, Installation and Structures on the Continental Shelf

In general, in both the exclusive economic zone and on the continental shelf beyond 200 nautical miles, unless otherwise noted, the coastal State has the exclusive right to construct and to authorize and regulate the construction, operation, and use of:

-artificial islands;

-installations and structures for the exploration, exploitation, conservation, and management of the natural resources, whether living or non-living of the waters superjacent to the seabed within 200 nautical miles, and of the seabed and its subsoil both within and beyond 200 nautical miles. Includes infrastructure related to other activities for the economic exploration and exploitation within 200 nautical miles, such as the production of energy from the water current and winds;

-installations and structures that may interfere with the exercise of the rights of the coastal State in the exclusive economic zone and the continental shelf.

The coastal State has exclusive jurisdiction over the previously mentioned artificial islands, installations, and structures, including jurisdiction with regard to customs, fiscal, health, safety, immigration laws, and regulations.

The coastal State has an obligation to give due notice of the construction of such artificial islands, installations, and structures, and must maintain permanent means for giving warning of their presence. Any installations or structures which are abandoned or disused shall be removed to ensure the safety of navigation, taking into account any generally accepted international standards established by the competent international organization. Such removal shall also have due regard to fishing, the protection of the marine environment and the rights and duties of other States. Appropriate publicity must also be given to the depth, position and dimensions of any installations and structures not entirely removed.

Safety zones, where necessary, may be established in which the coastal State may take appropriate measures to ensure the safety both of navigation and of the artificial islands, installations and structures. The breadth of the safety zones shall be determined by the coastal State, taking into account applicable international standards. The zones shall not exceed a distance of 500 meters measured from each point of their outer edge, except as authorized by generally accepted international standards or as recommended by the competent international organization. Due notice must be given of the extent of the safety zones.

All ships must respect these safety zones and comply with generally accepted standards regarding navigation in the vicinity of the artificial islands, installations, structures, and safety zones. However, the artificial islands, installations, structures, and safety zones may not be established where interference might be caused to the use of recognized sea lanes essential to international navigation.

8.2.9. Drilling on the Continental Shelf

The coastal State has the exclusive right to authorize and regulate drilling on the continental shelf for all purposes.

8.2.10. Payments and Contributions with Respect to the Exploitation of the Continental Shelf Beyond 200 Nautical Miles

The Conference faced the objective of broad margin States, those States with continental shelves extending beyond 200 nautical miles, to maximize the geographic limits of their sovereign rights. In contrast, the conference faced the competing objective of land locked and so-called geographically disadvantaged States to restrict such limits and thereby maximize the geographical sweep of the jurisdiction of International Seabed Authority. The latter's jurisdiction is limited to non-living resources in the seabed and subsoil beyond 200 nautical miles or the limit of the continental shelf beyond, as defined by Article 76, whichever is farther seaward.

The essence of the compromise was a broad continental shelf possibility for broad margin States and revenue sharing by the coastal State with respect to exploitation of non-living resources of the legal continental shelf beyond 200 nautical miles. The coastal State is required to make payments or contributions in kind with respect to such exploitation. No payments or contributions are required for any exploitation within 200 nautical miles or for exploitation of living resources of sedentary species beyond 200 nautical miles.

The payments and contributions are to be made annually with respect to all production at a site after the first five years of production at that site. For the sixth year, the rate of payment or contribution shall be 1 percent of the value or volume of production at that site. The rate shall increase by 1 percent for each subsequent year until the twelfth year and shall remain at 7 percent thereafter. Production does not include resources used in connection with exploitation.

The coastal State has discretion to define a site, which will be a function, among other things, of its domestic practice, technology and economic factors. As a general proposition, sites should be defined narrowly. Thus, for example, if leases are made in blocks, the site can be a subset of a block. The value of production at a site will also take account of domestic practice, and market factors.

In any event, the obligation to share revenues is that of the coastal State and not that of the licensee or lessee. The relationship of the former and latter in regard to sorting out the payment or contribution due is not the concern of the international community. Rather it is a domestic policy and legal matter.

In addition, a developing State, which is a net importer of a mineral resource produced from its continental shelf, is exempt from making such payments or contributions with respect to that mineral resource. The term "developing State" is not defined in the Convention.

The payments or contributions are made "through" the International Seabed Authority, which shall distribute them to States Parties to the Convention, on the basis of equitable sharing criteria, taking into account the interests and needs of developing States, particularly the least developed and the land-locked among them. The payments are not made to the International Seabed Authority for carrying out its functions regarding its jurisdiction over the non-living resources of the deep seabed. Its function regarding the payments and contributions is to distribute them based upon yet to be considered unspecified equitable sharing criteria.

8.2.11. Delimitation of the Continental Shelf Between States with Opposite or Adjacent Coasts

The delimitation of the continental shelf between States with opposite or adjacent coasts is to be effected by agreement on the basis of international law in order to achieve an equitable solution. The actions of the Commission shall not prejudice matters relating to such delimitation. Thus, in the absence of an agreed boundary, manifested by, for example, an agreement in force, the

Commission may not make recommendations regarding an outer limit in the relevant area of the continental shelf. Nevertheless, two or more States, with an unresolved common offshore boundary, may decide to make jointly a submission of the outer limit of the continental shelf to the Commission with a view to the Commission making a recommendation regarding the outer limit in that contested area. In that case, the Commission's recommendation would not be prejudicial to an ultimate resolution of a bilateral or multilateral delimitation. In certain circumstances, such an approach might facilitate resolution of a contested boundary.

8.2.12. Tunneling

The Convention also provides that it will not prejudice the right of the coastal State to exploit the subsoil of the continental shelf by means of tunneling, irrespective of the depth of water above the subsoil. The provision was hardly necessary, in light of the broad sweep of coastal State sovereign rights over the natural resources of the continental shelf. It was seemingly included out of an abundance of caution and taken from the 1958 Geneva Convention on the Continental Shelf. This provision is of no practical importance, in light of the regime of the continental shelf both within and beyond 200 nautical miles. It cannot be read as derogating from the regime of the deep seabed and the jurisdiction of the International Seabed Authority.

8.3. STATEMENT OF UNDERSTANDING CONCERNING A SPECIFIC METHOD TO BE USED IN ESTABLISHING THE OUTER EDGE OF THE CONTINENTAL MARGIN

During the negotiation of the Convention, the Government of Sri Lanka argued that Article 76, in its application to the area off its coast, would be inequitable. Essentially, it maintained that the Bengal Fan was, in part, the natural prolongation of its land territory. Article 76 provides no rules specific to the situation off Sri Lanka. Nevertheless, the Final Act of the Conference, Annex II, includes a 'Statement of Understanding Concerning a Specific Method to be Used in Establishing the Outer Edge of the Continental Margin.' Furthermore, the Annex to Convention relating to the Commission, includes as a function to make recommendations in accordance with Article 76 and the Statement of Understanding. While the Final Act gives no rights and imposes no obligations on States, the incorporation of a reference to the Statement of Understanding in the Annex on the Commission seems to raise the possibility of it having legal implications.

First, the Statement considers the special characteristic of a State's continental margin where: 1) the average distance at which the 200-meter isobath occurs is not more than 20 nautical miles; and 2) the greater proportion of the sedimentary rock of the continental margin lies beneath the rise. Second, it takes into account the inequity that would result to that State in applying

Article 76, in that the mathematical average of the thickness of sedimentary rock along a line established at the maximum distance permissible under paragraph 4(a)(i) and (ii) would not be less than 3.5 kilometers, and that more then half the margin would thereby be excluded. Third, it recognizes that such a State may, notwithstanding Article 76, establish the outer edge of its continental margin by straight lines not exceeding 60 nautical miles in length connecting fixed points, at each of which the thickness of sedimentary rock is not less than one kilometer. Should a State establish the outer edge of its continental shelf as noted, clearly in derogation of Article 76, a neighboring State may similarly delineate its outer edge on a common geological feature. Finally, the Statement requests that the Commission be governed by this Statement in making recommendations related to the establishment of the outer edge of the continental margins of these States in the southern part of the Bay of Bengal. The Statement does not mention Sri Lanka or India explicitly.

The Statement of Understanding does not address the natural prolongation of the landmass, a central feature of the legal continental shelf, defined in Article 76. Rather, it focuses on paragraphs 4(a)(i) and (ii), and the inequities inherent therein if applied in the southern part of the Bay of Bengal.

However, the Bengal Fan, the extensive "continental margin" in the southern part of the Bay of Bengal, does not appear to be the natural prolongation of the land territory of Sri Lanka. If that is so, the Statement of Understanding has no application in the southern part of the Bay of Bengal. It does not purport to apply anywhere else. Therefore the legal peculiarities of the structure of the Convention in that case, i.e., the relationship of Article 76, the Annex on the Commission, and the Final Act, are of only theoretical interest.

8.4. THE AREA BEYOND THE LIMITS OF NATIONAL JURISDICTION

The Area is defined in the Convention as the "seabed and ocean floor and subsoil thereof, beyond the limits of natural jurisdiction." It is beyond the 200 nautical mile exclusive economic zone or the continental shelf beyond the 200 nautical mile limit established in accordance with Article 76, whichever is further seaward. One of the major sea areas that probably hosts hydrate in deep marine sediments beyond the 200-nmi limit is the Arctic Ocean (Max and Lowrie, 1993). Even here, however, and in most other places where continental margin sediments extend beyond the 200 nmi limit, methane-enriched sediments are mostly base of slope and rise sediments, which are unlikely to host high-grade (Chapter 4) economic deposits of gas hydrate.

The Area and its non-living resources are the common heritage of mankind. Thus, no State may claim or exercise sovereignty or sovereign rights over any part of the Area or its resources. No State or any natural or juridical person may appropriate any part of the Area. All rights in the resources of the Area are vested in mankind as a whole, on whose behalf the International

Seabed Authority shall act. Furthermore, no State or natural or juridical person may claim, acquire, or exercise rights with respect to minerals recovered from the Area, except in accordance with Part XI of the Convention and the Agreement. The authority of the International Seabed Authority relates only to non-living resources, meaning all solid, liquid or gaseous mineral resources.

Activities in the Area with respect to resource deposits which lie across limits of national jurisdiction, shall be conducted with due regard to the rights and legitimate interests of the relevant coastal State. Consultations, including a system of prior notification, shall be maintained with the State concerned, with a view to avoiding infringement of such rights and interests. In cases where activities in the Area may result in the exploitation of resources lying within national jurisdiction, the prior consent of the coastal State concerned is required.

The United States sought a regime for the administration of the Area relating to non-living resources that included guaranteed access, under reasonable conditions, together with security of tenure for States and their nationals. Unfortunately, Part XI, as negotiated at the Conference, was deemed to be deeply flawed by the United States, a view shared by other industrialized nations. In 1990 the United Nations Secretary-General initiated informal consultations that led to the Agreement, which was adopted by the United Nations General Assembly in 1994.

The objections to Part XI, included: (1) a structure which did not accord industrialized States influence in the regime commensurate with their interests; (2) principles at odds with a free market approach; and (3) provisions that impeded access by industrialized States. The Agreement met fully those objections. The Agreement and Part XI are to be interpreted and applied together as a single instrument. In the event of any inconsistency between the two, the provisions of the Agreement shall prevail.

The main function of the Authority to administer the exploration for and exploitation of, the non-living resources of the Area, shall be carried out by: (1) an Assembly (composed of all members of the Authority), (b) a Council-the executive organ (composed of 36 members subject to important criteria), and (c) a secretariat. There are also subsidiary bodies, including (1) a Legal and Technical Commission (composed of 15 technical experts, elected by the Council), (2) an Economic Planning Commission (composed of 15 technical experts elected by the Council), and (3) a Finance Committee composed of 15 experts elected by the assembly. The Economic Planning Commission's establishment is subject to a future decision of the Council, or upon the approval of the first plan of work; significant functions will be performed by the Legal and Technical Commission for the time being. In this regard, the Agreement provides that the setting up and the functioning of the organs and subsidiary bodies of the Authority shall be based on an evolutionary approach. This approach recognizes that activities relating to the exploration and exploitation of the non-living resources of the Area may not be imminent.

Prospecting in the Area does not require prior approval of the Authority, although a prospector must submit a written undertaking to comply with the Convention. Different prospectors may prospect the same parts of the Area. Prospecting does not convey any rights regarding resources. Exploration and exploitation activities may be conducted by States, Parties sponsored by States and Parties are subject to approval by the Council, explicitly or implicitly. The so-called Enterprise, an operating arm of the Authority, may only become operational following a decision by the Council.

Procedurally, an applicant submits a plan of work that, if approved, constitutes a contract between the applicant and the Authority. The Legal and Technical Commission reviews applications and makes recommendations to the Council on the approval of plans of work. The Commission is required to base its recommendations on (1) whether the applicant meets financial and technical qualifications, (2) whether the proposed plan of work otherwise meets the rules and regulations adopted by the Council and (3) whether the applicant has included undertakings to comply with the Convention and with rules, regulations and procedures adopted pursuant thereto. Plans of work shall be reviewed and approved on a first come, first served basis. Decisions of the Commission are to be taken by a majority of its members present and voting. If the Commission recommends against approval of an application, it can nevertheless be approved by the Council.

There are application fees of $250,000 for the exploration phase and $250,000 for the exploitation phase. Should these fees exceed the costs of processing the applications, the applicant will receive the difference. Financial arrangements shall be established in the future, including an annual fee during commercial production. This fee can be credited against payments under a royalty, profit sharing, or other arrangement.

Plans of work for exploration are for 15 years, at the end of which an applicant must apply for a plan of work for the exploitation phase. However the applicant may seek not to proceed to the exploitation phase for a number of reasons, in which event the work plan for exploration will be extended in 5-year increments by the Authority if certain conditions are met. A contractor, whose plan of work has been approved, has exclusive rights in the area covered with respect to a specific category of resource.

The Agreement provides that the four-member consumer chamber of the Council shall include the State that, upon entry into force of the Convention, has the largest economy in terms of gross domestic product. Generally, decisions are to be taken by consensus in the Council (as well as the other organs of the Authority). However, if all efforts to reach a decision by consensus have been exhausted, and the Convention does not specifically provide for certain decisions to be taken by consensus, the Council shall take decisions by a two-thirds vote of the members present and voting, provided that such decisions are not opposed by a majority in any chamber, i.e., the four member consumer, investor or producer chambers, or by the two other groups which shall be treated as one chamber, i.e., the six members from among developing States subject to certain criteria and the eighteen States to insure equitable geographical representation. Thus, industrialized countries have significant influence in blocking decisions.

There are specific provisions for grandfathering into the regime the mining sites of entities that have engaged in substantial activities before the Convention entered into force. There are also provisions for areas to be set aside for possible use of the Enterprise that may undertake mining activities under certain conditions, initially through joint ventures.

Moreover, rules, regulations, and procedures for administration of the mining regime must be adopted by the Council by consensus. With a guaranteed seat for the United States, once it becomes party to the Convention, its consent, or at least absence of objection, must be obtained before such rules, regulations, and procedures can be adopted or applied.

The Authority, of course, must over time adopt rules for different categories of minerals. This will be a time consuming process, as the particular needs of the interested miners may not be clearly known before the activity takes place. Rules, regulations, and procedures for the exploration and exploitation of any resource, other than polymetallic nodules, shall be adopted pursuant to a request to the Authority by any of its members.

Suffice it to say, the development of an approach to the implementation of a regime for the exploration and exploitation of gas hydrate in the Area, will be a significant endeavor. However, should this resource be found in the Area, and if the United States is Party to the Convention and Agreement, the specific rules of the game ultimately adopted should not be adverse to US interests.

8.5. THE RELATIONSHIP OF THE CENTRAL GOVERNMENT TO LOCAL AUTHORITIES

In a federal system, such as in the United States, the relationship regarding resource rights between the State and Federal government is not affected by the Convention. Thus, should hydrate be found within the offshore jurisdiction of a State of the Union, relevant domestic law, regarding the regulatory regime would apply. In other federal systems, a different result between domestic authorities may be found.

8.6. CONCLUSION

The international legal elements described provide a framework, which is conducive to the exploration for, and exploitation of, oceanic gas hydrate. The regime applicable to each offshore area promotes the certainty and stability expectations essential to such commercial undertakings.

AUTHOR COMMENT

The regulatory environment of individual countries with respect to exploration and exploitation of natural gas hydrate in some States may concern the issue of whether gas hydrate should be considered under petroleum or mining regulations. This varies from country to country in those few countries that have established policy.

Chapter 9

Conclusions and Summary

9.1. CONCEPTUALIZATION OF THE HYDRATE GAS RESOURCE

Gas hydrate and gas concentrations associated with hydrate appear to be widespread on oceanic continental margins and in some permafrost regions. Kvenvolden's (1988) estimate of extremely large volumes of natural gas sequestered in hydrate is being essentially substantiated by exploration. Concentrations of hydrate have been identified in both oceanic and permafrost regions, and drilling evaluation and hydrate tests for conversion and production have been achieved. This is rapid progress in developing a new energy resource play considering that hydrate was first notice in the general scientific literature related to single locality discovery as recently as 1971 (Stoll et al., 1971) Earlier Russian hydrates were recognized sources of natural gas (Makogon, 1965; Makogon et al., 1971), before the hydrate energy resource potential was recognized in North America, but these publications were not available in the West at that time.

A commercially successful strategy for gas hydrate production does not yet exist and hydrate has yet to be proven to be an economically viable unconventional gas resource. Indeed, a number of oceanic hydrate resource evaluations have been negative with respect to the eventual recovery of methane from hydrate deposits because of low gas hydrate content in low permeability sediments (e.g., Hovland et al., 1997). One of the present difficulties in making assessments concerning the economic viability of hydrate, however, has been the lack of systematic distinction between high- and low-grade deposits and their characteristics and economic potential, even though commercial interest has inherently sought what we define as deposits of high-grade type. Although it is conjectural as to whether low-grade hydrate deposits will ever be 'mined' for their natural gas, observations on the commerciality of low-grade deposits may not be relevant for the commercial exploitation of hydrate deposits. High-grade deposits, of Class 1 and 2 character (Chapter 7) that have the greatest potential to concentrate hydrate, almost certainly will be the first to be exploited, although Class 2 deposits will require substantial reservoir management. In these high-grade deposits, of which the Mallik (5.2) and Nankai (5.3) discoveries are good examples, hydrate may exceed 30% of bulk volume in permeable sands (Max, 2000) and prospects of recovery of at least some hydrate deposits are good.

Modern drilling, sensor, and control technologies that have been developed of conventional hydrocarbon exploration and exploitation appear to

Michael D. Max et al. (eds.), Economic Geology of Natural Gas Hydrate, 289–296.

us to be generally of a sufficient level to allow recovery of natural gas from hydrate deposits. In some respects, some of these technologies, such as coiled tube drilling, have been proposed for development specifically for hydrates. The main element to the successful application of this technology will be the proper modeling of geologic reservoir and extraction procedures.

In any field, one standard of measure of its maturity is the number and subject of patents. In the field of exploration and recovery of natural gas from conventional gas deposits, there are thousands of patents, many of which are current as new inventions are continually being made. In contrast, there appear to be very few patents that bear on the recovery of natural gas from hydrate. Agee et al. (1999) envisage harvesting seafloor and near-seafloor hydrate using partly mechanical means and conversion of gas to liquids after recovery and delivery to a ship for transport (Christensen, 1994). Chatterji & Griffith (1998) suggest another method of causing in-situ dissociation (Chapter 5). Matsuo et al. 2004) use fluid injection to dissociate hydrate and recover the fluid-gas mixture, which requires an additional separation step. Lyon (2004) uses the injection of CO_2 and a catalyst to release methane. Heinemann et al. (2001) dissociate hydrate through the use of microwave heating without a focusing effect. Related patents, such as Guo et al. (1995), Yosikawa et al. (2003), are mainly in the field of using hydrate as a safe transport and storage media. But these methods inject water into a gaseous atmosphere but do not dissociate hydrate in a novel manner.

Many of the new techniques necessary for the identification of hydrate will be patentable. Because the development of the hydrate resource appears to be on a fast track, a considerable body of patents that control the identification and extraction of hydrate could suddenly develop. This rapid development would be in contrast to patent activity in the more general field of conventional hydrocarbon energy resources where most of the patents have lapsed and now constitute prior art. It is to be hoped that the development of a large body of active patents does not retard the development of the hydrate resource.

Perhaps one of the primary impediments to the development of oceanic hydrate resources could be identified as psychological rather than technical. For instance, in a special technology issue of Geotimes, Lubick (2004), shows “traditional drill-well technology”, and suggests that any solutions (to recovery of natural gas from oceanic hydrate) remain futuristic. This attitude ignores existing drilling technology innovations such as coiled tube drilling, multi-lateral collectors, well head drilling tools, computer aided and assisted drilling and monitoring using MEMS down-hole sensors, and high pressure jet drilling technologies in the place of rotary drilling. Many of these technologies are already being used for conventional hydrocarbon exploration and exploitation and application to hydrate exploration and recovery would not be futuristic, requiring new inventions, but rather would be adaptive.

Efficient recovery of natural gas from oceanic hydrate will depend on developing detailed extraction models and procedures. This development, in turn, depends on understanding (1) how natural gas comes to be sequestered in hydrate (Chapter 2), (2) the form and characteristics that individual deposits may take, and (3) where the sweet spots or hydrate concentrations may be (Chapters 3, 4 & 6). Modeling of hydrate extraction (Chapters 5 & 7) is presently incomplete. Containment of converted gas where there is no natural trap other than within the hydrate itself, for instance, is a major issue that has emerged from the experiences of early production testing that urgently requires resolution.

9.2. GAS HYDRATE: A NEW HYDROCARBON RESOURCE AT THE RIGHT TIME

For virtually the whole of the 20th century, the United States controlled the market for world energy demand, and for much of the century controlled directly or indirectly the world hydrocarbon supply. This control allowed the United States to promote a policy of inexpensive energy that fueled the economic dynamo and imparted a natural advantage over countries with higher energy costs for industrialization and production. Because of increasing demands for energy among countries that are industrializing at a rapid pace, principally China (which is now the second-largest oil consumer after the United States) and India, the world supply of both gas and oil is stretching thin and prices are rising accordingly. The International Monetary Fund (Dougherty, 2005) has now joined other financial organizations in warning about the economic effect of global energy superdemand (1.11). Even if there is no unexpected disruption of energy supplies, which could possibly cause energy prices to spike to double the pre-disruption price, it is likely that the era of inexpensive energy is gone forever.

Demand has the potential to outstrip energy production and keep energy costs high. For instance, if India and China reach the stage where their per capita oil consumption equals that of South Korea, a prosperous, industrialized Asian country that offers a good economic model for them, those two nations alone will require 104 million barrels per day. Because the entire world now uses 84 million barrels per day, it is likely that emerging demands can never be met with the present world energy supply. The only real economic limit to consumption is likely to be the price of energy, and strongly rising prices that plateau at high levels will have an effect on economics that could alter large-scale societal assumptions. Recovery of natural gas from hydrate has the potential to mitigate this energy shortage through the provision of additional gas supplies, particularly where they are adjacent to the nations that require the energy.

9.3. GAS HYDRATE CHARACTERIZATION

Few developed nations are self-sufficient in energy, although larger geographic areas may provide their own energy supplies. For example, Russia is self-sufficient in gas supplies and is a major supplier of gas to Western Europe. Norway is a net energy exporter, as are many of the small West African and SW Asian countries with major oil deposits. A number of Middle Eastern countries, of course, hold a major portion of the world's oil and gas.

Hydrate deposits, like conventional hydrocarbon deposits, are found in those permafrost and oceanic areas where conditions are suitable for their formation. In some places, there is a strong overlap between the location of potential hydrate deposits and both known and potential conventional hydrocarbon resources. This overlap may be particularly true of deep water hydrocarbon deposits. In other areas, however, hydrate may be a unique store of hydrocarbon energy resource. The deep continental shelf and slope of Japan, is the best example known today of an area in which hydrate may constitute the principal hydrocarbon energy supply. The continental slopes of the Indian Ocean and the western margin of South America may also provide indigenous energy supplies to countries once thought to have limited energy resources. Hydrate formation in a GHSZ is a natural process that is capable of concentrating gaseous and dissolved natural gas over a very long period of time. Hydrate constitutes a huge reservoir from which natural gas may be recovered. How much of this resource will ever become a commercial reserve, however, is unknown (Preface).

9.3.1. Permafrost Hydrate

In addition to very substantial supplies of conventional gas, extraction from permafrost hydrates in both North America and Russia would have the effect of providing significantly larger gas reserves. Permafrost hydrate deposits, are amenable to near-term development, amongst other reasons, because:

1. They are predominantly high-grade deposits.
2. The converted gas will generally remain trapped,
3. They are all near-surface and relatively easy drilling targets,
4. Gathering and processing facilities, pipelines, and infrastructure on land are less expensive than for oceanic hydrate deposits.
5. Some permafrost hydrate deposits have already been identified, quantified, and valued in at least a preliminary fashion. Extraction models (Chapter 6) indicate that a steady state conversion of these hydrate deposits can result in adequate gas flows.
6. They commonly are associated with conventional gas or petroleum deposits rather than being independent recovery targets.

9.3.2. Oceanic Hydrate

A number of oceanic hydrate deposits have been recognized that are immediately adjacent to industrial countries, which are not self-sufficient in fuel, particularly in gas. The most prominent of these is the Nankai deposit in the deep continental shelf off SE Japan (Chapters 3 & 5). Other deposits have been recognized offshore India, but their commerciality has not yet reached the level of the Nankai hydrate 'field'. Japan, among major industrial nations (Chapter 1), is particularly short of indigenous energy.

A number of recent developments in the last five years have both broadened and deepened interest in developing hydrates as a source of methane gas.

1. Oceanic hydrate holds the largest portion (about 95%) of the world's hydrate resources, and it is possible that super giant-sized hydrate deposits exist. These will be the big prizes.

2. Commercial exploration for conventional hydrocarbon deposits has moved into deep water and the technology required for exploitation of hydrate is largely available or is in the late stages of development. Drilling is taking place in water depths where hydrate may be encountered at and beneath the seafloor. Industry is learning to cope with the drilling safety aspects of hydrate, which do not occur in the shallower waters in which industry had previously been restricted. Infrastructure for moving gas from deep water wellheads to market is being planned and financed for conventional deepwater hydrocarbon deposits. Because the development costs of hydrate exploration and exploitation, being largely amortized under the conventional hydrocarbon accounting subhead as a byproduct of deeper water drilling for conventional hydrocarbon deposits, the conventional hydrocarbon infrastructure can be shared. Hydrate work may be directly piggybacked and the coincident recovery of both conventional and unconventional deposits can be contemplated.

3. Government agencies (Preface) have begun to develop National hydrate research programs focused on the recovery of natural gas from oceanic hydrate, especially in energy-poor coastal States.

4. Some energy companies are beginning to regard investment in hydrate resources as advisable, especially where these coincide with conventional offshore developments.

5. Oceanic hydrate will constitute 'new' energy, in contrast to permafrost hydrate, which thus far has only been found in close association with existing conventional hydrocarbon deposits and can be regarded as only adding to those resources.

9.3.3. Hydrate Natural Gas Quality

Natural gas produced from hydrate, especially where no chemical additives are absorbed from groundwater prior to production, is often very pure methane. Higher density hydrocarbon gases often occur where gas is of a thermogenic origin. Other gases, particularly SOx, nitrogen, and carbon dioxide, which are often constituents of conventional natural gas, are not typically found in hydrate natural gas. Because the process of forming gas hydrate is capable of purifying a gas mixture, the overall quality of hydrate natural gas will be extremely high. In a situation where conventional natural gas requires processing to remove contaminant gases, especially where nitrogen or carbon dioxide must be removed to bring the btu value of the gas above 750-800 btu/ft^3, blending the conventional gas with hydrate natural gas will dilute the contaminant gases while raising btu values to a marketable level, without further processing. In addition, hydrates, especially oceanic hydrates are often not associated with any oil or hydrocarbon condensates. Thus, their production has no danger of pollution commonly associated with oil spills.

9.4. HYDRATE EXPLORATION AND RECOVERY

Hydrate delineation and provincing, and the application of semi-quantitative assessments of the likelihood of the presence of hydrate will take into account all existing information for any particular area. As with other exploration scenarios, more general methods will be used at first, with higher precision methods then yielding to drilling for final evaluation.

The presence of substantial indications of oceanic hydrate was first revealed by reflection seismic survey, although hydrate in the Black sea was observed in cores (K. Kvenvolden, pers. com.) It is likely that this exploration technique will continue to be important for the first order identification of the presence of the general conditions required for the formation of hydrate deposits. Although the identification of BSR cannot be used as a detailed exploration technique, its presence is a good indicator (1) of substantial gas generation, (2) porosity suitable for fluid and gas migration, and (3) the existence of pressure and temperature conditions suitable for spontaneous formation of hydrate. For identification of individual deposits, however, other seismic analysis techniques that will reveal range-independent velocities, attenuation, and sediment type and porosities, among other characteristics, are likely to be more important. Other techniques, such as electromagnetics may prove to be of greater importance in the identification of hydrate resources.

Drilling, probably with some adaptation of coiled tube technology, with MEMS sensor arrays deployed in the drill string, will probably be required for the in-situ evaluation of potential hydrate deposits. Ideally, the exploration holes will also become part of the stimulation and extractive system. In detail, the drilling strategy will almost certainly differ considerably from that used for conventional hydrocarbons.

In order to recover the natural gas, conversion conditions have to be maintained in the entire extraction system. The water that will be derived from the hydrate will be gas saturated, the gas will be water saturated and the general conditions surrounding the conversion volume will be suitable for the spontaneous formation of hydrate from either an aqueous or gaseous precursor. Therefore, attention will have to be given to maintenance of increased heat or lower pressures in order to preclude the unwanted formation of hydrate. In contrast, conventional gas deposits are usually drawn from greater depths and are often very warm. These conditions usually require that the fluids recovered before they can cool to temperatures where hydrate may form. Alternatively, new down-hole technology for gas drying may provide an alternative solution that would preclude the formation of unwanted hydrate.

The commercial development of the hydrate resource is much closer to being a reality today than it has ever been. Companies are hanging back while the governments of energy-poor countries are taking the lead, even though some of these countries are awash with money because of the increased sale price of their hydrocarbon fuels. It is presently not possible to say exactly where or when the first hydrate well or field will begin producing commercial natural gas. It is possible to say, however, that immediately following the opening of the first commercial field, investment in and development of, the hydrate resource will become a much more attractive energy option.

9.5. COMMERCIAL HYDRATE NATURAL GAS DEVELOPMENT

The major commercial concerns active in hydrate energy developments are the national energy and natural resources companies of Japan and India, where there is government influence or control of energy. Issues of national energy security primarily drive these government companies. In the United States, and in many other countries, however, energy companies function outside of direct government control. Although the perceived higher risk of investing in hydrate recovery is often used as a reason for not investing, the simple reason could be that any investment would diminish their immediate bottom line profitability on a short horizon. This outlook is demonstrated by the fact that these same companies, or their pre-merger predecessors used to lead the world in geological and geophysical research bearing on the identification of new energy resources and in increasing resource base by improving production. These major research assets have largely been divested and these energy companies have a greatly diminished research capability.

The major multinational energy companies are unlikely to become entrepreneurial in their choosing of short-term profitability over improving longer term company value through the identification of new energy resources. At present, the perceptions of these major energy companies appears so accord with the conclusion of Bil (2003) on behalf of Shell International Exploration and Production B.V. that identified no show-stoppers to hydrate development but had no enthusiasm for near-term investment for hydrate development. Thus, it is likely that real entrepreneurialism in the field of hydrate development will be made by small companies or consortia of small companies willing to accept the risk of developing hydrate natural gas resources.

The national gas hydrate research program of the United States is presently in review following a report by the National Research Council (NRC, 2004). In this, the research that was funded under the program since its inception in 2000 is listed. It is possible that a national gas hydrate research program that attempts to provide relatively small levels of funding to a large number, and possibly geographically balanced, of essentially unrelated research projects is not the best vehicle for developing commercial hydrate natural gas. There appears to have been no clearly identified objective to develop exploration, evaluation, and recovery methods for hydrate gas deposits because other issues, such as seafloor stability, global carbon/climate, and drilling and seafloor safety, as well as other issues not directly related to recovery of hydrate natural gas have shared the modest level of funding. This is in contrast to the apparently well-defined hydrate programs of Japan and India and the European hydrate research program, which is mainly focused on basic research and global climate.

We would suggest that the emphasis on broad hydrate research programs is beginning to impede development of hydrate natural gas exploitation. A great deal of biogeological, climate, and other basic science issues are simply peripheral to the development path that will lead to commercial hydrate natural gas extraction. We need to learn how to reliably identify concentrations of natural gas hydrate and how to extract it. A critical path needs to be identified so that peripheral hydrate research, no matter how interesting or important for other reasons, can be excluded. Existing technical expertise needs to be co-opted. For instance, the water and gas management in an oceanic hydrate natural gas reservoir and recovery system will almost certainly require coalbed methane technology and practices in some form. Experience from heavy oil - tar sand recovery will prove useful in conversion of hydrate and tight sand drilling experience can be almost directly applied. If a concerted program for commercial recovery of hydrate is enacted, possibly by a consortium of small, entrepreneurial companies, the successful of exploitation of hydrate natural gas may not lie far in the future.

Glossary of Terms

Hydrate science is rapidly evolving. Both new terms and different terminology for the same gas hydrate related objects and mechanisms are commonly found in the literature and in discussions. Commercial terms may differ somewhat from scientific usage.for descriptions of the same objects or processes. Geophysical and gas hydrate terminology is current usage, mining terms are common usage and taken largely from AGI (1997) and Jackson (1997). Also see the on-line Schumberger glossary <http://www.glossary.oilfield.slb.com/> for more detailed terms relating to hydrocarbon exploration and exploitation. American usage (meter, kilometer) is used.

AAPG: American Association of Petroleum Geologists.

AGC: Automatic gain control: A system to automatically control the increase in amplitude of an electrical signal from the original input to the amplified output. AGC is commonly used in seismic processing to improve visibility of late-arriving events in which attenuation or wavefront divergence has caused amplitude decay.

API: American Petroleum Institute.

Aquifer: An underground formation of permeable rock or sediment that can hold water in pore space. Aquifers can be limited in aerial extent or very large, underlying thousands of square kilometers of the earth's surface. With relation to hydrate and related gas deposits, those marine sediments which can either host hydrate and related gas or can provide the hydrate forming gas by flow or diffusion through the pores. A water-bearing layer of rock or sediment capable of yielding water flow.

(Confined) Aquifer: An aquifer whose upper, and perhaps lower, boundary is defined by a layer of natural material that does not transmit water readily. Often a gas or petroleum reservoir.

(Unconfined) Aquifer: An aquifer which is not bounded by an impermeable layer in at least one, and possibly two directions. See Chapter 7. The terms open or communicating aquifer may also be used.

Atm, atm: Pressure measured in atmospheres (1 bar or 14.7 lb/in^2)

Attribute: A measurable property of seismic data, such as amplitude, dip, frequency, phase and polarity. Attribute analysis includes assessment of various reservoir parameters, including hydrocarbon indicators, using techniques such as amplitude variation with offset (AVO) analysis.

Authigenic: Rock or sediment constituents and minerals that have formed place and have not been transported form elsewhere. Applies specifically to minerals that have crystallized locally at the spot where they are now found. The term often refers to minerals formed during diagenesis following deposition of the original sediment.

AVO: Amplitude Variation with Offset is a seismic attribute for the variation in seismic reflection amplitude with change in distance between shotpoint and receiver. It indicates differences in lithology and fluid content in rocks above and below a reflector.

Barrel: One petroleum barrel = 159 liters or 42 U.S. Gallons.

Bcf: Billion cubic feet (of gas).

BGR: Base of Gas Reflector. This is sometimes a strong positive impedance contrast at the base of a gas-rich zone that may be trapped below hydrate. Where a free gas zone occupies pore space, there may be a sharp contact with subjacent water flooded sediments.

BHT: Bottom Hole Temperature.

Biogenic methane: Methane formed by microbial processes. With respect to oceanic hydrate, the source is deep biosphere acting upon buried organic material within the marine sediments.

Blanking: Blanking refers to a decreased reflection strength on reflection seismic records that display true reflection amplitude. It is often located in the lower third of the GHSZ. Blanking is attributed to: hydrate filling of pore spaces replacing water. This reduces the acoustic impedance contrasts found in the original sedimentary bedding. Some apparent blanking may have causes that are unrelated to gas hydrate accumulations.

Bm^3 Billion cubic meters

BGHS: Base of Gas Hydrate Stability. Also: **BOH**, Base of hydrate. This refers to the actual base of naturally occurring hydrate stability. Normally, this occurs more or less uniformly at some depth in the sediments depending on water depth, seafloor temperature, geothermal gradient and for any particular HFG or mixture of HFGs (which have different stability fields). In abnormal circumstances, for instance where there is a local high heatflow (see 3.2.1; 4.7.2; 4.8), the BGHS may be pushed up to the seafloor. In this instance, cooling of the HFG-enriched vent water by seafloor water has the potential to produce highly HFG-saturated water that will cause rapid growth of seafloor hydrate near the vent. The BGHS may be difficult to identify when there is no BSR, especially where marine sediment bedding is parallel with the seafloor.

BOP: Blow out preventer. Usually installed at the wellhead and part of a stack of BOPs. These are optimized to provide maximum pressure control in the event of a well control incident.

BSR: Bottom-Simulating Reflector. Seismic reflection from a negative impedance contrast at the contact between higher Vp hydrate-rich sediment zone and sediments containing gas in pores below which have much slower acoustic velocities. The BSR is commonly taken as is a remotely sensed reflection that marks the base of the hydrate stability zone and is used as a first-order indicator of the presence of gas hydrate.

BSR push-up: Occurs where the BGHSZ is pushed up toward the seafloor by, 1. An anomalous subjacent heatflow, such as over a salt plug, 2. In the vicinity of a usually near vertical fluid-gas flow passage through the HSZ where lateral heat flow results in sub-hydrate stability conditions extending laterally from the passage. Often associated with seafloor vents and seafloor hydrate. Push rather than pull is used because the effect is caused by warmer fluids moving up.

btu or (BTU): British thermal unit, A measure of heat energy required to raise the temperature of one pound of water by one degree Fahrenheit. 1 btu = 1055J, $2.973x1^{-4}$ kWh.

Compound hydrate: A species of gas hydrate with more than one type of HFG occupying guest sites within the same crystal lattice. For example, natural gas hydrate with methane and ethane present.

Condensate: Light hydrocarbons that condense when cooled to surface temperature. Also known as natural gas liquids.

Connate water: Water that has been entrapped during sedimentation along with the sediments. During compaction of marine sediments, connate water may be expelled into porous strata or secondary porosity from which it may migrate upward into the sea.

Darcy: A measurement of the ability of a fluid to pass through a porous material. The permeability that will allow a fluid of 1 centipoise (cP) viscosity to flow at a 1 cm/s velocity for a pressure drop of 1 atm/cm. The millidarcy (md) is the common unit for hydrocarbon reservoirs.

Depressurization: The gas hydrate gasification technique whereby pressure in the immediate vicinity of hydrate is lowered, causing ambient conditions of hydrate instability.

Diagenesis: The sum of all chemical and physical changes in minerals during and after their initial accumulation. Diagenesis involves addition and removal of material, transformation by dissolution and recrystallization or replacement, or both, and by phase changes. Authigenic refers to minerals formed in place. It is also applied to minerals that are clearly the result of new crystal growth on older crystals. Diagenesis embraces those processes such as compaction, cementation, authigenesis, replacement, crystallization, hydration, etc., that occur under conditions of pressure up to 1 kb and temperature in a range that encompasses gas hydrate stability conditions on Earth.

Diffusional pressure: The tendency for a material to be transported through a given medium. A strong tendency, or high diffusional pressure, leads to faster diffusion. A weak tendency, or low diffusional pressure, leads to slower diffusion.

Dissociation. The breakdown of the hydrate crystal structure or 'melting' of hydrate into its water and gas components owing to rising temperature and/or diminishing pressure (2.6).

Dissociation feedback: Gas hydrate contains within it a self-moderating or feedback system which act as a rate-controlling mechanism during dissociation. Whatever the cause of dissociation, heat is absorbed. This cools the system and may cause dissociation to slow or cease. Run-away dissociation feedback may take place when hydrate is brought to near-atmospheric pressures which usually will initiate dissociation; the system may absorb enough heat under this condition to freeze the local water. When a sample becomes coated with ice, the rate of dissociation slows (see Self-preservation).

Dissociation point: The temperature, at a constant pressure, or the pressure, at a common temperature, at which a compound or mineral breaks up reversibly to form two or more other substances. For instance, $CaCO_3$ becomes CaO plus CO_2 and gas hydrate becomes water and gas.

Dissolution: The process of dissolving into a homogeneous solution, as when a gas disperses into its component molecules in a fluid or other gas or where a hydrate structure becomes unstable owing to the diffusion of gas molecules from the hydrate into surrounding gas or fluid. Maximum concentrations of dissolved solids (TDS) are governed by the solubility for each species.

Dry gas: Gas with no condensate, generally almost pure thermogenic methane.

Effective porosity: That portion of the porosity that is available for storing recoverable gas or fluid.

Endothermic: Gas hydrate dissociation is endothermic. It absorbs heat when it dissociates about equal to the heat produced during crystallization.

Exothermic: Gas hydrate formation is exothermic. It produces heat when it forms.

Fraccing (Frac): Industrial term for the artificial fracturing of rock in an oil or gas reservoir for the purpose of increasing permeability. Fracturing can be caused in a number of ways in the vicinity of a drilled hole but hydraulic fracturing is most common. In this method, high pressure fluid is forced into the reservoir materials so that cracks propagate, often along bedding planes. Dispersed sand and may be introduced into the fractures to hold the fractures open when the pressure is reduced in order to recover the gas or oil. The term spelled as 'fracking' is also used, but not as commonly, and at the time of writing appears to be associated with environmental legislation.

Gas: One of the three common states of matter along with solid and liquid. As used in this book, the word gas usually refers to the hydrocarbon natural gases, principally methane. Where other gases such as CO_2 and SO_x are mentioned they are identified.

Gas drive: Energy within a fluid resulting from the expansion of gas where pressure on the fluid is lowered.

Gas hydrate: A solid crystalline material formed from cages of water molecules containing voids or guest sites that are occupied by gas molecules whose presence stabilizes the crystal structure through van der Waals bonding. In the case of oceanic and permafrost hydrates the gas molecules are mostly hydrocarbon gases, principally methane, although other gases such as hydrogen sulfide and carbon dioxide may also form natural gas hydrate.

Gas Hydrate Stability Zone (GHSZ): The region within permafrost and oceanic sediments and rocks in which natural gas hydrate is stable. The GHSZ in oceanic sediments extends downward from the seafloor to some pressure-depth at which temperature is too high for hydrate to be stable. In permafrost regions, the GHSZ begins at some distance below the surface of frozen ground where temperature is low enough and pressure is high enough for hydrate to be stable and extends downward to some pressure-depth at which temperature is too high for hydrate to be stable.

Geothermal gradient The rate in temperature increase per depth in the solid Earth. The gradient may vary depending on the thermal characteristics of geological materials.

Geothermal profile: A graphical representation of temperature and depth.

Groundwater: Generally, all water contained in the ground. In the case of hydrate, the principal mineralizing solution.

(Principal) Gas laws:

- **Boyle's Law:** Boyle's Law defines the relationship between pressure and volume. It states that at a constant temperature, the volume of a given mass of gas varies inversely with pressure, or $P_1V_1 = P_2V_2$.
- **Charles' Law:** Charles' Law concerns the relationship between temperature, volume and pressure. It states that 'if the pressure remains constant, the volume of a given amount of gas is directly proportional to temperature'. Consequently, if pressure is increased, temperature also increases.
- **Dalton's Law:** Dalton's Law concerns the composition of mixed gas. It states that 'the total pressure exerted by a mixture of gases is the sum of the pressures that would be exerted by each of the gases if it alone occupied the total volume'. The total pressure is the sum of the partial pressures of the gases present.

➢ **Graham's Law:** Graham's Law concerns the rate at which different gases diffuse in a fluid. It states that "the rate of diffusion of a gas, in relation to another, is inversely proportional to the square roots of the gas densities (or molecular weights).

➢ **Henry's Law:** Henry's Law relates to absorption of gas in liquids. It states that 'the amount of gas that will dissolve in a liquid at a given temperature is directly proportional to the partial pressure of that gas over the liquid'. At increased pressures, increased volumes of gas dissolves in liquid although the ratio of liquid to gas molecules may not remain constant. The amount of gas that will dissolve in a liquid increases as the temperature decreases.

➢ **Ideal gas law (PV=nRT):** Equation of state of an ideal gas and used as an approximation for the behavior of a real diffuse gas. Pressure is proportional to temperature and inversely proportional to volume for 1 mol of gas. P is the pressure, V is the volume, n is the number of moles of gas, T is the absolute temperature, and R is the Universal Gas Constant. Accurate to about 99% in the P – T range for natural gas hydrate and gas recovery scenarios but becomes less accurate at higher and lower temperatures.

➢ **Raoult's Law:** Raoult's law states that the vapor pressure of a solvent over a solution is equal to the mol fraction of the solvent in the solution. This law predicts the vapor pressure of water over dilute solutions containing HFG. The vapor pressure of a solvent decreases as more substances are dissolved in it. This is an opposite effect to increasing the pressure on a solvent, which raises the vapor pressure.

GOC: Gas – oil contact. A bounding surface in a reservoir above which predominantly gas occurs and below which predominantly oil occurs. The boundary may be transitional, with a mixed gas and oil zone.

HFG: Hydrate forming gas. In this book primarily methane or methane-based mixed hydrocarbon gas such as ethane, propane and butane. Other gases such as sulfur compounds and CO2 may rarely occur naturally. HFR, hydrate forming reactant, may also be used, especially in a chemical discussion.

HEZ: Hydrate economic zone. This refers to the entire thickness of sediments that contains solid hydrate and gas that is trapped is association with hydrate.

GHSZ: Gas Hydrate Stability Zone. This is the zone in which hydrates are expected to be stable based on calculations of pressure and temperature with depth in the seafloor. Also known as Hydrate Stability Zone (**HSZ**). It extends from the seafloor downward in the marine sediments to some depth determined by rising temperature. Hydrate in the GHSZ is most stable near the surface and progressively less stable downward.

Hydrate, gas hydrate, natural gas hydrate: Hydrate or gas hydrate (singular) is commonly used where one gas species is dominant, such as in naturally occurring natural gas hydrate where methane hydrate is the dominant species. Where more than one gas is present in the hydrate lattice and where more than one hydrate structure is present, hydrates (plural) is correct usage. Hydrate deposits, which may include more than one type of hydrate is correct, as is, 'deposits of hydrates', where more than one species of hydrate is commonly present.

HZ: Hydrate Zone. This is the zone within the HSZ in which hydrate is actually present. The upper surface of the HZ may extend to the surface or it may be depressed below a sulfate reduction, or other chemically-active zone. The base of the HZ may occur either higher or lower, or be imbricated and complex, depending on local chemical, structural, and thermal deviations from a simple theoretical model.

IAEA: International Atomic Energy Agency

IHO: International Hydrographic Organization

IMO: International Maritime Organization

Inhibition: The gas hydrate gasification technique whereby an antifreeze with respect to gas hydrate is introduced, causing ambient conditions of hydrate instability. The presence of the fluid (e.g. brines, methanol, etc.) changes the position of the phase boundary in pressure-temperature space and causes hydrate lying between the original and the inhibitor-determined position of the phase boundary to dissociate.

IOC: Intergovernmental Oceanographic Commission (UNESCO)

IUCN: International Union for the Conservation of Nature and Natural Resources - World Conservation Union

K: Notation for the Kelvin absolute temperature scale. Usually shown as K following the absolute temperature without a degree (°) symbol.

Kerogen: Chemically complex organic debris and accumulations. From the Greek *keros*, meaning wax. Source of most of the biogenic methane and other hydrocarbons.

LNG: Liquified natural gas. Natural gas, mainly methane and ethane, which has been liquefied at cryogenic temperatures. Principal means of transporting natural gas across oceans in LNG large tankers.

LPG: Liquefied Petroleum Gas; Gas mainly composed of propane and butane, which has been liquefied at low temperatures and moderate pressures.

Mallik: Test well site in the Mackenzie Delta of the Canadian Arctic Ocean (cover photo).

Mcf: Thousand cubic feet (of gas).

md: Millidarcy; a measure of permeability.

MEMS: Also Mems. Micro-electronic mechanical systems. General term for small, mechanically tough instruments that often include elements of nano-technology (very small). These include fiber-optic and other small, hardended sensors.

MCF: Thousand standard ft^3. M is Roman number for 1,000.

MMSCFD: Million standard ft^3/day.

Mineral: **(a)** A naturally occurring inorganic element or compound having a periodically repeating arrangement of atoms (or molecules) and a characteristic chemical composition, which results in distinctive physical properties. **(b)** An element or chemical compound that is crystalline and that has formed from naturally occurring materials as a result of geologic processes. Water-ice and gas hydrate are minerals.

Mineralization: A general term for the process or processes by which a mineral or minerals are introduced or impregnated into a rock or sediment that results in a potentially valuable deposit.

Mineralizer: A gas or fluid that dissolves, receives by fractionation, transports, and precipitates minerals. A mineralizer is typically aqueous, with various hyperfusible gases including CO_2 and CH_4.

Mineral assessment: (a) The process of appraisal of identified and undiscovered mineral resources within some specified region, and the product of that appraisal (Barton, 1995) (b) The estimation of mineral endowment, meaning the number of deposits or tonnage or volume that occurs in the region, given some minimum size of accumulation (deposit), minimum concentration (grade), and maximum depth of occurrence.

Mineral belt: An elongated region of mineralization; an area containing several mineral deposits.

Mineral deposit: A mass of naturally occurring mineral material which is either metal ores or nonmetallic minerals, usually of economic value, without regard to mode of origin.

Mineral deposit model: The systematically arranged information describing the essential attributes (properties) of a class of mineral deposits. The model may be empirical (descriptive), in which instance the various attributes are recognized as essential even though their relationships are unknown; or it may be theoretical (genetic), in which instance the attributes are interrelated through some fundamental concept.

Mineral economics: Study and application of the technical and administrative processes used in management, control, and finance connected with the discovery, development, exploitation, and marketing of minerals.

Mineral occurrence: Any economic mineral that occurs in any concentration or volume sufficient to justify further exploration.

Mixed hydrate: Two or more different pure hydrate species mixed together. For example, a mixture of SI methane hydrate and SII propane hydrate, each occupying different crystal lattices which are adjacent or intermingled.

mM: One thousandth of a Mol (Mole).

MMSCFD: Million standard cubic feet

NGL: Natural gas liquids (propane, butante, etc.). Components of natural gas that may naturally be liquid in field facilities or in gas-processing plants. Natural gas liquids include propane, butane, pentane, hexane and heptane, but not methane and ethane, which require refrigeration to be liquefied.

nmi: nautical mile (also naut mi, n mile, or NM). The nautical mile is defined to be the average distance on the Earth's surface represented by one minute of latitude, which, because the Earth is not a perfect sphere, varies from 59.7 to 60.3 nautical miles. In 1929 an international conference in Monaco defined the nautical mile to be exactly 1852 meters or 6076.115 49 feet, a distance known as the international nautical mile, although 6080 feet (1853.18 meters) established by the British, is still most commonly used.

OECD: Organization for Economic Co-operation and Development.

Ore: The naturally occurring material from which a mineral or minerals of economic value can be extracted profitably or to satisfy social or political objectives. The term is generally but not exclusively used to refer to metaliferous material.

Orebody, ore deposit: A continuous, well-defined mass of material of sufficient ore content to make extraction economically feasible. A mineral deposit that has been tested and is known to be of sufficient size, grade, and accessibility to be producible to yield a profit. Often synonymous with mineral deposit.

Ore-forming fluid: A gas or fluid that dissolves, receives by fractionation, transports and precipitates ore minerals. A mineralizer is typically aqueous, with various hyperfusible gases including CO_2 and CH_4.

OBS: Ocean Bottom Seismometer. This device senses and records pressure, shear, and boundary wave acoustic energy on 3-axes.

Paragenesis: A characteristic association or occurrence of minerals or mineral assemblages in ore deposits, connoting contemporaneous formation.

Pay zone: Productive interval. Zone of mineral enrichment above the cut-off grade.

Percolation: Water that moves through an aquifer by natural processes.

Phase boundary: The interface in temperature/pressure space where a phase change occurs. In the case of gas hydrate the phase change is from stability of solid gas hydrate to stability of gas + water.

Play: Hydrocarbon traps of a particular genetic type. Examples of oil plays are reef plays, growth fault plays, stratigraphic plays, etc. Gas hydrate and related gas deposits (hydrate system) are a type of unconventional gas play.

Production: The process of extracting and recovering natural gas. The solid, crystalline gas hydrate must first induced to dissociate. The resulting gas then can be treated as conventional natural gas and recovered to the surface for use. An unconventional production method that is possible is to dissolve the hydrate and produce gas from the water through natural depressurization.

Prospect: A location where a well could be drilled to locate an oil or gas deposit.

Pure hydrate: A single species of gas hydrate with (essentially) a single type of HFG occupying guest sites within contiguous crystal lattices. For example, methane hydrate.

Resedimentation: Sedimentation of material derived from a pre-existing sedimentary material, usually by mechanical processes driven by pressure or gravity. Sediment collapse may be caused by hydrate dissociation and sediment mass movements may be caused by the presence of hydrate and related gas deposits.

Reserve: An estimate within specified accuracy limits of the amount or value of a naturally occurring deposit that could be extracted or produced with existing technology under the economic conditions at the time of the determination. Reserves are the recoverable fraction of the resource base.

Reservoir: Any porous and permeable rock that yields oil or gas. Hydrate is unique among gas deposits in forming part of the solid fraction, which may decrease porosity and permeability to nearly zero.

Resource: A concentration of naturally occurring materials from which a commodity is potentially economically recoverable. Resources are speculative, unproven, or hypothetical until proven by exploration methods.

Ryukyu Trough: Also spelled Ryokyo. Subduction zone margin of the eastern Eurasian plate that passes from north of Taiwan to offshore SE Japan. Synonymous in its northern end with the Nankai Trough.

Saturation: The state of equilibrium in which no further gas is taken up in solution; the vapor pressure of gas is about equal to the vapor pressure of the solute.

Self-preservation effect: When dissociating hydrate becomes coated with water ice that forms because of heat absorption of the endothermic reaction, the rate of dissociation slows (2.6.4).

Seismic reflection profiling: This is the standard method for carrying out reflection seismic profiles from ships at sea. A source of energy, such as an airgun, watergun, transducer (boomer), or a sparker, is used to provide for regular, short bursts of acoustic energy. This pressure wave energy is detected by a horizontal array. Reflections from the seabottom and from reflectors within the seafloor are 'stacked' and a vertical cross-section of the impedance structure of the subbottom is assembled.

Show: Any indication of oil/gas hydrate mineralization during drilling or testing.

Sour gas: Natural gas containing hydrogen sulfide.

Source, Source rock: Sedimentary rock or sediments in which organic material under pressure, heat, and time may be transformed to liquid or gaseous hydrocarbons.

SPAR Platform: A stationary floating device. Not an acronymn. Derives from nautical name for any rounded, elongate buoy or rigging spar floating with the long axis vertical owing to weighting at the lower end. Used for buoys and lights in order to have the top of the SPAR buoy standing well above the sea surface. Developed as a deepwater platform as a generally circular-sectioned, very large semi-submersible that extends down for a considerable distance into the water and also stands will off the water surface. Platforms contain gas/oil/water processing and storage zones and sometimes offload directly to tankers after processing. Where a SPAR platform is teathered to the seafloor with tensioned moorings, it is referred to as a 'tension leg platform'.

SPE: Society of Petroleum Engineers.

STATOIL: Semi-state oil company of Norway.

STP: Standard temperature and pressure, about 20 °C and 1 atmosphere (14.7 lbs/in^2 or 1 bar).

Superdemand: An economic condition of energy demand where the United States no longer controls the price of energy. Caused because of rising demand elsewhere in the world, principally in China and India.

TCF: Trillion cubic feet (of gas).

TD: Total depth.

TGR: Top of Gas Reflector. This is a strong negative impedance contrast that marks the top of a gas-rich zone. Where this occurs immediately below the HSZ, it may form the BSR. It may occur within an HSZ where gas has expelled pore fluids and has precluded further formation of hydrate. Neither the top nor the base of gas reflectors necessarily mark a thermodynamic phase boundary like that defined by the BSR.

Thermal gradient: Temperature – depth plot of temperature in the Earth or ocean water. In open oceans in summer, temperature decreases from a maximum at the surface to a minimum at the seafloor. In the winter the surface layers may be cold; temperature – density stratification may also occur.

Thermogenic methane: Methane formed from organic matter by heating (Chapters 3 & 4).

Thermal stimulation: Term for artificial dissociation process where warm fluids are introduced in a hydrate section and the rise in temperature and the availability of heat allows the hydrate to 'melt'.

Total pressure: Hydrostatic plus lithostatic pressure.

Transmissivity: A measure of the ability of an aquifer to transmit water

Trap: Any barrier to the upward or lateral movement of oil or gas, which cause them to accumulate. A trap occurs structurally, hydrologically, or geomorphically above a reservoir rock in which the accumulation can take place.

Unconventional gas: Natural gas deposits that are fundamentally different from conventional deposits in which gas in a geological trap flows at economic rates to the surface under its own pressure. In a conventional gas deposit, a discrete gas zone occurs above water or oil, which are also in the trap, and pressurize the gas zone. May also be applied to a conventional gas deposit in which conventional (normal industry practices) secondary recovery techniques are used. Unconventional gas deposits presently are divided into; coalbed methane, tight gas sands, shale gas, and gas hydrate.

UNDP: United Nations Development Programme

UNEP: United Nations Environment Programme

Vertical Array: A vertical array of hydrophones used in conjunction with stepped-out sources to acquire an acoustic dataset used primarily for determining physical properties of the uppermost seafloor structure and for acoustic propagation modeling.

Vp: Notation for pressure wave velocity.

Vs: Notation for shear wave velocity.

Void ratio: Ratio of voids to solid rock. Can also be expressed as a percentage.

VSP: A seismic record that is collected by firing a seismic source, commonly an airgun near the sea surface while pulling a hydrophone upward through a nearby borehole to record the direct arrival of seismic pulses. This provides excellent sediment velocity data.

Wet gas: Gas with condensate fraction, usually ethane, propane, and butane.

Wipe-Out: The common term for obliteration of seismic structure in a reflection seismic profile by free gas in sediment pore space owing to compressional wave (Vp) attenuation.

UNESCO: United Nations Educational, Scientific and Cultural Organization

Selected References

This reference list is an update of the reference list in Max (Ed.) 2000 and 2003. It is not exhaustive and does not include all references in the gas hydrate literature since that time. A number of references from non-U.S.-based scientists are included for completeness, although they may not be refereced specifically in the text. This list is focused on economic applications for gas hydrate, particularly in the field of hydrate as a diagenetic mineral constituent of marine sediments and as a potential energy resource. There are more references than are referred to in text. Also see: [*http://walrus.wr.usgs.gov/globalhydrate/References.pdf*] [*http://www.nap.edu/openbook/0309092922/html/105.html#pagetop*] [*http://walrus.wr.usgs.gov/globalhydrate/*].

ABS. 2004. American Bureau of Shipping, Two if by Sea, http://www.eagle. org/prodserv/EnergyNews /meche213336.pdf

Adler, P.M. & Thovert, J.-F. 1999. *Fractures and fracture networks. Theory andapplications of transport in porous media.* Kluwer Acacemic Publishers, Dordrecht.

Agee, M.A., Weick, L.J. & Agee, K.L. 1999. System and method for hydrate recovery. U.S. Patent 5,950,732 (September 14, 1999), 19pp.

AGI. 1997. *Dictionary of Mining, Mineral, and Related Terms.* American Geological Institute w/ the Society for Mining, Metallurgy, and Exploration, Inc., 646pp.

AP. 2004. Associated Press. Mississippi Chemical announces closure of Louisiana plants, March 25, 2004.

Akhmanov G.G. & Woodside J.M. 1998. Mud Volcanic Samples in the Content of the Mediterranean Ridge Mud Diapiric Belt. In: Robertson, A.H.F., Emeis, K.-C., Richter, C., & Camerlenghi, A. (Eds.), Proceedings ODP, Scientific Results 160, College Station, TX (Ocean Drilling Program), 597-605.

Andreassen, K., Mienert, J., Bryn, P. & Singh, S.C. 2000. A double gas-hydrate related bottom simulating reflector at the Norwegian continental margin, Annals of the N.Y. Academy of Science 912, 126-135.

Anonymous. 1964. Searchers "Sniffing" for North Sea gas. Oil and Gas 62, 93.

Aoki, Y., Shimizu, S., Yamane, T., Tanaka, T., Nakayama, K., Hayashi, T. & Okuda, Y. 2000. Methane hydrate accumulation along the western Nankai Trough. Annals of the New York Acamey of Sciences 912 (ISSN: 0077-8923), 136-145.

Askari, F. & Max, M.D. 1998. Executive Summary: GiS applications for Rapid Environmental Assessment. In: Pouliquen, E., Kirwan, A.D. Jr. & Pearson, R.T. (Eds.), Rapid Environmental Assessment, SACLANTCEN Conference Proceedings CP-44, March 10-13 1997, Lerici (La Spezia), Italy, xiv-xvii.

Austing, J.A.Jr. 2004. The Integrated Ocean drilling Program: Utilizing new drilling platforms for ocean research. Marine Technological Society Journal 38(2), 44-54.

Auzende, J-M., Dickens, G.R., van de Buequue, S., Exon, N.F., François, C., Lafoy, Y. & Voutay, O., 2000. Thinned crust in southwest Pacific may harbor gas hydrate. EOS, Transactions of American Geophysical Union, 81(17), 182-185.

Auzende, J.M., van de Beuque, S., Dickens, G., François, C., Lafoy, Y., Voutay, O. & Exon, N. 2000. Deep sea diapirs and bottom simulating reflector in Fairway Basin (SW Pacific). Marine Geophysical Researches. 21(6), 579-587.

Aya, I., Yamane K. & Nariai H. 1997. Solubility of CO_2 and density of CO_2 hydrate at 30 MPa. Energy 22, 263–271.

Baba, K. & Yamada, Y. 2004. BSRs and associated reflections as an indicator of gas hydrate and free gas accumulation: An example of accretionary prism and forearc basin system along the Nankai Trough, off central Japan, Resource Geology 54(1), 11-24.

Baker, K., McLing, T., Street, L., Schafer, A., Ansley, S., Hull, L & Holt. R. 2005. A research park for studying processes in unsaturated fractured media. EOS 82(2), 15, 20.

Bangs, N.L.B., Musgrave, R.J. & Tréhu, A.M. 2005. Upward shifts in the southern Hydrate Ridge gas hydrate stability zone following postglacial warming, offshore Oregon. Journal of Geophysical Research 110, B03102,doi:10.1029/2004/JB003293,2005.

Barnes, R.O. & Goldberg, E.D. 1976. Methane production and consumption in anoxic marine sediments. Geology 4, 297-300.

Ben-Avraham, Z, Smith, G., Reshef, M. & Jungslager, E. 2002. Gas hydrate and mud volcanoes on the southwest African continental margin off South Africa. Geology 30, 927-930.

Bleeker, W. 2002. Archaean tectonics – a review, with illustrations from the Slave craton. Geological Society London Special Publication 199, 151-181.

Berndt, C., Mienert, J., Vanneste, M. & Bünz, S. 2004. Gas hydrate dissociation and seafloor collapse in the wake of the Storegga Slide, Norway. In: Wandås, B.T.G., Eide, E., Gradstein, F. and Nystuen, J.P. (eds): Onshore-offshore relationships on the North Atlantic Margin. Norwegian Petroleum Society (NPF) Special Publication, Elsevier, Amserdam, 285-292.

Berndt, C., Bünz, S., Clayton, T., Mienert, J. & Saunders, M. 2004. Seismic character of bottom simulating reflectors: Examples from the mid-Norwegian margin. Marine and Petroleum Geology 21(6), 723-733.

Berndt, C., Bünz, S. & Mienert, J. 2003. Polygonal fault systems on the mid-Norwegian margin: A long term source for fluid flow. In: van Rensbergen, P., Hillis, R.R., Maltman, A.J. & Morley, C. (Eds.) Subsurface Sediment Mobilization. Geological Society of London Special Publication 216, 283-290.

Bil, K. 2003. Economic Perspective of Methane from Hydrate. In: Max, M.D. 2003. (Ed.), Natural Gas Hydrate in Oceanic and Permafrost Environments. Kluwer Academic Publishers, The Netherlands, 349-360.

Bishnoi, P.R. & Natarajan, V. 2003. Formation and Decomposition of Gas Hydrate. Fluid Phase Equlibria 117, 168-177.

Broadbent, S.R. & Hammersley, J.M. 1957. Percolation processes, I. Crystals and mazes. Proceedings Cambridge Philisophical Society 53, 629-645.

Bünz, S., Mienert, J. & Berndt, C. 2003. Geological controls on the Storegga gas-hydrate system of the mid-Norwegian continental margin. Earth and Planetary Science Letters 209 (3-4), 291-307.

Boetius, A.K., Ravenschlag, C.J., Schubert, et al. 2001. A marine microbial consortium apparently mediating anaerobic oxidation of methane. Nature 407, 623-626

Bolton, A. & Maltman, A. 1998. Fluid-flow pathways in actively deforming sediments: the role of pore fluid pressures and volume change. Marine and Petroleum Geology 13, 281-297.

Booth, J.S. & O'Leary, D.W. 1991. A statistical overview of mass movement characteristics on the North American Atlantic outer continental margin. Marine Geotechnology 10, 1-18.

BP. 2004. Statistical Review of World Energy. <http://www.BP.com /subsection.do?categoryId=95& contentId=2006480>

Bohrmann, G,., Greinert, J., Suess, E. & Torres, M. 1998. Authigenic carbonates from the Cascadia subduction zone and their relation to gas hydrate stability. Geology 26, 647-650.

Bohrmann G, Ivanov M, Foucher JP, Spiess V, Bialas J, Weinrebe W, Abegg F, Aloisi G, Artemov Y, Blinova V, Drews M, Greinert J, Heidersdorf F, Krastel S, Krabbenhöft A, Polikarpov I, Saburova M, Schmale O,Seifert R, Volkonskaya A, Zillmer, M. 2003. Mud volcanoes and gas hydrates in the Black Sea - new data from Dvurechenskii and Odessa mud volcanoes. Geo-Marine Letters 23(3-4), 239-249.

Bohrmann G, Tréhu A, Rack F, Torres M & ODP Leg 204 Shipboard Scientific Party. 2003. Drilling gas hydrates on Hydrate Ridge, Cascadia Continental Margin. Energy, exploration and exploitation 21(4), 333-334.

Borowski, W.S. 2004. A review of methane and gas hydrates in the dynamic, stratified system of the Blake Ridge region, offshore southeastern North America. Chemical Geology 205(3-4), 311-346.

Borowski, W.S., Paull, C.K. & Ussler, W. III. 1996. Marine pore water sulfate profiles indicate in situ methane flux from underlying gas hydrate. Geology 24, 655-658.

Borowski, W.S., Hoehler, T.M., Alperin, M.J., Rodriguez, N.M. & Paul, C.K. 1997. Significance of anaerobic methane oxidation in methane-rich sediments overlying the Blake Ridge gas hydrates. Proceedings Ocean Drilling Program Scientific Results 164, 87-99.

Borowski, W. S., Paull, C.K. & Ussler, W. III. 1999. Global and local variation of interstitial sulfate gradients in the deep-water, continental margin sediments. Sensitivity to underlying methane and gas hydrates. Marine Geology 159, 131-154.

Bouriak S., Vanneste M. & Saoutkine A. 2000. Inferred gas hydrates and clay diapirs near the Storegga Slide on the southern edge of the Voring Plateau, offshore Norway. Marine Geology, 163. 125-148.

Bouriak, S.V. & Akhmetjanov, A.M. 1998. Origin of gas hydrate accumulations on the continental slope of the Crimea from geophysical studies. In: Henriet, J.P. & Mienert, J. (Eds.), Gas Hydrates: Relevance to World Margin Stability and Climatic Change. Geological Society London Special Publlication 137, 215-222.

Brekke, H., Dahlgren, S., Nyland, B. & Magnus, C. 1999. The prospectivity of the Vøring and Møre basins on the Norwegian Sea continental margin. In: A.J. & Boldy, S.A.R. (Eds.), Petroleum Geology NW Europe Procedings 5th Conference. Special Publication, Geological Society of London, 261–274.

Brewer, P.G., Friederich, G.E., Peltzer, E.T. & Orr, F.M. Jr. 1999. Direct experiments on the ocean disposal of fossil fuel CO_2. Science 284, 943–945.

Brewer, P.G., Paull, C.K., Peltzer, E.T., Ussler, W. III, Rehder, G. & Friederich, G.E. 2002a. Measurements of the fate of gas hydrates during transit through the ocean water column. Geophysical Research Letters, 29, doi10.1029/2002GL014727.

Brewer, P.G., Peltzer, E.T., Friederich, G., Aya, G.I., & Yamane, K. 2000. Experiments on the ocean sequestration of fossil fuel CO_2: pH measurements and hydrate formation. Marine Chemistry 72, 83-93.

Brewer, P.G., Peltzer, E.T., Friederich, G.E. & G. Rehder, G. 2002b. Experimental determination of the fate of rising CO_2 droplets in seawater. Environmental Science & Technology, 36: 5,441–5,446.

Briaud, J.L. & Chaouch, A. 1997. Hydrate melting in soil around hot conductor. Journal Geotechnical and Geoenvironmental Engineering 123, 645-653.

Broek, D. 1984. *Elementary engineering fracture mechanics*. Martinus Nijhoff, Boston MA, 469pp.

Brooks, J.M., Bryant, W.R., Bernard, B.B. & Cameron, N.R. 1999. The nature of gas hydrates on the Nigerian continental slope. In: Annals of the New York Adcademy of Sciences, Third International Conference on Gas Hydrates, July 18-22, Park City, Utah,

Brown, K.M., Bangs, N.L., Froelich, P.N. & Kvenvolden, K.A. 1996. The nature, distribution, and origin of gas hdyrate in the Chile tripple junction region. Earth Planetary Science Letters 139, 471-483.

Buffet, B. & Archer, D. 2004. Global inventory of methane clathrate: sensitivity to changes in the deep ocean. Earth and Planetary Sciences Letters 222, 185-199.

Bünz, S. Mienert, J. & Berndt, C. 2003. Geological controls on the Storegga gas-hydrate system of the mid-Norwegian continental margin. Earth and Planetary Science Letters 209, 291-307.

Cameron, N. & Ziegler, T. 1997. Probing the lower limits of a fairway; Further pro Permian potential in the southern North Sea. In; Ziegler, K, Turner, P. & Daines, S.R. (Ed.), Petroleum geology of the southern North Sea: future potential. Geological Society Special Publication 123, 123-141.

Carbó, A., Muñoz, A, Druet, M., Llanes, P. & Álvarez, J. 2004. New gravity map of the western Galacia Margin: The Spanish Exclusive Economic Zone Project. EOS 85(52), 565, 568.

Campbell, C.J. 1997. *The Coming Oil Crisis.* Multi-Science Publishing Co. Ltd, Brentwood, Essex, England, 210pp.

CGA. 2004. Canadian Gas Association. Natural Gas Statistics. <http://www.cga. ca/publications/gasstats.htm>

Chapman, L.H. 2004. Geology and Mineralization Styles of the George Fisher Zn-Pb-Ag Deposit, Mount Isa, Australia. Economic Geology 99, 233-255.

Chapman, R, Gettrust, J, Walia, R., Wood, W., Hannay, D., Spence, G., & Hyndman, R. 2002. High resolution deep-towed multichannel seismic survey of deep sea gas hydrates off western Canada, Geophysics 67, 1038-1047.

Chapman, R., Pohlman, J., Coffin, R., Chanton, J. & Lapham, L. 2004. Thermogenic gas hydrates in the northern Cascadia margin. EOS 85(38), 361, 365.

Chatterji, J. & Griffith, J.E. 1998. Methods of decomposing gas hydrates. U.S. Patent 5,713,416 (February 3, 19998), 6pp.

Chen, D.F. & Cathles, L.M. III. 2003. A kinetic model for the pattern and amounts of hydrate precipitated from a gas stream: Application to the Bush Hill vent site, Green Canyon Block 185, Gulf of Mexico. Journal of Geophysical Research 108, NO. B1, 2058, doi:101029/2001JB001597.

Chen, F. & Yapa, P.D, 2001. Estimating hydrate formation and decomposition of gases released in a deepwater ocean plume, Journal of Marine Systems 30, 21-32.

Chen, Z., Osadetz, K.G., Gao, H., Hannigan, P.K. & Watson, J.C. 2001. Improving exploration success through uncertainty mapping, Keg River Reef play, Western Canada Sedimentary Basin. Bulletin of Canadian Petroleum Geology 49(3), 367-375.

Cherkis, N.Z., Max, M.D., Vogt, P.R., Crane, K., Midthassel, A. & Sundvor, E. 1999. Large-scale mass wasting on the north Spitsbergen continental margin, Arctic Ocean in: Gardiner, J., Vogt, P. & Crane, K. (Eds). Mass wasting in the Arctic. Geomarine Letters Special Issue 19, 131-142.

Chow, R., Blindt, R., Kamp, A., Grocutt. P. & Chivers, R. 2004. The microscopic visualisation of the sonocrystallisation of ice using a novel ultrasonic cold stage. Ultrasound Sonochemistry 11(3-4), 245-50.

Christensen, T.S. 1994. Improved production using autothermal reforming. Hydrocarbon Processing, March, 39-46.

Chuvilin E.M., Ebinuma T., Kamata Y., Uchida T., Takeya S., Nagao J., Narita H. 2003. Effects of temperature cycling on the phase transition of water in gas-saturated sediments. Canadian Journal of Physics 81, 1-8.

Chuvilin, E.M., Ebinuma, T., Kamata, Y., Uchida, T., Takeya, S., Nagao, J. & Narita H. 2002. An Experimental Study of Gas Hydrate and Ice Accumulation in Sediments. Proceedings of the Twelfth international offshore and polar engineering conference. Kitakyushu, Japan, 429-433.

Chuvilin, E.M. & Kozlova, E.V. 2004. Experimental estimation of gas hydrates accumulation and existence conditions in sediments, In: Collett, T. & Johnson, A. (Eds.), 2004. Proceedings of Natural Gas Hydrates: Energy Resource Potential and associated Geologic Hazards. A.A.P.G. Hedberg Research Conference, September 12-16, 2004, Vancouver, BC, Canada, 3pp.

Chuvilin, E.M., Kozlova, E.V., Makhonina, N.A., Yakushev, V.S. & Dubinyak D.V. 2002. Peculiarities of methane hydrate formation/dissociation P/T conditions in sediments of different composition. Proceeding of the Fourth International Conference on Gas Hydrate. Yokohama, 433-438.

Chuvilin, E.M., Makhonina, N.A., Titenskaya, O.A. & Boldina, O.M. 2002. Petrophysical investigations on frozen sediments artificially saturated by hydrate. Proceeding of the Fourth International Conference on Gas Hydrate. Yokohama, 734-739.

Chuvilin, E.M. & Perlova E.V. 1999. Forms and formation conditions of gas component of frozen sediments (in Russian). Vestnik of Moscow University, ser. 4 Geology, N5, 57-59.

Chuvilin, E.M., Perlova, E.V. & Dubinyak, D.V. 2001. Experimental modeling of a gas hydrates existence conditions within marine sediments of the Yamal peninsula (*in Russian*) Proceeding of the 2nd Russian conference on Geocryology 1, 169-173.

Chuvilin, E.M., Perlova, E.V., Makhonina, N.A. & Yakushev V.S. 2000. Research of hydrate and ice formation in soils during cyclic fluctuations of temperature. In: Thimus, J.-F. (Ed), Ground Freezing. Balkema, Rotterdam, 9-14.

Chuvilin, E.M., Yakushev, V.S. & Perlova, E.V. 1998 (published 2000). Gas and possible gas hydrate in the permafrost of Bovanenkovo gas field, Yamal peninsula, West Siberia. Polarforschung 68, 215-219.

Chuvilin, E.M., Yakushev, V.S. & Perlova, E.V. 2002. Natural gas and gas hydrate association in permafrost of Yamal peninsula (West Siberia). Proceeding of the Fourth International Conference on Gas Hydrate. Yokohama, 216-221.

Chuvilin, E.M., Yakushev, V.S., Perlova, E.V. & Kondakov, V.V. 1999. Gas component of permafrost within Bovanenkovo gas field (Yamal peninsula) (*in Russian*). Doklady Akademii Nauk SSSR 369 (N4), 522-524.

Chuvilin, E.M., Perlova, E.V., Makhonina, N.A., Kozlova, E.V. & Yakushev V.S. 2001. Gas hydrate formation and dissociation conditions in clay sediments of different composition. Proceedings of the 7th International Symposium on Thermal Engineering and Sciences for Cold Regions, Seoul, Korea, 231-235.

Chuvilin, E.M., Perlova, E.V., Makhomina, N.A. & Yakushev, V.S. 2002. Phase transitions of water in gas-saturated sediments. In: Proceedings of the VI International Conference on "Gas in Marine Sediments". Special Issue, Geologiya I Geofizika 43(7), Siberian Branch, Russian Academy of Sciences, 689-697.

Circone, S., Stern, L.A. & Kirby, S.H. 2004. The Effect of Elevated Methane Pressure on Methane Hydrate Dissociation, American Mineralogist 89, 1192-1201.

Circone, S., Stern, L.A. & Kirby, S.H. 2004. The role of water in gas hydrate dissociation. Journal Physical Chemistry B 108, 5747-5755.

Clarke, M. & Bishnoi, P.R. 2001. Determination of Activation Energy and Intrinsic Rate Constant of Methane Gas Hydrate Decomposition, Canadian Journal of Chemical Engineering 79(1), 143-147.

Clarke, R.A. 2004. *Against All Enemies: Inside America's War on Terror*. Simon and Schuster, New York, 320pp.

Clennell, M.B. 2000. Submarine gas hydrates: nature, occurrence & perspectives for exploration in the brazilian continental margin. ISSN 0102-261X. Review Brasil Geofisica 18(3), 397-409.

Clift, P.D. & Molnar, P. Drilling of submarine fans in the Indian Ocean. Eos 84(42), 442-443.

Cloos, M. 1982. Flow melanges: Numerical modeling and geologic constraints on their origin in the Franciscan subduction complex, California. Geological Society of America 93, 330-345.

Coffin, R.B., Grabowski, K.S. & Chanton, J.P. 2003. The role of methane hydrate in ocean carbon chemistry and biogechemical cycling. In: Max, M.D. 2003. (Ed.), Natural Gas Hydrate in Oceanic and Permafrost Environments. Kluwer Academic Publishers, The Netherlands, 77-90.

Coffin, R.B., Pohlman, J., Grabowski, K., Knies, D. & Downer, R. 2005. Keathley Canyon, August 2003 Gulf of Mexico, Piston Coring Tested for Prediction of Hydrate Exploration. NRL/MR/6110--05-? (as of 16 Mar 05).

Collett, T.S. 2001. Quantitative well-log analysis of in-situ gas-hydrate-bearing sediments, In: Geophysical Monograph 124, Paull, C.K & Dillon, W.P. (Eds.), American Geophysical Union, 189-210.

Collett, T.S. 2003. Natural Gas Hydrate as a Potential Energy Resource. In: Max, M.D. (Ed.), *Natural Gas Hydrate in Oceanic and Permafrost Environments*. Kluwer Academic Publishers, Dordrecht, 123-136.

Collett, T.S., 2004, Testimony before the House Committee on Resources, Subcommittee on Energy and Mineral Resources, Washington, DC, July 15, 2004.

Collett, T.S. 2004. Gas Hydrates as a Future Energy Resource, Geotimes, November, 2004, 24-27.

Collett, T.S., Bird, K.J., Burruss, R.C. & Lee, M.W. 1992. Gas hydrate reservoir characteristics and economics. U.S. DOE Report DOE/MC/20422-92/C0027 (DE92014351)), 13pp.

Collett, T.S. & Dallimore, S.R. 2002. Integrated well Log and Reflection Seismic Analysis of Gas Hydrate Accumulations on Richards Island in eth Mackenzie Delta, N.W.T., Canada, CSEG Recorder, October, 2002, 25-40.

Collett, T.S., Dallimore, S.R., Lewis, R.E., Lee, M.W. Ochiai, K. & Sawamura, F. 2003. Downhole log evaluation of gas hydrate occurrences in the Mallik 5L-38 gas hydrate production research well. In: Program Abstracts, From Mallik to the Future. Mallik International Symposium in Makuhari, Japan. December 8-10, 4-5.

Collett, T.S., Goldberg, D.S., Aleksandra J. & Guerin, G. 2004. Occurrence of Gas Hydrate and Associated Free-Gas Accumulations on Hydrate Ridge, Offshore Oregon, USA (Abs.) American Association of Petroleum Geologists Annual Meeting.

Conti, S. & Fontana, D. 2002. Sediment instability related to fluid venting in Miocene authigenic carbonated deposits of the northern Appennines (Italy). International Journal of Earth Sciences 91(6), 1030-1040.

Cronin, B.T., Ivanov, M.K., Limonov, A.F., Egorov, A.V., Akhmanov, G.G., Akhmetjanov, A.M., & Kozlova, E.V. 1997. New discoveries of mud volcanoes on the Eastern Mediterranean Ridge. Journal Geological Society London 154, 173-182.

Curt, B. 2004. Marine transportation of LNG, Intertanko Conference, http://www.intertanko.com/pubup load/curt.pdf.

Curray, J. R. 1989. The Sunda Arc: a model for oblique plate convergence. Netherlands Journal of Sea Research. 24, 131-140.

Curry, C.A. Wang, K., Hyndman, R.D. & He, J. 2004. Ther thermal effects of slab-driven mantle flow above a subducting plate: The Cascadia subduction zone and backarc. (doi:10.1026/j.epsl.2004.04.020) Earth and Planetary Science Letters 223, 35-48.

Dahlen, F.A., Suppe, J. & Davis, D. 1984. Mechanics of fold-and-thrust belts and accretionary wedges: Cohesive Coulomb theory. Journal of Geophysical Research 89, 10,087-10,101.

Dallimore S., Chuvilin E.M., Yakushev V.S., Grechischev S. E. Ponomarev V., Pavlov A. 1996. Field and laboratory characterization of intrapermafrost gas hydrates Mackenzie Delta, NWT, Canada. 2th International Conference on Natural Gas Hydrates, Tolouse France, 525-531.

Dallimore, S.R., Uchida, T. & Collett, T.S. (Eds.) 1999. Scientific Results from JAPEX/JNOC/GSC Mallik 2L-38 Gas Hydrate Research Well, Mackenzie Delta, Northwest Territories, Canada. Geological Survey of Canada Bulletin 544.

Darcy, H. 1856. Les fontaines publiques de la Ville de Dijon. V. Dalmont, Paris.

Das, S.K. 2004. Indo-Russian collaboration in gas hydrate research. In: Report on the Regional Workshop, CODATA task group on natural gas hydrates, Indian National Science Academy, New Delhi, http://www.codata.org/taskgroups/TGhydrates/NewDelhiFeb04.pdf, 9-10.

Davie, M.K. & Buffett, B.A. 2001. A numerical model for the formation of gas hydrate below the seafloor. Journal Geophysical Research 106, 497–514.

Davie, M.K. & Buffett, B.A. 2003. A steady state model for marine hydrate formation: Constraints on methane supply from pore water sulfate profiles. Journal Geophysical Research 108(B10), 2495, doi:10.1029/2002 JB002300.

Davie, M.K. & Buffett, B.A. 2003. Sources of methane for marine gas hydrate: inferences from a comparison of observations and numerical models. Earth and Planetary Science Letters 206, 51 63.

Davie, M.K., Zatsepina, O.Y. & Buffett, B.A. 2004. Methane solubility in marine hydrate environments. Marine Geology 203(1-2), 177-184.

Davis, D., Suppe, J. & Dahlen, F.A. 1983. Mechanics of fold-and-thrust belts and accretionary wedges, Journal of Geophysical Research 88, 1153-1172.

Davis, E.E., Hyndman, R.D. & Villinger, H. 1990. Rates of fluid expulsion across the northern Cascadia accretionary prism: Constraints from new heat flow and multichannel seismic reflection data, Journal of Geophysical Research 95, 8869-8889.

De Capitani, L. & Cita, M.B. 1996. The "marker-bed" of the Mediterranean Ridge diapiric belt; geochemical characteristics. Marine Geology 132, 215-225.

Deffeyes, K.S. 2003. *Hubbert's Peak: The Impending World Oil Shortage.* Princeton University Press, 224pp.

D'Hondt, S. Jergensen, B.B., Miller, D.J., Batzke, A., Blake, R., Cragg, B.A., Cyplonka, H., Dickens, G.R., Ferdelman, T., Hinrichs, J.-U., Holm, N.G., Mitterer, R. Spivack, A., Wang, G., Berkins, B., Engelen, B., Ford, D., Gettemy, G., Rutherford, S.D., Sass, H., Skilbeck, C.G., Aiello, I.W., Guerin, G., House, C.H., Inagalki, F., Meister, P., Naeher, T., Niitsuma, S., Parkes, R.J., Schippers, A., Smith, D.C., Teske, A., Wiegel, J. Padilla, C.N. & Acosta, J.L.S. 2004. Distributions of microbial activities in deep subseafloor sediments. Science 306, 2216-2221.

Delong, E.F. 2004. Microbial life breathes deep. Science 306, 2198-2200.

Devarakonda, S., Evans, J.M.B. & Myerson, A.S. 2003. Impact of Ultrasonic Energy on the Crystallization of Dextrose Monohydrate. Crystal Growth and Design 3(5), 741-746.

Dholabhai, P.D., Englezos, P., Kalogerakis, N. & Bishnoi, P.R. 1991. Equilibrium conditions for methane hydrate formation in aqueous mixed electrolyte solutions, Canadian Journal of Chemical Engineering 69, 800-805.

Dickens, G.R. 2001. The potential volume of oceanic methane hydrates with variable external conditions. Organic Geochemistry 32, 1179-1193.

Dickens, G.R. 2001. Sulfate profiles and barium fronts in sediment on the Blake Ridge: Present and past methane fluxes through a large gas hydrate reservoir, Geochimica et Cosmochimica Acta 65(4), 529-543.

Dickens, G.R. 2001. Modeling the global carbon cycle with a gas hydrate capacitor: Significance for the latest Paleocene thermal maximum. In: Natural Gas Hydrates: Occurrence, Distribution, and Detection, Paull. C.K & Dillon, W.P. (Eds.), AGU. Geophysical Monograph 124, Washington, DC, 19-38.

Dickens, G., Exon, N.F., Holdway, D., Lafoy, Y., Auzende, J., Dunbar, G., Summons, R. 2001. Quaternary sediment cores from the Southern Fairway Basin on the northern Lord Howe Rise (Tasman Sea). GA - Geoscience Australia, Record, 2001/31 ISBN: 0642467110, CD-ROM. 0642467110

Dickens, G.R., Paull, C.K., Wallace, P. & ODP Leg 164 Scientific Party. 1997. Direct measurement of in situ methane quantities in a large gas-hydrate reservoir. Nature 385, 426-428.

Dieckmann, V., Ondrak, R., Mangelsdorf, K. & Horsfield, B. 2003. Origin of hydrocarbon gasses in the Mallik 5L-38 gas hydrate: new insights from kinetic modeling. In: Program Abstracts, From Mallik to the Future. Mallik International Symposium in Makuhari, Japan. December 8-10, 8-9.

Dillon, W.P., Austin, J.A., Jr., Scanlon, K.M., Edgar, N.T. & Parson, L.M. 1992. Accretionary margin of north-western Hispaniola: morphology, structure and development of part of the northern Caribbean plate boundary, Marine and Petroleum Geology 9(1), 70-88.

Dillon, W.P., Holbrook, W.S., Drury, R., Gettrust, J., Hutchinson, D., Booth, J. & Taylor, M., 1997. Faulting of gas hydrate-bearing marine sediments - contribution to permeability. Paper # 8296. Offshore Technology Conference Houston, Texas, 5-8 May, 1997, 201-209.

Dillon, W.P., Hutchinson, D.R. & Drury, M. 1996. 4. Seismic reflection profiles on the Blake Ridge near sites 994, 995, and 997. In: Paull, C.K., Matsumoto, R. & Wallace, P.J., et al. (Eds.), Proceedings of the ocean Drilling Program, Initial reports 164, 47-56.

Dillon, W.P., Lee, M.W. & Coleman, D.F. 1994. Identification of marine hydrates in situ and their distribution off the Atlantic Coast of the United States. In: Sloan, E.D. Jr., Happel, J. & Hnatow, M.A. (Eds.), Natural Gas Hydrates, Annals of the New York Academy of Sciences 714, 364-380.

Dillon, W.P. & Max, M.D. 2001. Gas hydrate in seafloor sediments: Impact on future resources and drilling safety. Technical Procedings of the Offshore Technology Conference Houston TX 30 April - 3 May 2001, CD ROM, OTC Paper #13034, 10pp.

Dillon, W.P. & Max, M.D. 2003. Chapter 6. Oceanic Gas Hydrate. In: Max, M.D. (Ed.), Natural Gas Hydrate in Oceanic and Permafrost Environments. Kluwer Academic Publishers, The Netherlands, 61-76.

Dillon, W.P., Nealon, J.W., Taylor, M.H., Lee, M.W., Drury, B.M. & Anton, C.H. 2001. Seafloor collapse and methane venting associated with gas hydrate on the Blake Ridge – Causes and implications to seafloor stability and methane release. In: Natural Gas Hydrate – Occurrence, Distribution, and Detection. Paull, C.K. & Dillon, W.P. (Eds.), AGU. Washington, DC, 211-233.

DOE. 2004. U. S. Department of Energy's Methane Hydrate R&D Multi-Year Program Plan: 2006-2010. Draft, 25pp.

Dogherty, C. 2005. World must live with high oil prices, IMR warns. International Herold Tribune, Saturday, April 9.

Duckworth, K. & O'Neill, D. 1989. Detection of a brine conductor under an oil field by means of a fixed, transmitter electromagnetic survey using a SQUID magnetometer. Canadian Journal of Exploration Geophysics 25, 61-73.

Dunlap, H.F. & Hutchinson, C.A. 1961. Marine seep detection. Offshore 14(4), 11-12.

Durham, W.B., Kirby, & Stern, L.A. 2003a. The strength and rheology of methane clathrate hydrate. Journal Geophysical Research 108(B4), 2182, doi:10.1029/2002JB001872.

Durham, W.B., Stern, L.A. & Kirby, S.H. 2003b. Ductle flow of methane hydrate. Canadian Journal Physics 81, 373-380.

Dvorkin, J., Helgerud, M.B., Waite, W.F., Kirby, S.H. & Nur, A. 2003. Introduction of physical propteries and elasticity. In: Max, M.D. (Ed.), Natural Gas Hydrate in Oceanic and Permafrost Environments. Kluwer Academic Publishers, The Netherlands, 245-260.

Economides, M. & Oligney, R. 2000. The Color of Oil, Round Oak Publishing Company, Katy, Texas, 10-11.

Egorov A.V.; Crane K.; Vogt P.R.; Rozhkov A.N. 1999. Gas hydrates that outcrop on the sea floor: Stability models. Geo-Marine Letters, 19(1-2), 68-75.

Egeberg, P.K. & Dickens, G.R. 1999. Thermodynamic and pore water halogen constraints on gas hydrate distribution at ODP Site 997 (Blake Ridge), Chemical Geology 153, 53-79.

Egorov, A.V. & Ivanov, M.K., 1998. Hydrocarbon gases in sediments and mud breccia from the central and eastern part of the Mediterranean Ridge. Geo-Marine Letters, 18, 127-138.

EIA (U.S. Energy Information Administration). 1998. Natural Gas 1998: Issues and Trends, Washington DC, 58.

EIA. 2003. The Global Liquefied Natural Gas Market: Status and Outlook, Report DOE/EIA-0637, December.

EIA. 2004. Annual Energy Outlook 2004 with Projections to 2025, Report #: DOE/EIA-0383, January.

Ershov, E.D, & Yakushev, V.S. 1992. Experimental research on gas hydrate decomposition in frozen rocks. Cold Regions Science Technology 20, 147-156.

Ershov, E.D, Lebedenko, Yu.P. & Chuvilin, E.M., et al. 1991. The peculiarities of the existence of gas hydrates in cryozone (*in Russian*). Doklady AN SSSR 321(4), 788-791.

Ershov, E.D., Lebedenko, Yu.P, Chuvilin, E.M., Yakushev, V.S. 1990. Experimental investigations of microstructure of ice–methane hydrate agglomerate (*in Russian*). Engineering Geology N3, 38-44.

Ershov, E.D., Lebedenko, Yu.P., Chuvilin, E.M., Yakushev, V.S. 1993. Peculiarities of gas hydrate formation in sands. Proceedings of the 6th Int'l. Conference on Permafrost, Beijing, China, 160-163.

ESA. 2004. Resources of Minerals and Energy. Earth Sciences Australia. Adopted to HTML from Prof. S. A. Nelson, Tulane University, U.S.A.

ESAS. 2005. Mission to Titan, Saturn's largest moon. European Space Agency [http://www.esa.int/esaCP/SEMQ1QQ3K3E_index_0.html]. Also NASA.

Eschard, R. 2001. Geological factors controlling sediment transport from platform to deep basin: a review. Marine & Petroleum Geology 18(4), 487-490.

Etiope, G. & Milkov, A.V. 2004. A new estimate of global methane flux from onshore and shallow submarine mud volcanoes to the atmosphere. Environmental Geology 46, 97-1002.

Exon, N.F., Dickens, G.R., Auzende, J-M., Lafoy, Y., Symonds, P.A. & van de Beuque S., 1998. Gas hydrates and free gas on the Lord Howe Rise, Tasman Sea. Petroleum Exploration Society of Australia Journal 26, 148-158.

Exon, N., HILL, P., Lafoy, Y., Fellows, M., Perry, K., Mitts, P., Howe, R., Chaproniere, G., Dickens, G., Ussler, W., & Paull, C., 2004. Geology of the Fairway and New Caledonia Basins in the Tasman Sea: sediment, pore water, diapirs and bottom simulating reflectors (Franklin Cruise FR9/01 and Geoscience Australia Survey 232). Geoscience Australia Record 2004/26, 112 pp. and CD-ROM

Davidson, D.W. 1973. Chapter 3, Clathrate Hydrates. In: Franks, F. Water, A Comprehensive Treatise, Volume 2, Water in Crystalline Hydrates; Aqueous Solutions of Simple Nonelectrolytes, Plenum Press, New York, 115-234 (Table, 123-127).

FERC. 2004. Federal Energy Regulatory Commission. Existing and Proposed North American LNG Terminals. http://www.ferc.gov/industries/gas/gen-info/horizon-lng.pdf.

Fink, C.R. & Spence, G.D., 1999. Hydrate distribution off Vancouver Island from multi-frequency single-channel seismic reflection data, Journal Geophysical Research 104, 2909–2922, 1999.

Fetterer, C.W. 2001. *Applied Hydrogeology (4th Ed.)*, Prentice Hall, ISBN 0-13-122687-8, 598pp.

Fisher, C.R., MacDonald, I.R., Sassen, R., Young, C.M., Macko, S.A., Hourdez, S., Carney, R.S., Joye, S. & McMullin, E. 2000. Methane ice worms. Hesiocaeca methanicola colonizing fossil fuel reserves: Naturwissenschaften 87, 184-187.

Fisher, D.M. 1996. Fabircs and veins in the forearc: a record of cyclic fluid flow at depths of < 15 km. Subduction: Top to Bottom. Geophysical Monograph 96, American Geophysical Union, 75-89.

Fleischer, P., Orsi, T.H., Richardson, M.D. & Anderson, A.L. 2001. Distribution of free gas in marine sediments: a global review. Geo-Marine Letters 21, 103-122.

Flemings, P.B., Liu, X. & Winters, W.J. 2003. Critical pressure and multiphase flow in Blake Ridge gas hydrates. Geology 31, 1057-1060.

Flemings, P.B., Stump, B.B., Finkbeiner, T. & Zoback, M. 2002. Flow focusing in overpressured standstones: Theory, observations and applications. American Journal of Science 302, 827-855.

Floriano, W.B. & Nascimento, M.A.C. 2004. Dielectric Constant and Density of Water as a Function of Pressure at Constant Temperature. Brazilian Journal of Physics 34(1), 38-41.

Fohrmann, F., Gorman, A.R. & Pecher, I.A. 2004. Examination of Marine Seismic Reflection Data for the Occurrence and Distribution of Gas Hydrate on the Continental Margin of Fiordland, New Zealand (Poster). In: Proceedings 85(47) AGU Fall Meeting, San Francisco, Ocean Sciences General Contributions: Gas Hydrates II, OS41C-484.

Formolo, M.J., T.W. Lyons, T.W., Zhang, C. Kelley, C., Sassen, R., Horita, J. & Cole, D.R. 2004. Quantifying the relative roles of methane and oil hydrocarbon oxidation in the formation of authigenic carbon carbonates at gas hydrate sites in the Gulf of Mexico. Chemical Geology 205(3-4), 253-264.

Franks, F. 1973. (Ed.), Water, A Comprehensive Treatise, v2, Water in Crystalline Hydrates; Aqueous Solutions of Simple Nonelectrolytes, Plenum Press, New York, 123-127.

Fu, X., Yang, M., Wen, P., et al. 2001. Seismic data processing and characteristics for gas hydrates in South China Sea, Geological Science and Technology Information 20(4), 33~40

Fultz, J.D. & Pittard, F.J. 1990. Openhole drilling using coiled tubing and a positive displacement motor," SPE 20459, presented at 65th Annual Technical Conference and Exhibition, New Orleans September 23-26, Society of Petroleum Engineers, Richardson Texas.

Ganguly, N., Spence, G.D., Chapman, N.R. & Hyndman, R.D. 2000. Heat flow variations from bottom simulating reflectors on the Cascadia margin, Marine Geology 164, 53–68.

Gardner, J.M., Diaz-Naveas, J., Pohlman, J.W., Hagen, R.A., Coffin, R.B. & Wood, W.T. 2004. Methane hydrates and fluid flow along the Chilean margin. In: Proceedings 85(47) AGU Fall Meeting, San Francisco, Ocean Sciences General Contributions: Gas Hydrates II OS41C-0496.

Gardner, J.M., Shor, A.N. & Jung, W.Y. 1998. Acoustic imagery evidence for methane hydrates in the Ulleung Basin. Marine Geophysical Researches 20(N6), 495-503.

Garven, G., Appold, M.S., Toptygina, V.I & Hazlett, T.J. 1999. Hydrogeologic modeling of the genesis of carbonate-hosted lead-zinc ores. Hydrogeology Journal 7(1), 108-126.

Gaynanov, V.G., Bouriak, S.V. & Ivanov, M.K., 1998. Seismic evidence for gas accumulation related to the area of mud volcanism in the deep Black Sea. Geo-Marine Letters 18, 139-145.

Geofagat, M.Y., Podrazansky, A.M., Basovich, V.S. & Surkov, D.V. 2004. Drilling technique for gas hydrate exploratory drilling. Not published. Paper available from Aquatic Company, Marine Surveys & Drilling Operations, 7/9 Letnikovskaya str. Moscow 113114, Russia.

Gering, K.L. 2003. Simulations of methane hydrate phenomena over geologic timescales. Part I: Effect of sediment compaction rates on methane hydrate and free gas accumulations. Earth and Planetary Science Letters 206, 65–81.

Germanovich, L. & Xu, W. 2004. Failure of marine sediments due to gas hydrate dissociation. In: Proceedings 85(47) AGU Fall Meeting, San Francisco, Ocean Sciences General Contributions: Gas Hydrates II, OS41C-0495.

Gerner, F., Svensson, B. & Djumena, S. 2004. Gas Flaring and Venting: A regulatory framework and incentives for gas utilization. Public Policy Journal 279, 4pp.

Germanovich, L. & Xu., W. 2004. Failure of marine sediments due to gas hydrate dissociation. In: Proceedings 85(47) AGU Fall Meeting, San Francisco, Ocean Sciences General Contributions: Gas Hydrates II OS41C-0495.

Gettrust, J.M., Grim, M., Madison, S & Rowe, M. 1988. Results of a deep-towed multichannel survey on the Bermuda Rise, Geophysical Research Lettters 15(12), 1413-1416.

Gettrust, J.F., Wood, W.T., Lindwall, D.A., Chapman, R., Walia, R., Hannay, C., Spence, G., Louden, K., MacDonald, R. &Hyndman, R.D. 1999. New Seismic Study of Deep Sea Gas Hydrates Results in Greatly Improved Resolution. Eos 80(38), 439-440.

Ginsburg G. D. 1993. Gas hydrates from the continental slope, offshore Sakhalin Island, Okhotsk Sea. Geo-Marine Letters 13, 41-48.

Ginsburg G.D, Soloviev V., Matveeva T. & Andreeva I. 2000. Sediment grain-size control on gas hydrate presence, sites 994, 995, and 997. Paull, C.K., Matsumoto, R., Wallace, P.J. & Dillon, W.P. (Eds.), Proceedings of the Ocean Drilling Program, Scientific Results 164, College Station, TX Ocean Drilling Program), 237-24.

Gilbert, L.Y., Waite, W.F., Winters, W.J. & Mason, D.H. 2004. Characterization of Gulf of Mexico sediment in hydrate and non-hydrate bearing cores using specific surface area. In: Proceedings 85(47) AGU Fall Meeting, San Francisco, Ocean Sciences General Contributions: Gas Hydrates II, OS41C-0491.

Goldberg, D. 1997. The role of downhole measurements in marine geology and geophysics. Reviews of Geophysics 35, 315-342.

Goldberg, D. & Saito, S. 1998. Detection of gas hydrates using downhole logs. Geological Society London special Publication 137, 129-132.

Gonzalez, R. 2004. GIS Approach to the Geology, Production, and Growth of the Barnett Shale Play in Newark East Field (Master's Thesis), University of Texas at Dallas, 3.

Gorman, A.R., Holbrook, W.S., Hornbach, M.J., Hackwith, K.L., Lizarralde, D., Pecher, I. 2002. Migration of methane gas through the hydrate stability zone in a low-flux hydrate province. Geology 30, 327-330.

Grauls, D. 2001. Gas hydrates: Importance and applications in petroleum exploration. Marine Petroleum Geology 18, 519-523

Greenspan, A. 2003. Natural gas supply, Testimony of Chairman Alan Greenspan before the Committee on Energy and Natural Resources, U.S. Senate, July 10, 2003.

Guerin, G., Goldberg, D. & Collett, T.S. 2004. Analysis of sonic velocity in an active gas hydrate system, Hydrate Ridge, offshore Oregon. In: Proceedings 85(47) AGU Fall Meeting, San Francisco, Ocean Sciences General Contributions: Gas Hydrates II, OS41C-0494.

Guo, B., Bretz, R.E. & Socorro, N.M. 1995. Method and apparatus for generating, transporting and dissociating gas hydrates. U.S. Patent 5,473,904 (December 12, 1995), 8pp.

Hagen, R.A., Gardner, J.M. & Wood, W.T. 2004. Thermal Measurements Over Hydrate Mounds in Atwater Valley, Gulf of Mexico (Poster). In: Proceedings 85(47). AGU Fall Meeting, San Francisco, Ocean Sciences General Contributions: Gas Hydrates II OS41C-0490.

Hamilton, E. 1980. Geoacoustic modeling of the seafloor. Journal Acoustic Society of America 68, 1313-1340.

Handa, Y.P. 1986a. Calorimetric determinations of the composition, enthalpies of dissociation, and heat capacities in the range 85 to 270K for clathrate hydrates of xenon and krypton, Journal of Chemical Thermodynamics 18, 891-902.

Handa, Y.P., 1986b, Compositions, enthalpies of dissociation, and heat capacities in the range of 85 to 270K for clathrate hydrates of methane, ethane, and propane, and enthalpy of dissociation of isobutane hydrate, as determined by heat flow calorimeter, Journal of Chemical Thermodynamics 18, 915-921.

Hannegan, D.M. 2005. Methane Hydrate Drilling Technology. Procedings Offshore Technology Conference (CD-ROM), Houston Texas, 2-5 May, 2005, (OTC 17448), 3pp.

Haq, B.U. 1999. Methane in the deep blue sea. Science 285, 543-544.

Haq, B.U. 2003. Climate Impact of Natural Gas Hydrate. In: Max, M.D. (Ed.), Natural Gas Hydrate in oceanic and permafrost environments. Kluwer Academic Publishers, Dordrecht, The Netherlands, 137-148.

Hardy, D.R., Rath, B.B., Hurdle, B.G., Carhart, H.W., Coffey, T. & Saalfeld, F.E. 2004. DOD Future Energy Resources. Special Report of the National Defense University, Washington, D.C., 84pp.

Harris, C., Murton, J. & Davies, C.R. 2000. Soft-sediment deformation during thawing of ice-rich frozen soils: results of scaled centrifuge modelling experiments. Sedimentology 47(3), doi:10.1046/j.1365-3091.2000.00322.x

Hato, M., Inamori, T., Bahar, A. & Matsuoka, T. 2004. Application of AVO analysis to seismic data for detection of gas below methane hydrate stability zone in Nankal Trough area, Resource Geology54(1), 105-113.

Head, I.M., Jones, D.M. & Larter, S.R. 2003. Bilogical activity in the deep subsurface and the origin of heavy oil. Nature 426, 344-352.

Heeschen, K.U. 2003. Distribution and height of methane bubble plumes on the Cascadia Margin characterized by acoustidc imaging. Geophsical Research lletters 30(12), 1243, doi:10.1029/2003GL01674, 45-1 to 45-4.

Hein, J.R., Scholl, D.W., Barron, J.A., Jones, M.G. & Miller, J.J. 1978. Diagenesis of Late Cenooic diatomaceous deposits and formation of the bottom simulating reflector in the southern Bering Sea, Sedimentology 25, 155-181.

Heinemann, Huang, D.D.-T., Long, J. & Saeger, R.B. 2001. Method for recovering gas from hydrates. U.S. Patent 6,214,175 B1 (April 10, 2001), 9pp.

Helgerud, M. B., Dvorkin, J., Nur, A. Sakai, A. & Collett, T. 1999. Elastic-wave velocity in marine sediments with gas hydrates: Effective medium modeling. Geophysical Research Letters 26, 2021–2024.

Helgerud, M.B., Circone, S., Stern, L.Al, Kirby, S.H. & Nur, A. 2002. Measured temperature and pressure dependence of compressional and shear wave speeds in polycrystalline sI methane hydrate and polycrystalline ice Ih. Proceedings of the 4th International Conference on Gas Hydrates, Yokohama Japan, 2002, 711-721.

Henriet, J.P., De Mol, B., Vanneste, M., Huvenne, V., Van Rooij, D. 2001. Carbonate mounds and slope failures in the Porcupine Basin: a development model involving past fluid venting. In: Shannon, P.M., Haughton, P., Corcoran, D. (Eds.), Petroleum exploration of Ireland's offshore basins. Special Publication 188, Geological Society of London, 375-383.

Henriet, J.-P., De Mol, B., Pillen, S., Vanneste, M., Van Rooij, D., Versteeg, W., Croker, P.F., Shannon, P.M., Unnithan, V., Bouriak, S., Chachkine, P. & the Porcupine-Belgica 97 shipboard party. 1998. Gas hydrate crystals may help build reefs. Nature 391, 648-649.

Henrys, S.A., Ellis, S., Uruski, C. 2003. Conductive heat flow variations from bottom-simulating reflectors on the Hikurangi margin, New Zealand. Geophysical Research Letters 30, 1065–1068.

Hiller, R.D. & Cosgrove, J.W. 2002. Core and seismic observations of overpressure-related defomation within Eocene sediments of the Outer Moray Firth, UKCS. Petroleum Geoscience 8(2), 141-149.

Hitoshi, S., Soloviev, V.A., Tatiana, M., et al. Hydrate bearing structures mapped in the Sea of Okhotsk (CHAOS project). Eos (in press).

Hjelstuen, B.O., Sejrup, H.P., Andreassen, K., Boe, R., Eldholm, O., Hovland, M., Knies, J., Kvalstad, T., Kvamme, B., Mienert, J. & Pedersen, R.B. 2004. Norwegian Research Strategies on gas Hydrates and

Natural Seeps in the Nordic Seas Region (GANS). In: Proceedings 85(47) AGU Fall Meeting, San Francisco, Ocean Sciences General Contributions: Gas Hydrates II, OS41C-0488.

Hoehler, T.M., Alperin, M.J., Albert, D.B. & Martens, C.S. 1994. Field and laboratory studies of methane oxidation in an anoxic marine sediment: Evidence for methanogen – sulfate reducer consortium. Global Biogeochemical Cycles 8, 451-463.

Holbrook, W.S. 2001. Seismic studies of the Blake Ridge. Implications for hydrate distribution, methane expulsion, and free gas dynamics, in Natural Gas Hydrates, Occurrence, Distribution, and Detection, American Geophysical Union Special Publication 124. Paull, C.K. & Dillon, W.P. (Eds.), 235 – 256.

Holbrook, W.S., Gorman, A.R., Hornbach, M., Hackwith, K.L., Nealon, J., Lizarralde, D. & Pecher, I.A. 2002a. Seismic detection of marine methane hydrate. The Leading Edge 21, 686-689.

Holbrook, W.S., Lizarralde, D.. Pecher, I.A., Gorman, A.R., Hackwith, K.L., Hornbach, M & Saffer, D. 2002b. Escape of methane gas through sediment waves in a large methane hydrate province. Geology 30, 467-470.

Holman, S., Max, M., & Kolle, J. 2003. Experimental Data on Ethane Solubility in Natural Seawater. Proceedings of the AIChE 2003 Spring National Meeting, New Orleans, LA, 30 March – 3 April 2003, 457-472.

Holman, S.A., Osegovic, J.P., Young, Jr., J.C. Max, M.D. & Ames, A.L. 2003. Nucleation and Growth of Gas Hydrate in Natural Seawater (Abs). Proceedings American Geophysical Union Fall Meeting San Francisco, 8-12 December.

Holman, S.A., Osegovic, J.P., Young Jr., J.C., Ames, A.L., Tatro, S.R. & Max, M.D. 2004. Controlled growth of polycrystalline ethane hydrate and inferences for gas recovery from hydrate deposits, Offshore Technology Conference, Houston, TX, May 3-6, 2004, 6pp.

Hornbach, M.J., Holbrook, W.S., Gorman, A.R., Hackwith, K.L., Lizarralde, D. & Pecher, I. 2003. Direct seismic detection of methane hydrate on the Blake Ridge. Geophysics 68, 92-100.

Hornbach, M.J., Holbrook, W.S., Saffer, D.M., Gorman, A.R., Van Avendonk, H., Lizarralde, D. & Pecher, I. 2004. 3D seismic imaging of the Blake Ridge gas hydrate province: Evidence for a dynamic gas system. In: Proceedings 85(47) AGU Fall Meeting, San Francisco, Ocean Sciences General Contributions: Gas Hydrates II OS41C-0493.

Hovland, M. & Gudmestad, O.T. 2001. Potential influence of gas hydrates on seabed installations. In Paull. C.K & Dillon, W.P. (Eds.), Natural Gas Hydrate – Occurrence, Distribution, and Detection. AGU. Washington, DC, 307-315.

Hovland, M. & Orange, D. 2001. Gas hydrate and seeps - Effects on slope stability: The "hydraulic model". Proceedings of the International Offshore and Polar Engineering Conference Stavanger, Norway Conference 1, 471-476.

Hoyle, F.1955. Frontiers of Astronomy, Heinemann, London, 366pp.

Hu, Z., Sun, J., & Zhao, Q. 2002. Identification of gas hydrate with seismic method, Progress in Exploration Geophysics 25(6), 23~27

Huang, Y. 2004. Progress of marine gas hydrate research in northern South China Sea in the past five years in China. Proceedings 32nd International Geological Congress, Florence, Italy, 1250.

Hue, A.Y. 1990. (Ed.), *Deposition of Organic Facies.* AAPG Studies in Geology #30. American Association of Petroleum Geologists, 234pp.

Hugman, R.H., Springer, P.S. & Vidas, E.H. 1993. Tight gas field, reservoir, and completion analysis of the United States, Vol. 1: Project summary: Arlington, Virginia, Energy and Environmental Analysis, topical report no. GRI-92/0226.1, prepared for the Gas Research Institute.

Hunt, J.M. 1996. *Petroleum Geochemistry and Geology.* W.H. Freeman, San Francisco, 720pp.

Hunt, J.M. 2005. (Minerals Management Service).Personal communication to W. Dillon (Chapter 3).

Huvenne, V.A.I., Blondel, Ph. & Henriet, J.-P. 2002. Textrual analyses of sidescan sonar imagery from two mound provinces in the Porcupine Seabight. Marine Geology 189, 323-341.

Huvenne, V.A.I., Croker, P.F. & Henriet, J.-P. 2002. A refreshing 3D view of an ancient sediment collapse and slope failure. Terra Nova 14(1), 33-40.

Hyndman, R.D., Spence, G.D., Chapman, R., Riedel, M. & Edwards, R.N. 2001. Geophysical studies of marine gas hydrates in northern Cascadia. In Paull. C.K & Dillon, W.P. (Eds.), Natural Gas Hydrate – Occurrence, Distribution, and Detection. AGU. Washington, DC, 273-295.

Hyndman, R.D., Currie, C.A. & Mazzotti, S.P. 2005. Subduction zone backarcs, mobile belts, and orogenic heat. (doi: 101130/1052-5173(2005)015 <4SZBMBA>2.0CO;2) GSA Today 15(2), 4-10.

Hyndman, R.D., Yuan, T. & Moran, K. 1999. The concentration of deep sea gas hydrates from downhole electrical resistivity logs and laboratory data, Publication 19980.10, Geological Survey of Canada, Ottawa, Ontario, Canada, 1998.fluid migration? Journal Geophysical Research 104, 15,351– 15,364.

IAGS (Institute for the Analysis of Global Security) 2004. The Future of Oil, <www.iags.org/futureofoil.html>.

IELE (Institute for Energy, Law, & Enterprise) 2002. "Brief History of LNG", University of Houston<http://www.energy.uh.edu/LNG/LNG_introduction _06.asp>

Inamori, T. & Hato, M. 2004. Detection of methane hydrate-bearing zones from seismic data, Resource Geology 54(1), 99-104.

Ion, D.C. 1979. World energy supplies. Proceedings Geological Association 90,193-203.

Irvin, G., Sichu, Li., Simmons, B., John, V.T, McPherson, G., Max, M.D. & Pellenbarg, R.E. 2000. Control of gas hydrate formation using surfactant systems. Underlying concepts and new applications. Paper 51.In: Holder, G.D. & Bishnoi, P.R. (Eds). Gas Hydrates Challenges for the Future. Anals of the New York Academy of Sciences 912, Special publication of papers from the Third International Conference on gas hydrates, 515-526.

Istomin, V.A. & Yakushev, V.S. 1992. Gas hydrates under natural conditions (*in Russian*) Nedra, Moscow, 326pp.

Ivanov, M.K., Limonov, A.F. & van Weering, T.C.E. 1996. Comparative characteristics of the Black Sea and Mediterranean Ridge mud volcanoes. Marine Geology 132, 253-271.

Ivanov. M.K., Limonov, A.F., & Woodside, J.M. 1998. Extensive deep fluid flux through the sea floor on the Crimean continental margin (Black Sea). In: J.-P. Henriet and J. Mienert (Eds.). Gas Hydrates: Relevance to World Margin Stability and Climatic Change. Geological Society London Special Publication 137, 195-214.

Jackson, J.A. 1997. *Glossary of Geology* (4th Ed.), American Geological Institute, 769pp.

Jager, M.D. & Sloan, E.D. 2001. The effect of pressure on methane hydration in pure water and sodium chloride solutions. Fluid Phase Equilibria, 185.

Jakobsen, M., Hudson, J.A., Minshull, T.A. & Singh, S.C. 2000. Elastic properties of hydrate-bearing sediments using effective medium theory, Journal Geophysical Research 105(B1), 561-577.

Jassim, S.Z., Raiswell, R. & Bottrell, S.H. 1999. Genesis of the Middle Miocene stratabound sulphur deposits of northern Iraq. Journal Geological Society London 156(1), 25-39.

Jennings, H.M. & Thomas, J.J. 2004. A discussion of the Paper "The BET-specific surface area of hydrated Portalnd cement and related materials" by Ivan Odler. Cement and Concrete Research 34, 1959-1950.

Ji, C., Ahmadi, G., & Smith, H.S. 2001. Natural gas production from hydrate decomposition by depressurization. Chemical Engineering Science, 56.

Ji, C., Ahmadi, G., & Smith, D.H. 2003. Constant rate natural gas production from a well in a hydrate reservoir. Energy Conversion and Management, 44, 2403–2423.

Jiang, S.-Y. 2004. Search for gas hydrates in the South China Sea. Proceedings 32nd International Geological Congress, Florence, Italy, 1250.

Johnson, A.H., Dillon, W.P. & Max, M.D. 2003. Strategies for gas hydrate exploration in the Gulf of Mexico. Geophysical Research Abstracts 5, 01237.

Johnson, B.D. Boudreau, B.P., Gardiner, B.S. & Maass, R. 2002. Mechanical response of sediments to bubble growth. Marine Geology 187, 347-363.

Jönsson, B., Karlström, Wennerström, H., Forsén, Roos, B. & Almlöf, J. 1977. *Ab initio* molecular orbital calculations on the water – carbon dioxide system. Reaction pathway for $H_2O + CO_2 \rightarrow H_2CO_3$. Journal American Chemical Society 99(14), 4628-4632.

Jones, V.T. III & Drozd, 1983. R.J. Predictions of oil or gas potential by near surface geochemistry. Bulletin American Association of Petroleum Geologists 67(6), 932-952.

Jones, V.T. III, Matthews, M.D. & Richers, D.M. 1999. Light hydrocarbons for petroleum and gas prospecting. In: Hale, M. (Ed.), Geochemical remote sensing of the subsurface. In:, Govett, G.J.S., (Ed.), Handbook of Exploration Geochemistry 7, Elsevier Science BV. The Netherlands, 133-212.

Judd, A.G., Hovland, M., Dimitrov, L.I. Garcia Gil, S. & Jukes, V. 2002. The geological methane budget at continental margins and its influence on climate change. Geofluids 2, 109-doi:10.1046/j,1468-8123.202.00027.x.

Kanninen, M.F. & Popelar, C.H. 1985. *Advanced fracture mechanics.* Oxford University Press, New York, NY, 563pp.

Kargel, J.S., Kaye, J.Z., Head III, J.W., Marion, G.M., Sassen, R., Crowley, J.K., Ballesteros, O.P., Grant, A. & Hogenboom, D.L. 2000) Europa's crust and ocean. Origin, composition, and the prospects for life: Icarus 148, 226-265.

Kashchiev, D. & Firoozabadi, A. 2002. Nucleation of Gas Hydrates. Journal of Crystal Growth 243, 476-489.

Kashchiev, D. & van Rosmalen, G.M. 2003. Review: Nucleation in solutions Revisited. Crystal Research Technology 38(7-8), 555-574.

Kastner, M. 2001. Gas hydrate in convergent margins: Formation, occurrence, geochemistry and global significance. In: Natural Gas Hydrate – Occurrence, Distribution, and Detection, Paull. C.K & Dillon, W.P. (Eds.), AGU, Geophysical Monograph 124. Washington, 67-86.

Katoh, A., Nakayama, K, Baba, K. & Uchida, T. 2000. Model simulation for generation and migration of methane hydrate. In: Lerche, I. (Ed.), Gas Hydrates Special Issue, Energy Exploration & Exploitation 18(4), 401-421.

Katsube, T.J., Connell, S., Uchida, T., Dallimore, S.R., Wright, F., Johasson, I.R. & Perez, N. 2003. Effect of porosity, permeability and adsorbed water on gas-hydrate content. In: Mallik 2002 Gas hydrate prodution research well program. Mallik International Symposium in Makuhari, December 8-10, 2003, 25-26.

Katz, H.R. 1981. Probable gas hydrate in continental slope east of the North Island, New Zealand, Journal of Petroleum Geology 3, 315-324.

Kennett, J.P., Cannariato, K.G., Hendy, I.L. & Behl, R.J. 2002. *Methane hydrates in Quaternary climate change; The clathrate gun hypothesis.* American Geophysical Union, Washington DC, 224pp.

Kennett, J.P. & Fackler-Adams, B.N. 2000. Relationship of clathrate instability to sediment deformation in the upper Neogene of California. Geology 28, 215-218.

Kerr, R. A. 2005. Titan, once a world apart, becomes eerily familiar. Science, 307(5708), 330-331.

Kim, H.C., Bishnoi, P.R., Heidemann, R.A. & Rizvi, S.S. 1987. Kinetics of methane hydrate decomposition, Chemical Engineering Science 42, 1645-1653.

Kirchner, M.T., Boese, R., Billups, W.E. & Norman, LR.J. 2004. American Chemical Society 126, 9407-9412.

Klauda, J.B. & Sandler, S.I. 2000. A fugacity model for gas hydrate phase equilibria. Industrial and Engineering Chemistry Research 39, 3377–3386.

Klauda, J.B. & Sandler, S.I. 2003. Predictions of gas hydrate phase equilibria and amounts in natural sediment porous media. Marine and Petroleum Geology 20, 459–470.

Kobayashi, H. 1996. Development of methane gas resources. Research and Development Activitieson Methane Hydrate for The Future Energy Resources. National Inst. Resources and Environment Shigen to Kankyo Journal of NIRE (National Institute for Resources and Environment) 5(4), 181-192.

Kolle, J.J. 1994. Low-cost drilling system for characterizing marine gas hydrate accumulations. Proceedings of the fourth (1994) International Offshore and Polar Engineering Conference, Osaka, Japan, April 10-15, 333-336.

Kolle, J.J. & Bibee, L.D. 1996. Deep ocean sensor implantation system. International Journal of Offshore and Polar Engineering 6(4), 255-260.

Kondakova, VV, Kondakova, O.A. & Nikolaeva, G.V. 1989. The dependence of the phase compostion of wter on froozen and freezing sediments on external pressure. Imzhenernaya Geologiya 5, 21-29.

Kopf, A. Deyhle, A. & Zuleger, E. 2000. Evidence for deep fluid circulation and gas hydrate dissociation using boron and boron isotopes of pore fluids in forearc sediments from Costa Rica (ODP Leg 170). Marine Geology 167, 1-28.

Kopf, A., Robertson, A.H.F., Clennell, M.B, Flecker, R., Techmer, K. & Heinrichs, T. 1998. Mechanisms of mud extrusion on the Mediterranean Ridge accretionary complex. Geo-Marine Letters 18, 97-114.

Kuhs, W.F., Klapproth, A. & Gotthardt, F. 2000. The formation of meso- and macroporous gas hydrate. Geophysical Research Letters 27(18), 2929-2932.

Kutas, R.I., Rusakov, O.M. & Kobolev, V.P. 2002. Gas seeps in the northwestern Black Sea: geological and geophysical studies. In: Proceedings of the VI International Conference on Gas in Marine Sediments. Special Issue, Geologiya I Geofizika 43(7), Siberian Brancch, Russian Academy of Sciences, 698-705.

Kvamme B. 2000. Phase transitions and stabilities in seawater containing carbon dioxide. In: Recent Advances in Marine Science and Technology, Honolulu, Hawaii USA, monograph, 279–289.

Kvenvolden, K.A. 1987. Gas hydrates offshore Alaska and western continental United States. Circum-Pacific council Energy & Mineral Resources. Earth Sciences Series 6, 581-593.

Kvenvolden, KA. 1993. Gas hydrate- Geological perspective and global change. Reviews of Geophysics 31, 173-187.

Kvenvolden, K.A. 2003. Natural gas hydrate: Background and history of discovery. In: Max, M.D. (Ed.), Natural Gas Hydrate in Oceanic and Permafrost Environments. Kluwer Academic Publishers, The Netherlands, 9-16.

Kvenvolden, K.A. & Lorenson, T.D. 2000. A global inventory of natural gas hydrate occurrence, website http://walrus.wr.usgs.gov/globalhydrate/index.html.

Kvenvolden, K.A., & Lorenson, T.D. 2001. The global occurrence of natural gas hydrate, in Natural Gas Hydrates: Occurrence, Distribution, and Dynamics, Paull, C.K., & Dillon, W.P. (Eds.), AGU Monograph No. 124, 3-18.

Kvenvolden, K.A. & Rogers, K.A. 2005. Gaia's breath – Global methane exhalations. Marine and Petroleum Geology ((Fig. 3.3), in press).

Laherrere, J. 2000. Oceanic hydrates: More questions than answers. Energy Exploration and Exploitation 18(4), 349-383.

Lanoil, B.D., Sassen, R., La Duc, M.T., Sweet, S.T. & Nealson, K.H. 2001. Bacteria and Archea Physically Associated with Gulf of Mexico Gas Hydrates. Applied and Environmental Microbiology 67, 5143-5153.

Large, D. & Walcher, E. 1999. The Rammelsberg massive sulphide Cu-Zn-Pb-Ba-deposit, Germany: an example of sediment-hosted, massive sulphide mineralisation. Mineralium Deposita 34(5-6), 522-538.

Lay, T., Kanamori, H., Ammon, C.J., Nettles, M., Ward, S.N., Aster, R.C., Beck, S.L., Bilek, S.L., Brudzinski, M.R., Butlet, R., DeShon, H.R., Ekström, G., Satake, K. & Sipkin, S. 2005. The great Sumatra-Andaman earthquake of 26 December 2004. Science 308, 1127-1133.

Lee, M.J. & Wilkinson, J.J. 2002. Cementation, hyrothermal alteration, and Zn-Pb minerralization of carbonate breccias in the Irish midlands: Textrual evidence from the Cooleen Zone, near Sivermines, County Tipperary. Economic Geology 97. 653-662.

Lee, M.W. & Dillon, W.P. 2001. Amplitude blanking related to the pore-filling of gas hydrate in sediments: Marine Geophysical Researches 22, 101-109.

Lee, T.H., Linden, H.R., Dreyfus, D.A. & Vasko, T. 1988. *The Methane Age.* Kluwer Academic Publishers, Dordrecht, Boston, London, 204pp.

Lekvam, K. & Ruoff, P. 1997. Kinetics and mechanism of methane hydrate formation and decomposition in liquid water: description of hysteresis, Journal of Crystal Growth 179, 618-624.

Lerche, I. 2000. Estimates of worldwide gas hydrate resources. Energy Exploration & Exploitation 18, 329-337.

Lerche, I. & Bagirov, E. 1998. Guide to gas hydrate stability in various geological settings. Marine Petroleum Geology 15, 427-437.

Levin, L.A., James, D.W., Martin, C.M., Rathburn, A.E., Harris, L.H. & Michener, R.H. 2000. Do methane seeps support distinct macrofaunal assemblages" Observations on community structure and nutrition from the northern California slope and shelf. Marine Ecology Progress Series 208, 21-39.

Liaghat, S., Moore, F. & Jami, M. 2000. The Kuh-e-Surmeh mineralization, a carbonate-hosted Zn-Pb deposit in the simply folded belt of the Zagros Mountains, SW Iran. Mineralium Deposita 35(1), 72-78.

Lide, D. E. 2002. (Ed.), *CRC Handbook of Chemistry and Physics, (83rd Ed.),* CRC Press, Boca Raton, Florida, 6-5.

Limonov, A.F., van Weering, T.C.E., Kenyon, N.H., Ivanov, M.K., & Meisner, L.B. 1997. Seabed morphology and gas venting in the Black Sea mud volcano area: observations with the MAK-1 deep-tow sidescan sonar and bottom profiler. Marine Geology 137, 121-136.

Limonov, A.F., Woodside, J.M., Cita, M.B. & Ivanov, M.K. 1996. The Mediterranean Ridge and related mud diapirism; a background. Marine Geology 132, 7-19.

Liu, C.-S., Schnurle, P., Hsiuan, T.-H., Yang, T., Lin, S., Wang, Y. & Chung, S.-H. 2004. Gas hydrate distribution offshore southwestern Taiwan and its geophysical and geochemical characteristics (Abs.) In: Collett, T. & Johnson, A. (Eds.) AAPG Hedberg Research Conference, Natural gas hydrates: Energy resource potential and associated geologic hazards, September 12-16, Vancouver BC, Canada, 2pp.

Liu, H. & Zhou, Z. 2002. Seismic data processing methods for the study of gas hydrates in the East China Sea, Journal Ocean University, Qingdao 32(3), 441-448.

Lohse, D. 2003. Bubble Puzzles. Physics Today (February), 36-41

Lombardi, T.E. & Lambert, S.W. 2001. Coalbed Methane in the Illinois Basin, SPE Paper 72367, Society of Petroleum Engineers Eastern Regional Meeting, Canton, OH.

Locat, J. 2001. Instabilities along ocean margins: a geomorphological and geotechnical perspective. Marine and Petroleum Geology 18(4). 487-490.

Lodolo, E. & Camerlenghi, A. 2003. The occurrence of BSRs on the Antarctic margin. In: Max, M.D. 2003. (Ed.), *Natural Gas Hydrate: In Oceanic and Permafrost environments* (2nd Ed, 1st Ed, 2000). Kluwer Academic Publishers, London, Boston, Dordrecht, 199-212.

Lorenson, T.D., Whiticar, M.J., Collett, T.S., Dallimore, S.R. & Dougherty, J.A. Gas composition and isotopic geochemistry of cuttings, core, gas hydrate, and production testing samples from the Mallik 5L-38 international gas hydrate research well. In: Mallik 2002 Gas hydrate prodution research well program. Mallik International Symposium in Makuhari, December 8-10 2003, 37.

Lubick, N. 2004. Detecting marine gas hydrates. Geotimes 49(11), 28-30.

Lüdmann, T., Wong, H.K., Konerding, P., Zillmer, M., Petersen, J. & Flüh, E. 2004. Heat flow and quantity of methane decuded from a gas hydrate field in the vicinity of the Dnieper Canyon, northwesternn Black Sea. Geo-Marine Letters 24, 182-193.

Lyon, R.K. 2004. Catalyst allowing conversion of natural gas hydrate and liquid CO_2 to CO_2 hydrate and natural gas. U.S. Patent 6,757,573 B2 (May, 2004), 10pp.

Ma, Z., Geng, J., Dong, L. et al. 2002. Seismic recognition studies on marine gas hydrate. Marine and Quaternary Geology 22(1), 1-8.

MacDonald, I.R., Guinasso, N.L. Jr., Sassen, R., Brooks, J.M., Lee, L. & Scott, K.T. 1994. Gas hydrate that breaches the sea floor on the continental slope of the Gulf of Mexico. Geology 22, 699-702.

MacDonald, I.R., Leifer, I., Sassen, R., Stine, P., Mitchell, R. & Guinasso, Jr. N.L. 2002. Transfer of hydrocarbons from natural seeps to the water column and atmosphere: Geofluids 2, 1-13.

MacKay, M.E., Jarrard, R.D., Westbrook, G.K., et al. 1994. Origin of bottom simulating reflectors: Geophysical evidence from the Cascadia accretionary prism. Geology 22, 459-462.

Maekawa, T. 2001. Equilibrium conditions for gas hydrates of methane and ethane mixtures in pure water and sodium chloride solution. Geochemical Journal 35, 59-66.

Magra, K. 1968. Compaction and migration of fluids in Miocene mudstone, Nagaoka plain, Japan. American Association of Petroleum Geologists Bulletin 52, 2466-2501.

Makogon, Y.F. 1965. Hydrate Formation in the Gas-Bearing Beds Under Permafrost Conditions, Gazovaia Promyshlennost, 5, 14-15.

Makogan, Yu. F. 1974. *Hydrates of Natural Gases* (in Russian). Nedra, Moscow, 208pp.

Makogon, Y.A., Holditch, S.A. & Makogon, T.Y. 2005. Russian field illustrates gas-hydrate production, Oil & Gas Journal, Feb. 7, 43-47

Makogon, Y.F., Trebin, F.A., Trofimuk, A.A., Tsarev, V.P. & Cherskii, N. 1971. Detection of A Pool of Natural Gas in a Solid (Hydrated Gas) State, Doklady Akademii Nauk SSSR 196, 203-206.

Mallik 2002 Partners, 2003, Press Release: Mallik 2002. Gas Hydrate Production Research Well Program, December 10, 2003.

Manheim, F.T. 2004. U.S. Offshore Oil Industry new perspectives on an old conflict. Geotimes 49(12), 26-28.

Majorowicz, J.A. & Osadetz, K.G. 2001. Gas hydrate distribution and volume in Canada. American Association of Petroleum Geologists 85(7), 1211-1230.

Marjorowicz, J.A. & Osadetz, K.G. 2003. Natural gas hydrate stability in the east coast offshore - Canada. Natural Resources Research 12(2), 93-104.

Martens, C.S. & Berner, R.A. 1974. Methane production in the interstitial waters of sulfate-depleted marine sediments. Science 185, 1167-1169.

Matsumoto, R. 1999. Gas Hydrate. New Natural Gas Resources for the Coming Generation. Geological and Geochemical Survey of Gas Hydrate Deposits. Present status and future problems of R/D program (ISSN NO: 0916-9997) (*In Japanese*). Shigen to Kankyo(Journal of NIRE (National Institute for Resources and Environment) 8(5), 285-295

Matsumoto, R. 2002. Comparison of marine and permafrost gas hydrates: Exmaples from Nankai Trough and Mackenzie Delta. Proceedings, 4th International Conference on Gas Hydrates, Yokohama, Japan,

Matsumoto, R., Tomaru, H. & Lu, H. 2004. Detection and evaluation of gas hydrates in the eastern Nakai Trough by geochemical and geophysical methods, Resource Geology 54(1), 53-67.

Matsumoto, R., Tomaro, H., Chen, Y., Lu, H. & Clark, I. 2003. Geochemistry of the Interstitial Waters of the Mallik 5L-38 Gas Hydrate Research Well, Mackenzie Delta, Canada. Mallik 2002 Gas hydrate prodution research well program. Mallik International Symposium in Makuhari, December 8-10 2003, 44.

Matsubayashi, O. 1998. Heat flow measurement as an exploration tool for subbottom methane hydrates (*In Japanese*). Bulletin Geological Survey Japan 49(10), 541-549.

Matsuo, K., Kurosaka, S., Yanagimori, Y., Asano, S. & Shinoda, J. 2004. Device and method for extracting a gas hydrate. U.S. Patent 6,817,427 (November 16, 2004), 10pp.

Matveeva, T.V., Mazurenko, L.L., Soloviev, V.A., Klerkx, J., Kaulio, V.V., Khlystov, O. & Prasolov, E.M. 2003. Filtrogenic gas hydrates in the subsurface sediments of Lake Baikal (Eastern Siberia). Geo-Marine Letters Special Issue 23(3/4), 289-299.

Matveeva, T. & Soloviev, V. 2002. Gas hydrates of the Okhotsk Sea: a review of available data and current state of research. Proceedings of the 4th International Conference on Gas Hydrates. Yokohama 1, 185-189.

Matveeva, T.V. & Soloviev, V.A. 2002. Geologically controlled gas hydrate accumulation on the Blake outer Ridge, western North Atlantic (from DSDP and ODP data). In: Proceedings of the VI International Conference on "Gas in Marine Sediments". Special Issue, Geologiya I Geofizika 43(7), Siberian Brancch, Russian Academy of Sciences, 662-671.

Matveeva, T., Soloviev, V., Wallmann, K., Obzhirov, A., Biebow, N., Poort, J., Salomatin, A. & Shoji, H. 2003. Geochemistry of gas hydrate accumulation offshore NE Sakhalin Island (Sea of Okhotsk): results from the KOMEX-2002 cruise. Geo-Marine Letters Special Issue 23(3/4), 278-288.

Max, M.D., Dillon, W.P. & Malone, R.D. 1991. Proceedings: Report on National Workshop on Gas Hydrates April 23 and 24, 1991, U.S. Geological Survey, Reston, VA, U.S. Department of Energy DOE/METC - 91/6/24, DE 91016654, 38pp.

Max, M.D. 2003. (Ed.), *Natural Gas Hydrate: In Oceanic and Permafrost environments* (2nd Ed, 1st Ed, 2000). Kluwer Academic Publishers, London, Boston, Dordrecht, 422pp.

Max, M.D. 2003a. Gas hydrate potential of the Indian sector of the NE Arabian Sea and northern Indian Ocean. In: Max, M.D. 2003. (Ed.), *Natural Gas Hydrate: In Oceanic and Permafrost environments* (2nd Ed, 1st Ed, 2000). Kluwer Academic Publishers, London, Boston, Dordrecht, 224.

Max, M.D., Holman, S. Osegovic, J., Tatro, S. & Hill, J. 2004. Use of gas hydrate for desalination and water purification. Special application to seawater. MDS Contractor Final Report 2004-01, DARPA contract NBCHC010003 (DITEC), 8 Appendices, 47pp.

Max, M.D. & Clifford, S. 2000. The state, potential distribution, and biological implications of methane in the Martian crust. Journal of Geophysical Research-Planets 105/E2, 4165-4171.

Max, M.D. & Clifford, S.M. 2001. Initiation of Martian outflow channels: Related to the Dissociation of Gas Hydrate. Geophysical Research Letters 28, 1787-1790.

Max, M.D. & Dillon, 2002. W.P. Comment on: Kennedy, M.J., Christie-Blick, N., Sohl, L.E., 2001, Are Proterozoic cap carbonates and isotopic excursions a record of gas hydrate destabilization following Earth's coldest intervals? Geology 30, 762.

Max, M.D. & Dillon, W.P. 2000. Natural gas hydrate: a frozen asset? Chemistry & Industry (U.K.) 1, January, 16-18.

Max, M.D. & Holman, S. 2003. Variable gas solubility effect of the gas hdyrate system and its influence on optimization of gas recovery. Proceedings in CD ROM, Offshore Technology Conference 5-8 May 2003, Houston Texas, Paper # 15123, 1-4.

Max, M.D., John, V.T. & Pellenbarg, R.E. 2000. Methane hydrate fuel storage for all-electric ships: An opportunity for technological innovation. In: Holder, G.D. & Bishnoi, P.R. (Eds.), Gas Hydrates Challenges for the Future. Anals of the New York Academy of Sciences 912, Special publication of papers from the Third International Conference on gas hydrates, 460-473.

Max, M.D.& Kolle, J.J. 2000. Seafloor drilling and monitoring of the hydrate economic zone for exploration and production of methane. Proceedings 10th International Offshore and Polar Engineering Conference & Exhibition (http://www.tempresstech.com).

Max, M.D., Pellanberg, R.E. & Hurdle, B.G. 1997. Methane Hydrate, A Special Clathrate: Its Attributes and Potential. Naval Research Laboratory Report NRL/MR/6101-97-7926, 74pp.

Max, M.D., Spina, F. Bovio, E. & Vasseur, N. 1998. The E-Map, Innovative COTS software for data fusion and interactive communication for Rapid Envrionmental Assessment. In: Pouliquen, E., Kirwan, A.D. Jr. & Pearson, R.T. (Eds.), Rapid Environmental Assessment, SACLANTCEN Conference Proceedings CP-44, March 10-13 1997, Lerici (La Spezia), Italy, 269-278.

Mazurenko L.L., Soloviev V,A., BelenkayaI., Ivanov M.K. & Pinheiro L.M. 2002. Mud volcano gas hydrates in the Gulf of Cadiz. Terra Nova 14, 321-329.

Mazurenko L.L. & Soloviev V.A. 2002. The nature of gas hydrate-forming mud volcano fluids. Proceedings of the Fourth International Conference on Gas Hydrates. Yokohama 1, 80-83.

Mazurenko, L.L., Soloviev, V.A., Gardner, J.M. & Ivanov, M.K. 2003. Gas hydrates in the Ginsburg and Yuma mud volcano sediments (Moroccan Margin): results of chemical and isotopic studies of pore water. Marine Geology 195, 201-210.

McCaffrey, R. 1996. Slip Partitioning at Convergent Plate Boundaries of SE Asia. Geological Society of London Special Publication 106, 3-18.

McDonnell, S.L., Max, M.D., Cherkis, N.Z. & Czarnecki, M.F. 2000. Tectono-sedimentary controls on the likelihood of gas hydrate occurrence near Taiwan, Marine and Petroleum Geology 17(8), 929-936.

McGee, T.M. 2004. Toward understanding the migration of hydrocarbons and brines through the gas hydrate stability zone of the northern Gulf of Mexico. Proceedings 32nd International Geological Congress, Florence, Italy, 1247.

McGee, T.M., Lutken, C.B., Rogers, R.E., Brunner, C.A., Dearman, J.S., Lynch, F.L. & Woolsey, J.R. 2005. Can fractures in soft sediments host significant quantities of gas hydrates. (Poster). International Gas Hydrate Workshop, Victoria, BC, Canada, 8-11 May, 2005.

McGuire, P.L. 1981. Methane hydrate gas production: An assessment of conventional production technology as applied to hydrate recovery, Report LA-9102-MS, Los Alamos National Laboratory, Los Alamos, NM.

McGuire, P.L., Chatham, J.R., Paskvan, F.K., Sommer, D.M. & Carini, F.H. Low salinity oil recovery: An exciting new EOR opportunity for Alaska's North Slope. Proceedings, Society of Petroleum Engineering SPE Western Regional Meeting, Irvine, CA, 30 March-1 April 2005, 15pp.

McLellan, P.J. 2005. Geomechanics of the Mallik Gas Hydrate Reservoir, Mackenzie Delta, NWT (Abs). Presentation to the Emerging Petroleum Systems Division of the Canadian Society of Petroleum Geologists, Calgary, Alberta, April 20, 1pp.

McLellan, P.J., Gillen, K., Podetz, C. & Dallimore, S. 2003. Characteristics of natural fractures in the Mallik 5L-38 gas hydrate reserch well, Mackenzie Delta, N.W.T. In: Program Abstracts, From Mallik to the Future. Mallik International Symposium in Makuhari, Japan. December 8-10, 47-48.

Melnikov, V.P. & Nesterov, A.N. 2001. Gas hydrate formation in mineralized pore water. Earth Cryosphere 5(1), 3-14.
Meng, X., Liu, B., Shi, X., et al. 2000. Possibility of gas hydrate occurring in the western lower slope of mid-Okinawa Trough, Acta Sedimentologica Sinica 18(4), 629~633.
Mervis, J. 2004. Risky business (U.S. Science Policy). Science 306, 220-221.
MHAC (Methane Hydrate Advisory Committee). 2002. Report of the Methane Hydrate Advisory Committee on Methane Hydrate Issues and Opportunities Including assessment of Uncertainty of the Impact of Methane Hydrate on Global Climate Change, U.S. department of Energy, 56pp.
Miles, P.R. 2003. Geophysical sensing and hydrate. In: Max, M.D. (Ed.), Natural Gas Hydrate in Oceanic and Permafrost Environments. Kluwer Academic Publishers, The Netherlands, 261-274.
Mikucki, J.A., Liu, Y., Delwiche, M, Colwell, F.S. & Boone, D.R. 2003. Isolation of a methanogen from deep marine sediments that contain methane hydrates, and descripton of Methanoculleus submarinus sp.nov. Applied and Environmental Microbiology 69(6), 3311-3316.
Milkov, A.V. 2000. Worldwide distribution of submarine mud volcanoes and associated gas hydrates. Marine Geology 167, 29-42.
Milkov, A.V., Claypool, G.E., Lee, Y.-J., Torres, M.E., Borowski, W.S., Tomaru, H., Sassen, R., Long, P. & ODP Leg 204 Scientific Party. 2004. Ethane enrichment, propane depletion, and other compositional variations in subsurface gases. a new indicator of gas hydrate occurrence in marine sediments: Organic Geochemistry 35, 1067-1080.
Milkov, A.V., Claypool, G.E., Lee, Y.-J., Xu, W., Dickens, G.R., Borowski, W.S. & ODP Leg 204 Scientific Party. 2003. *In-situ* methane concentrations at Hydrate Ridge, offshore Oregon: New constraints on the global gas hydrate inventory from an active margin. Geology 31(10), 833–836.
Milkov, A.V., Dickens, G.R., Claypool, G.E., Lee, Y.-J., Borowski, W.S., Torres, M.E., Xu, W., Tomaru, H., Tréhu, A.M. & Schultheiss, P. 2004. Co-existence of gas hydrate, free gas, and brine within the gas hydrate stability zone at the southern summit of Hydrate Ridge (Oregon margin): Evidence from prolonged degassing of a pressurized core. Earth and Planetary Science Letters 222, 829-862.
Milkov, A.V., & Sassen, R. 2000. Thickness of the gas hydrate stability zone, Gulf of Mexico continental slope: Marine and Petroleum Geology 17, 981-991.
Milkov, A.V., & Sassen, R. 2001. Economic geology of the Gulf of Mexico and the Blake Ridge gas hydrate provinces: Trans. Gulf Coast Association Geological Societies 51, 219-228.
Milkov, A.V., & Sassen, R. 2001. Estimate of gas hydrate resource, northwestern Gulf of Mexico continental slope: Marine Geology 179, 71-83.
Milkov, A.V., & Sassen, R. 2001. Structurally focused gas hydrate, a future northwest US Gulf resource. Offshore Magazine, September, 92-93.
Milkov, A.V., & Sassen, R. 2002. Economic geology of offshore gas hydrate accumulations and provinces: Marine and Petroleum Geology 19, 1-11.
Milkov, A.V. & Sassen, R. 2003. Preliminary assessment of resources and economic potential of individual gas hydrate accumulations in the Gulf of Mexico continental slope. Marine and Petroleum Geology 20. 111-128.
Milkov, A.V. & Sassen, R. 2003. Two-dimensional modeling of gas hydrate decomposition in the northwestern Gulf of Mexico: Significance to global change assessment. Global and Planetary Change 36, 31-46.
Milkov, A.V, Sassen, R., Apanosovich, T.V. & Dadashev, F.G. 2003. Global gas flux from mud volcanoes; a significant source of fossil methane in the atmosphere and ocean. Geophysical Research Letters 30(2), 9-1 to 9-4.
Milkov, A.V., Sassen, R., Novikova, I. & Mikhailov, E. 2000. Gas hydrates at minimum stability water depths in the Gulf of Mexico: Significance to geohazard assessment. Transactions Gulf Coast Association Geological Socities 50, 217-224.
Milkov, A.V., P.R. Vogt, K. Crane, A.Yu. Lein, Sassen, R., & Cherkashev, G.A. 2004. Geological, geochemical, and microbial processes at the hydrate-bearing Håkon Mosby mud volcano. a review: Chemical Geology 205(3-4), 347-366.
Miller, J.J., Lee, M.W. & von Huene, R. 1991. An analysis of a seismic reflection from the base of a gas hydrate zone. American Association Petroleum Geologists 75, 910-924.
Mitts, P.J., Paull, C.K., Ussler III, W., Dickens, G.R. & Perry, K.L. 2004. Pore water sulfate, chloride, and methane concentrations in sediment gravity cores from the Fairway and New Caledonia Basins: results from Franklin Cruise FR9/01. In: Exon, N., Hill, P., Lafoy, Y., Fellows, M., Perry, K., Mitts, P., Howe, R., Chaproniere, G., Dickens, G., Ussler, W., & Paull, C., 2004. Geology of the Fairway and New Caledonia Basins in the Tasman Sea: sediment, pore water, diapirs and bottom simulating reflectors (Franklin Cruise FR9/01 and Geoscience Australia Survey 232). Geoscience Australia Record 2004/26, 70-92.

MMS. 1997. Report of the Subcommittee on Environmental Information for Select OCS Areas under Moratoria, United States Minerals Management Service, 41pp.
MMS. 2004a. Shallow-Water Gas Discovery, MMS Ocean Science, September/October 2004, United States Minerals Management Service, 11.
MMS. 2004b. Offshore Stats and Facts, United States Minerals Management Service http://www.mms.gov/offshore/.
MMS. 2004c. Deepwater Gulf of Mexico 2004: America's Expanding Frontier, OCS Report MMS 2004-021.
Moon, C., Tayor, P.C. & Rodger, P.M.J. 2003. American Chemical Society 125, 4706-4707.
Moore, G.F., Taira, A. & Klaus, A. 2000. Tectonics and hydrogeology of the Nankai Trough accretionary prism: Preliminary results of ODP Leg 190. (Abs. T52D-09), EOS, Transactions American Geophysical Union. 81(48).
Moridis, G.J. 2003. Numerical Studies of Gas Production From Methane Hydrates, SPE Journal, 32(8), 359-370.
Moridis, G.J. 2004. Numerical Studies of Gas Production from Class 2 and Class 3 Hydrate Accumulations at the Mallik Site, Mackenzie Delta, Canada, SPE Reservoir Evaluation and Engineering 7(3), 175-183.
Moridis, G.J., Kowalsky, M. & Pruess, K. TOUGH-Fx/HYDRATE: A code for the simulation of system behavior in hydrate-bearing geologic media, Lawrence Berkeley National Laboratory, 2005 (LBNL Report No. pending [28 Feb. 05]).
Moridis, G. & Collett, T.S. 2003. Strategies for gas production from hydrate accumulations under various geologic conditions, Lawrence Berkeley National Laboratory. Paper LBNL-52568.
Moridis, G.J. & Collett, T.S. 2004. Gas Production from Class 1 Hydrate Accumulations, in Recent Advances in the Study of Gas Hydrates, C. Taylor and J. Qwan, Editors, Kluwer Academic/Plenum Publishers (Section I, Chapter 6), 75-88.
Moridis, G.J., Collett, T.S. Dallimore, S.R., T. Inoue, T. & Mroz, T. 2005. Analysis and Interpretation of the Thermal Test of Gas Hydrate Dissociation in the Mallik 5L-38 Research Well, Mackenzie Delta, Canada; In: Scientific Results from the Mallik 2002 Gas Hydrate Production Research Well Program, Mackenzie Delta, Northwest Territories, Canada, Dallimore, S.R. & Collett, T.S. (Eds.), Geological Survey of Canada, Bulletin 585.
Moridis, G.J., Collett, T., Dallimore, S.R., Satoh, T., Hancock, S. & Weatherhill, B. 2004. Numerical Studies Of Gas Production From Several Methane Hydrate Zones at the Mallik Site, Mackenzie Delta, Canada. Journal of Petroleum Science and Engineering 43, 219-239.
Moridis, G.J., Seol, Y., Collett, T.S., Dallimore, S.R., Inoue, T., Mroz, T.H. & Henninges, J. 2003. Thermal properties of hydrates from temperature data analysis of an isolated formation interval in the 5L-38 Mallik Research Well (abs.). Mallik International Symposium (Chiba, Japan), Japan National Oil Corporation, Technology Research Center, 58
Morita, S., Ashi, J., Kuramoto, S., Taira, A., Bangs, N.L., Shipley, T.H. & Moore, G.F. 2000. Interaction between geologic structure and distribution of BSR in the western Nankai accretionary prism. (Abs. T51A-14), EOS, Transactions American Geophysical Union 81(48).
Moore, J.N. 2003. Personal communication May 8, National Defense University presentation on Law of the Sea.
Moridis, G.J. & Collett, T.S. 2003. Strategies for gas production from hydrate accumulations under various geological and reservoir conditions. Proceedings of the TOUGH Symposium, Lawrence Berkeley National Laboratory, Berkeley, California, U.S.A.
Moridis, G.J., Collett, T.S., Dallimore, S.R., Satoh, T., Hancock, S. & Weatherill, B. 2002. Numerical studies of gas production from several CH4-hydrate zones at the Mallik site, MacKenzie Delta, Canada. LBNL-50527, Lawrence Berkeley National Laboratory, Berkeley, California, U.S.A.
Mosher, D.C., Louden, K., LeBlanc, C., Shimeld, J., & Osadetz, K. 2005. Gas hydrates offshore eastern Canada: fuel for the future? Proceedings Offshore Technology Conference, (CD-ROM), Houston Texas, 2-5 May, 2005, (OTC 17588), 9pp.
Myerson, A.S. 2002. (Ed.), *Handbook of Industrial Crystallization*, 2nd. Ed. Butterworth-Heinemann, Boston, 313pp.
MWD. Merriam-Webster Collegiate Dictionary (on-line, < http://www.m-w.com/cgi-bin/dictionary>
Naudts, L., Greinert, J., Artemov, Yu.G & De Batist, M. 2005. Subsurface and morphologic setting of 2778 methane seeps in the Dnieper paleo-delta in the northwestrn Black Sea. (Poster). International Gas Hydrate Workshop, Victoria, BC, Canada, 8-11 May, 2005.
NEB. 2004. Canada's Conventional Natural Gas Resources, National Energy Board (Canada), 38pp.
Neil, C. 2004. 2003 shows spot cargoes, tankers to dictate US LNG supplies, not terminal capacities, Oil & Gas Journal, April 12, 2004, 70.
Nelson, K. 2004. BP Sets Record. (www.petroleumnews.com) Petroleum News 9(41), 3pp.

NETL. 2005. The DOE/JIP Gulf of Mexico Hydrate Research Cruise, http://www.netl.doe.gov/scngo/Natural%20Gas/hydrates/rd-program/GOM_JIP/index.html

Newberry, C.J., Webster, G., Cragg, B.A., Parkes, R.J., Andrew J. Weightman, A.J. & Fry, J.C. 2004. Diversity of prokaryotes and methanogenesis in deep subsurface sediments from the Nankai Trough, Ocean Drilling Program Leg 190. Environmental Microbiology 6(3), 274-doi:101111/j.1462-2920.2004.00568.x.

Normile, D. 2005. Japan's new ship sets standard as modern, floating laboratory. Science 307, 1552-1553.

NPC. 2003. Balancing Natural Gas Policy – Fueling the demands of a Growing Economy, Vol. 1 – Summary of Findings and recommendations, National Petroleum Council, 109pp.

NRC, 2004. Charting the Fugure of Methane Hydrate Research in the United States. National Research Council Report of the Committee to Review the Activites Authorized undeer the Methane Hydrate Resarch and Development Act of 2000. Ocean Studies Board, Board on Earth Sciences and Resources, Division on Earth and Life Studies. The National Academies Press, Washington, D.C., 192pp.

Ohmura, R., Matsuda, S., Uchida, T., Ebinuma, T. & Narita, H. 2004. Chathrate hydrate crystal growth in liquid water saturated with a guest substance: Observations in a methane + water system. Crystal Growth & Design 5(3), 953-957.

Ohmura, R., Ogawa, M., Yasuoka, K. & Mori, Y.H. 2003. Statistical study of the clathrate-hydrate nucleation in a water/hydrochlorofluorocarbon system: Search for the nature of the "Memory Effect", Journal of Physical Chemistry (B) 107, 5289-5293.

ON&T. 2005. Deepwater GoM sees major advances in 2004. Ocean News & Technology 11(2), 31.

Orphan, V.J., House, C.H., Hinrichs, K.-U., McKeegan, K.D., DeLong, E.F., 2001. Methane-consuming Archaea revealed by directly coupled isotopic and phylogenetic analysis. Science, 293, 484-487.

Osadetz, K.G. & Chen, Z. 2005. A re-examination of Beaufort Sea-Mackenzie delta basin gas hydrate resource potential using a petroleum play approach. Proceedings 5th International Confrerence on Gas Hydrates, 13-16 June, 2005, Trondheim, Norway, 11pp.

OTC, 2004. General communication in discussion at the 2004 Gas Hydrate session. 2004 Offshore Technology Conference 3-6 May 2004, Houston, Texas U.S.A.

Oxman, B.H. 1980. The United Nations Conference on the Law of the Sea: The Ninth Session. 75 American Journal Intertional Law 211, 229.

Packham, G.H. & van der Lingen, G.J. 1973. Progressive carbonate diagenesis at Deep Sea Drilling Sites 206, 207, 208, and 210 in the Southwest Pacific and its relationship to sediment physical properties and seismic reflectors. In: Burns, R.E., Andrews, J.E. et al., 1973. Initial Reports of the Deep Sea Drilling Project, 21, U.S. Government Printing Office, Washington D.C., 495-521.

Parker, J.C., Lenhard, R.J. & T. Kuppusamy, T. 1987. A Parametric Model for Constitutive Properties Governing Multiphase Flow in Porous Media, Water Resources Research 23(4), 618-624.

Parkes, R.J., Wellsbury, P., Mather, I.D. & Maxwell, J.R. 2000. Possible deep biosphere energy sources: Where does bacterial diagenesis end and thermal alteration begin. (Abs. B72D-03), EOS, Transactions American Geophysical Union 81(48).

Patin, S. 1999. (Translated from: *Russian Envrionmental Impact of the Offshore Oil and Gas Industry.* EcoMonitor Publishing, East Northport, NY.

Paull, C.K., Borowski, W.S. & Rodriguez, N.M. 1998. Marine gas hydrate inventory: preliminary results of ODP Leg 164 and implications for gas venting and slumping associated with the Blake Ridge gas hydrate field. Geological Society Special Publication 137, 153-160.

Paull, C.K. & Dillon, W.P. (Eds.) 2001. *Natural Gas Hydrate – Occurrence, Distribution, and Detection. American Geophysical Union.* Washington, DC. 315pp.

Paull, C.K., Spiess, F.N., Ussler, W. & Borowski, W.S. 1995. Methane-rich plumes on the Carolina continental rise: associations with gas hydrates, Geology 23, 89-92.

Paull, C.K. Ussler, W. 2001. History and significance of gas samplind during DSDP and ODP drilling associated with gas hydrates. In: Paull, C.K. & Dillon, W.P. (Eds), Natural Gas Hydrates. Occurrence, Distribution, and Dynamics. Geophysical Monograph 124, AGU, 53-65.

Paull, C.K., Ussler, W., and Borowski, W.S. 1994. Sources of biogenic methane to form as hydrates, in situ production or upward migration, International Conference on Natural Gas Hydrates, Annals NY Academy of Sciences 715, Sloan, E.D., Happel, J. & Hnatow, M.A. (Eds.), 392-409.

Paull, C.K., Ussler, W. III & Dillon, W.P. 2003. Potential role of gas hydrate decomposition in generating submraine slope failures. In: Max, M.D. (Ed.), Natural Gas Hydrate in Oceanic and Permafrost Environments. Kluwer Academic Publishers, The Netherlands, 149-156.

Pecher, I.A. 2002. Gas hydrates on the brink, Nature (news and views), 420, 622-623.

Pecher, I.A., Henrys, S.A., Bannister, S., Davey, F., Nishimura, Y., Yamada, A., Shimamura, H., NIGHT Working Group. 2002. North Island Geophysical Transect (NIGHT)—first results from a combined

ocean bottom seismometer and multichannel seismic reflection survey. EOS, Western Pacific AGU Meeting Supplement 763. 85,763.
Pecher, I.A., Henrys, S.A. & Zhu, H. 2004. Seismic images of gas conduits beneath vents and gas hydrates on Ritchie Banks, Hikurangi Margin, New Zealand, N. Z. Journal Geology Geophysics 47, 275-279.
Pecher, I.A. & Holbrook, W.S. 2000, 2003. Seismic Methods for Detecting and Quantifying Marine Gas Hydrate/Free Gas Reservoirs: In: Oceanic and Permafrost environments. Max, M.D. (Ed.), Kluwer Academic Publishers, London, Boston, Dordrecht, 286-294.
Pecher, I.A., Holbrook, W.S., Sen, M.K., Lizarralde, D., Wood, W.T., Hutchinson, D.R., Dillon, W.P., Hoskins, H. & Stephen, R.A. 2003. Seismic anisotropy in gas-hydrate- and gas-bearing sediments on the Blake Ridge, from a walkaway vertical seismic profile, Geophysical Research Letters 30(14), 1733, 4, doi:10.1029/2003GL017477.
Pecher, I.A., Kukowski, N. Ranero, C.R. & von Huene, R. 2001b. Gas hydrates along the Peru and Middle America trench systems, In: Paull, C.K. & Dillon, W.P. (Eds), Natural Gas Hydrates. Occurrence, Distribution, and Dynamics. Geophysical Monograph 124, AGU, 257-271.
Pecher, I.A., Kukowski, N., Huebscher, C., Greinert, J. & Bialas, J. 2000. Evidence for a link between bottom simulating reflections and elevated fluid flow, Lima basin, Peru margin, (Abs. T61A-16), EOS, Transactions American Geophysical Union v. 81(48).
Pecher, I.A., Kukowski, N., Huebscher, C., Greinert, J, Bialas, J., & GEOPECO Working Group, 2001b. The link between bottom simulating reflections and methane flux into the gas hydrate stability zone -- new evidence from Lima Basin, Peru Margin, Earth Planetary Science Letters 185, 343-354.
Pellenbarg, R.E. & Max, M.D. 2001. Gas Hydrates: From Laboratory Curiosity to Potential Global Powerhouse. Journal of Chemical Education 78, 896-900.
Pellenbarg, R.E., Max, M.D. & Clifford, S.M. 2003. Methane and carbon dioxide hydrates on Mars: Potential origins, distribution, detection and imlications for future in-situ resource utilization. Journal of Geophysical Researches – Planets 108 E4, 8042, doi:10.1029/2002JE001901.
Perry, K.F., Cleary, M.P., & Curtis, J.B. 2004. New Technology for Tight Gas Sands, World Energy Council. http://www.worldenergy.org/wecgeis/publications/default/tech_papers/17th_congress/2_1_16.asp# Heading6.
Peltzer, E.T. & Brewer, G. 2003. Practical physical chemistry and empirical predictions of methane hydrate stability. In: Max, M.D. (Ed.), 2003. Natural Gas Hydrate in Oceanic and Permafrost Environments, Kluwer Academic Publishers, The Netherlands, 17-28.
Peltzer, E.T., White, S.N., Dunk, R.M., Brewer, P.G., Sherman, A.D., Schmidt, K., Hester, K.C., Sloan, E.D. 2004. (in-situ) Raman analyses of natural gas and gas hydrates at Hydrate Ridge, Oregon. In: Proceedings 85(47) AGU Fall Meeting, San Francisco, Ocean Sciences General Contributions: Gas Hydrates I OS34B-01.
Personal Communications allowing use of unpublished information.
Warren T. Wood & Clifford I. Voss. Thermal modeling of low-temperature seafloor seeps and implications for methane release. E-mail. December 23 2004.
Rick Hagan. Calculations and modeling relating to thermal modeling of low-temperature seafloor seeps and implications for methane release. E-mail. December 23 2004.
Pignot, P. & Chapman, N.R. 2001. Tomographic inversion of geoacoustic properties in a range-dependent shallow water environment, Journal Acoustic Society America 110, 1338–1348.
Pike, W. 2004. LNG takes off?, Hart's E&P, November, 2004, 5.
Piñero, E., Gràcia, E., Martínez-Ruiz, F., Larrasoaña, J.D., Vizeaino, A. & Ercilla, G. 2005. Sedimentological and mineralogical characterization of gas hydrate bearing sediments from ODP-Leg 204, southern Hydrate Ridge (Cascadia Margin). Geophysical Research Abstracts 7, 07704.
Piñheiro, L.M., Magalhães, V., Somoza, L., V. Diaz del Rio, V., Gardner, J. & Ivanov, M. Gas hydrates, mud volcanoes and carbonate chimneys in the Gulf of Cadiz. 2003. Geophysical Research Abstracts 5, 13693.
Pohlman, J.W., Wood, W., Boyd, T.J, Coffin, R.B., Ruppel, C., Hart, P., Dugan, B. & Hutchinson, D. 2004. Interstitial sulfate gradients as indicators of gas hydrates: How reliable are they? 32nd International Geological Conference, Florence Italy, August 20-28.
Pohlman, J., Camuel, E.A., Chapman, N.R., Spence, G.D., Whiticar, M.J. & Coffin, R.B. 2004. Origin, structure and composition of thermogenic gas hydrates on the northern Cascadia margin. (submitted to J. Petroleum Geology).
Poruban, S. 2004. Ziff: Only some LNG sites will be built in North America, Oil & Gas Journal, May, 3, 42.
Posewang, J., Mienert, J. 1999. The engima of double BSRs: Indicators for changes in the hydrate stability field? Geo-Marine Letters 19, 157-163.
Potter P.E., Maynard, J.B. & Depetris, P.J. 2005. *Mud and Mudstones, Introduction and Overview.* Springer, Berlin, Heidelberg, New York, 297pp.

Premoli Silva, I., Erba, E., Spezzaferri, S. and Cita, M.B., 1996. Age variation in the source of the diapiric mud breccia along and across the axis of the Mediterranean Ridge accretionary complex. Marine Geology 132, 175-202.

Priestley, J. 1790. Section III. Observations on the Freezing of Water, impregnated with Flou Acid Air, and with Vitriolic Acid Air. In: Experiments and Observations on Different Kinds of Air and Other Natural Branches of Natural Philosophy. Vol. II. (3 Vols.) Thomas Pearson, Birmingham. (Krauss Reprint Co. LC-71-125366, New York, 1970), 359-363.

Prindle, D.F. 1981. Petroleum Politics and the Texas Railroad Commission. Technical report. Austin, Tex.: University of Texas Press, 230pp.

Puesy, W.C. 1973. How to evaluate potential gas and oil source rocks. World Oil 176(5), 71-75.

Qian, J., Zhen, J. 2002. Identity method of gas hydrate in the seismic profile, Marine Oil 3, 54-59

Quirk, D.G. & Aitken, J.F. 1997. The strcture of the Westphalian in the northrn part of the southern North Sea. In: Ziegler, K, Turner, P. & Daines, S.R. (Ed.), Petroleum geology of the southern North Sea: future potential. Geological Society Special Publication 123, 143-152.

RCM. 2003. (*in Russian*). Russian Chemical Journal. D.M.Mendeleev Russian Chemical Society. (Rossiiskii Khimicheskii Zhurnal. Rossiiskoe Khimicheskoe obshchestvo imeni D.M.Mendeleeva. Moskva). Moscow. vol. XLVII (3).

Inerbaev, T.M., Subbotin, O.S., Belosludov, V.P., Kawazoe, E. & Kudo, D.I. Dynamic, Permodynamic and mechanical properties of gas hydrate structures I and II, 19-27.

Kuznezov, & V.A.Istomin, T.V. Rodionova- Gas hydrate: historical summary, current status, and prospects for study, 5-18.

Makogon, Yu.F. & Cholsti, D.S. Whisker hydrate crystals, 43-48.

Manakov, A.Yu. & Dyadin, Yu.A. Gas hydrate at high pressure, 28-42.

Matveeva T.V. & Solov'ev, V.A. Gas hydrate in the Sea of Ochotsk: Regularity of formation and dispostion, 100-111.

Yakushev, V.S., Perlova, E.V., Machonina, N.S., Chuvilin, E.M. & Kozlova, E.V. Gas hydrate in sediments of continents and islands, 80-90.

Reeburgh, W.S. 1976. Methane consumption in the Cariaco Trench waters sediments. Earth Planetary Science Letters 28, 337-344.

Reeburgh, W.S., 1982. A major sink and flux control for methane in marine sediments: Anaerobic Consumption. In: Fanning, K.A. & Manheim, F.T, (Eds.), The Dynamic Environment of the Ocean Floor. Lexington Books, Lexington, MA, 203-218.

Reed, D.L., Silver, E.A., Tagudin, J.E., Shipley, T.H., & Vrolijk, P. 1990. Relations between mud volcanoes, thrust deformation, slope sedimentation, and gas hydrate, offshore north Panama: Marine and Petroleum Geology 7, 44–54.

Rehder, G., Brewer, P.W., Peltzer, E.T. & Friederich, G. 2002. Enhanced lifetime of methane bubble streams within the deep ocean. Geophysical Research Letters 29, 1731, doi:10.1029/2001GL013966.

Rehder, G., Kirby, S.H., Durham, W.B., Stern, L.A., Peltzer, E.T., Pinkston, J. & Brewer, P.W. 2004. Dissolution rates of pure methane hydrate and carbon-dioxide in undersaturated seawater at 1000-m depth. Geochimica et Cosmochimica Acta 68, 285–292.

Riedel, M., Novosel, I., Bpence, G.D., Hyndman, R.D., Chapman, R.N., Solem, R.C., Lewis, T. & Zuelsdorff, L. 2005. Geophysical and geochemical signatures associated with gas hydrtate related venting at the north Cascadia margin. Geological Society of America Bulletin (accepted for publication).

Renssen, H., Beets, C.J., Fichefet, T., Goosse, H. & Kroon, D. 2004. Modeling the climate response to a massive methane release from gas hydrates. Paleoceanography 19, PA2010, doi: 10.1029/2003PA000968.

Reuters. 2004. Exxon forecasts LNG volumes growing 300 pct by '30'. http://www.reuters.com/financeQuoteCompanyNewsArticle.jhtml?duid=MTFH53413_2004-11-03_23-35 05_N03203350_NEWSML.

Riedel, M. & Theilen, F. 2001. AVO investigations of shallow marine sediments. Geophysical Prospecting 49, 198– 212.

Riedel, M., G.D. Spence, N.R. Chapman & R.D. Hyndman, 2002. 3D Seismic investigation of a vent field associated with gas hydrates offshore Vancouver Island. Journal Geophysical Research 107 (B9), EPM 5, 1-16.

Riedel, M., Spence, G.D., Chapman, N.R. & Hyndman, R.D. 2001. Deep sea gas hydrates on the northern Cascadia margin, The Leading Edge, January, 87-92.

Ripmeester, J.A. 2000. Hydrate Research – From Correlations to a Knowledge-based Discipline. In: Holder, G.D. & Bishnoi, P.R. Gas Hydrates, Challenges for the Fugure. Annals of the New York Academy of Sciences 912, 1-16.

Robb, L. 2004. *Introduction to Ore-Forming Processes*. Blackwell, Oxford, UK, Madon, MA, USA, 384pp.

Roberts, H.H. 2001. Fluid and gas expulsion on the northern Gulf of Mexico continental slope: Mud prone to mineral-prone responses, In: Natural Gas Hydrates: Occurrence, Distribution, and Detection. Paull, C.K. & Dillon, W.P. (Eds.), AGU Monogrraph 124, 145-161.

Roberts, H., Coleman, J., Hunt, J.L. Jr. & Shedd, W. 2000. Improving Interpretation of Seafloor Geology, Biology from Remote Sensing. Offshore Magazine, September 2000, 126.

Roberts, H.H., Hunt, J.L., Shedd, W.W. & Sassen, R. 2002. Surficial gas hydrates, part of the fluid and gas expulsion response spectrum: Identification from 3D seismic data. 2002 Offshore Technology Conference. OTC 14033, 10pp.

Roberts, H.H, Coleman, J.M., Hunt, J.L., Shedd, W.W., Sassen, R. & Milkov, A.V. 2002. Gas hydrate deposits in a complex geologic setting, northern Gulf of Mexico Slope: Transactions, Gulf Coast Association of Geological Societies52, 860.

Roberts, P. 2004. *The End of Oil: On the Edge of a Perilous New World.* Haughton Mifflin, 389pp.

Romero, S. 2004. Options exhausted, oil firms turn to tar. International Herald Tribune, September 1.

Rossetti, D.D.F. 1999. Soft-sediment deformation structures in late Albian to Cenomanian deposits, São Luís Basin, northern Brazil: evidence for paaeoseismicity. Sedimentology 46(6), 1065-1081.

Ruppel, C., Dickens, G., Castellini, D.G., Gilhooly, W. & Lizarralde, D. 2005. Heat and salt inhibition of gas hydrates in the northern Gulf of Mexico, Geophysical Research Letters 32, L04605

Ruppel, C. Kinoshita, M. 2000. Heat, methane, and energy flux in an active margin gas hydrate province, offshore Costa Rica, Earth Planetary Science Letters 179, 153-165.

Sad, A.R.E., Silveira, D.P., Sergio, R.P., Maciel, R.R. & Machado, M.A.P. 1998. Marine gas hydrates along the Brazilian margin. In: Extended Abstracts, AAPG International Conference and Exhibition, 146.

Sain, K., Minshull T.A., Singh, S.C., et al. 2000. Evidence for a thick free gas layer beneath the bottom simulating reflector in the Makran accretionary prism. Marine Geology 164, 3-12.

Sassen, R. 2000. Gulf of Mexico, West Africa Hydrates. Offshore Magazine, July, 164.

Sassen, R., & MacDonald, I.R. 1997. Thermogenic gas hydrates, Gulf of Mexico Continental Slope: Preprints, 213th ACS National Mtg., San Francisco, CA, 42(2), 472-474.

Sassen, R., Sweet, S.T., Milkov, A.V., DeFreitas, D.A. & Kennicutt, M.C. II. 2001. Stability of thermogenic gas hydrate in the Gulf of Mexico: Constraints on models of climate change. In: Natural gas hydrates occurrence, distribution, and detection, Paull, C.K. & Dillon, W.P. (Eds.), American Geophysical Union Geophysical Monograph, 124, 131-143.

Sassen, R., J.S. Watkins, C. Decker, Sweet, S.T., DeFreitas, D.A., D.A. &Losh, S. 2001b. High-maturity gas in the Main Pass area: Comparison to the central Gulf of Mexico slope: Transactions Gulf Coast Association Geological Societies 51, 285-291.

Sassen, R., Joye, S., Sweet, S.T., DeFreitas, D.A., Milkov, A.V. & Macdonald, I.R. 1999. Thermogenic gas hydrates and hydrocarbon gases in complex chemosynthetic communities, Gulf of Mexico continental slope. Organic Geochemistry 30, 485-497.

Sassen, R., Losh, S.L., Cathles III, Roberts, H.H., Whelan, J.K., Milkov, A.V., Sweet, S.T., & DeFreitas, D.A. 2001. Massive vein-filling gas hydrate: relation to ongoing gas migration from the deep subsurface in the Gulf of Mexico. Marine and Petroleum Geology 18, 551-560.

Sassen, R., Milkov, A.V., DeFreitas, D.A., Sweet, S.T., & Roberts, H.H. 2002. Gas hydrate in the Gulf of Mexico. Significance to sediment deformation. 2002 Offshore Technology Conference. 14034, 7pp.

Sassen, R., Milkov, A.V., Ozgul, E., Roberts, H.H., Hunt, J., Beeunas, M.A., Chanton, J. Sweet, S.T. & DeFreitas, D.A. 2003a. Gas venting and subsurface charge in the Green Canyon area, Gulf of Mexico continental slope: evidence of a deep bacterial methane source? Organic Geochemistry 34, 1555-1464.

Sassen, R., Milkov, A.V., Roberts, H.H., Sweet, S.T. & DeFreitas, D.A. 2003. Geochemical evidence of rapid hydrocarbon venting from a seafloor-piercing mud diapir, Gulf of Mexico continental shelf. Marine Geology 198, 319-329.

Sassen, R., Post, P.J., DeFreitas, D.A. & Sweet, S.T. 2003. Geochemistry of an oil seep and gas hydrate on the lower slope, central Gulf of Mexico: Significance to paleogeography and possible source rocks. Transactions Gulf Coast Association of Geological Societies 53, 741-754.

Sassen, R., Roberts, H.H., Carney, R., Milkov, A.V., DeFreitas, D.A., Lanoil, B. & Zhang, C. 2004. Free hydrocarbon gas, gas hydrate, and authigenic minerals in chemosynthetic communities of the northern Gulf of Mexico continental slope: relation to microbial processes. Chemical Geology 205(3-4), 195-217.

Sassen, R., Sweet, S.T., DeFreitas, D.A., & Milkov, A.V. 2000. Exclusion of 2-methylbutane (isopentane) during crystallization of structure II gas hydrate in sea-floor sediment, Gulf of Mexico. Organic Geochemistry 31, 1257-1262.

Sassen, R., Sweet, S.T., DeFreitas, D.A., Milkov, A.V., Salata, G.G. & McDade, E.C. 1999. Geology and Geochemistry of gas hydrates, central Gulf of Mexico continental slope. Transactions Gulf Coast Association Geological Societies 49, 462-468.

Sassen, R., Sweet, S.T., DeFreitas, D.A., Morelos, J.A. & Milkov, A.V. 2001. Gas hydrate and crude oil from the Mississippi Fan Foldbelt, downdip Gulf of Mexico Salt Basin. Significance to petroleum system: Organic Geochemistry 32, 999-1008.

Sassen, R., Sweet, S.T., Milkov, A.V., DeFreitas, D.A., & Kennicutt II, M.C. 2001c. Thermogenic vent gas and gas hydrate in the Gulf of Mexico slope: Is gas hydrate decomposition significant? Geology 29, 107-110.

Sassen, R., Sweet, S.T., Milkov, A.V., DeFreitas, D.A., Kennicutt II, M.C. & Roberts, H.H. 2001. Stability of thermogenic gas hydrate in the Gulf of Mexico: Constraints on models of climate change. In: Natural Gas Hydrates: Occurrence, Distribution, and Detection. In: Paull C.K. & Dillon, W.P. (Eds.), AGU Geophysical Monograph 124, 131-143.

Satoh, T., Dallimore, S., Collett, T., Inoue, T., S. Hancock, S., G. Moridis, G. & Weatherill, B. 2003. Production Test Planning for the Japex/JNOC/GSC Mallik 5L-38 Gas Hydrate Research Well. Program Abstracts, From Mallik to the Future. Mallik International Symposium in Makuhari, Japan. December 8-10, 71.

Sawai, Y. Satake, K., Kamataki, T., Nasu, H., Shishikura, M., Atwater. B.R., Horton, B.P., Kelsey, H.M., Nagumo, T. & Yamaguchi, M. 2004. Tranient uplift after a 17-th centruy earthquake along the Kuril subduction zone. Science 306, 1918-1920.

Schempf, F.J. 2004. The History of Offshore: Developing the E&P Infrastructure, Offshore, January, 20-35.

Schicks, J.M. & Ripmeester, J.A. 2004. The coexistence of two different methane hydrate phases under moderate pressure and temperature conditions: kinetic versus thermodynamic products" Angew. Chem. Int. Ed. 43, 3310-3313.

Schmidt, G.A. & Shindell, D.T. 2003. Atmospheric composition, radiative forcing, and climate change as a consequence of massive methane release from gas hydrates. Paleoceanography 18, No. 1004, doi: 10.1029/2002 PA000757.

Schmitz, B. & Pujalte, V. 2003. Sea-level, humidity, and land-erosion records across the initial Eocene thermal maximum from a continental-marine transect in northern Spain. Geology 31, 689–692.

Segall, M.P., Figueiredo, A.G. & Silva, C.G. 2000. Stability of near-surface gassy sediments, Amazon continental shelf. Procedings Offshore Technology Conference 1-4 May, Houston TX, OTC 12106, 5pp.

Selley, R.C. 1998. *Elements of Petroleum Geology. (2nd Ed.),* Academic Press, London, New York, Toronto, 470pp.

S&E. 2004. (*in Russian with Abstracts in English*). Science and Engineering in natural gas industry 1-2, (Nauka i technika v gozovoi promishlennosti, tom.1-2) (in order but page #s not available).

- Soloviev, V.A. & Mazurenko, L.L. Gas hydrate accumulation in fluid discharge centers as the subject of research and development, 14-19.
- Obzhirov, A.I., Salyuk, A.N., Shakirov, R.B., Druzhinin, V.V., Mishukova, G.I., Ageev, A.A., Solomatin, A.S., Petrikova, N.L., Veselov, O.V. & Koudelkin, V.V. Methane flows and gas hydrates in the Sea of Okhotsk, 20-25.
- Agalokov, S.E. & Kourchikov, A.R. Gas resources in the gas hydrate stability areas in the West Siberia(n) northern area, 26-35.
- Yakushev, V.S., Perlova, E.V. & Makhonina, N.A. Internally frozen accumulations of natural gas and gas hydrates, their origin and the development prospects, 36-42.
- Makhonina, N.A., Perlova, E.V., Yakushev, V.S. & Akhmedsafin, S.K. Gas accumulations in the Zapolyarnoye oil, gas, and gas condensate field's permafrost, 43-46.
- Narita, H. & Ebinuma, T. Research programme for developing methane hydrate resources in Japan, 47-55.
- Basniev, K.S. & Shchebetov, A.V. Natural hydrate formation development outlook, 90-95.
- Istomin, V.A., Derevyagin, A.M., Seleznev, S.V., Agaltsov, A.G. & Mikhailov, Yu.V. The dew point method for studying 2-phase equilibrium of gas hydrates, 63-70.
- Tohidi, B, Arhmandi, M., Anderson, R. & Yan, C. Study of low-dosing hydrate inhibitors at Heriot-Watt University, 81-89.
- Basniev, K.S. & Nifantov, A.N. 3D mathematical model of gas hydrate heat-based dissolution in porous media, 90-95.

Servio, P. & Englezos, P. 2001. Effect of temperature and pressure on the solubility of carbon dioxide in water in the presence of gas hydrate. Fluid Phase Equilibria 190, 127-134.

Servio, P. & Englezos, P. 2002. Measurement of dissolved methane in water in equilibrium with its hydrate, Journal of Chemical Engineering Data 47, 87-90.

Sherlock, R.L., Lee, J.K.W. & Cousens, B.L. 2004. Geologic and Geochronologic Constraints on the Timing of Mineralization at the Nanisivik Zinc-Lead Mississippi Valley-Type Deposit, Northern Baffin Island, Nunavut, Canada. Economic Geology 99, 279-293.

Shirota, H., Aya, I., Namie, S., Bollavaram, P., Turner, D., & Sloan, D.E. 2002. Measurement of Methane Hydrate Dissociation for Application to Natural Gas Storage and Transportation. Procedings 4th International Conference on Gas Hydrate, Yokahama, Japan, 972-977.

Shreve, R.L. & Cloos, M. 1986. Dynamics of sediment subduction, melange formation, and prism accretion. Journal of Geophysical Research 91, 10,229-10,245.

Shoji, H., Soloviev, V., Mateeva, T., Mazurenko, L., Minami, H., Hachkubo, A., Sakagami, H., Hyakutake, K., Kaulio, V., Gladysch, V., Logvina, E., Obzhirov, A., Baranov, B., Khlystov, O., Biebow, N., Poort, J., Jin, Y.K. & Kim, Y. 2005. Hydrate-bearing structures in the Sea of Okhotsk. Eos 86(2), 13, 18.

Singh, S. C., Minshull, T.A. & Spence, G.D. 1993. Velocity structure of a gas hydrate reflector. Science 260 , 204–207.

Singh, U.K., Pandey, D. Kurian, J., Vohat, P., Vijayalaxmi, S.N. & Rajan, S. 2004. Application of marine geophysical data for delineating the Indian legal continental shelf and hyrocarbon resources. In: Report on the Regional Workshop, CODATA task group on natural gas hydrates, Indian National Science Academy, New Delhi, http://www.codata.org/taskgroups/TGhydrates/NewDelhiFeb04.pdf, 12-15.

Skislock, K.B. 2005. Crisis on the China Rim: An economic, crude oil, and military analysis. Laguna Research Partners. https://www.researchconnect.com/reports/CrisisOnTheChinaRim_2005-04-14_LRP_FINAL.pdf

Sloan, E.D. 1998. *Clathrate hydrates of natural gases* (2nd Ed.). Marcel Dekker. Inc. New York, 628pp.

Sloan, E.D. 2000. Clathrate hydrates: The other common solid water phase. Industrial Engineering Chemical Research 39, 3123-3129.

Smil, V. 2003. *Energy at the Crossroads: Global Perspectives and Uncertainties.* MIT Press, 427pp.

Smith, J.D., Cappa, C.D., Wilson, D.R., Messer, B.M., Cohen, R.C. & Saykally, R.J. 2004. Energetics of hydrogen bond network rearrangements in liquid water. Science 306(29), 851-853.

Soloviev V.A. 2002a. Global estimation of gas content in submarine gas hydrate accumulations. Russian Geology and Geophysics.43(7), 648-661.

Soloviev, V.A. 2002b. Global estimation of gas content in submarine gas hydrate accumulations. In: Proceedings of the VI International Conference on "Gas in Marine Sediments". Special Issue, Geologiya I Geofizika 43(7), Siberian Brancch, Russian Academy of Sciences, 648-661.

Soloviev, V.A. Natural gas hydrate as potential mineral wealth, 59-69.

Soloviev, V.A. & Ginsburg, G.D. 1997. Water segregation in the course of gas hydrate formation and accumulation in submarine gas-seepage fields, Marine Geology 137, 59-68.

Somoza, L., Diaz-del-Rio, V., Hernández-Molina, F.J., León, R., Lobato, A., F. Lobo, F., Rodero & TASYO team. 2000. New discovery of a mud-volcanoe field related to gas venting in the Gulf of Cadiz: Imagery of multibeam data and ultra-high resolution seismic. In: Proceedings, 3th SYMPOSIUM on the IBERIAN ATLANTIC MARGIN, Faro (Algarve, Portugal), Sept. 2000 (Abs.), 397-398.

Song, H. & Geng, J. 2003. Geophysical evidence of gas hydrates existence in Dongsha region north of South China Sea. EOS, AGU, 2000, 81(48), F635.

Song, H., Geng, J., Wong, H., et al. 2001. A preliminary study of gas hydrates in Dongsha region north of South China Sea. Chinese Journal of Geophysics 44, 684-691.

Song, H., Jiang, W. & Zhang, L. 2002. Geophysical researches on marine gas hydrates (IV): Double bottom simulating reflections. Progress in Geophysics 18(3), 497-502

Song, H., Jiang, W. & Zhang, W., et al. 2002. Progress on marine geophysical studies of gas hydrates. Progress in Geophysics 17(2), 224-229

Song, H., Matsubayashi, O. & Kuramoto S. 2003. Full waveform inversion of gas hydrate-related bottom simulating reflectors. Chinese Journal of Geophysics 46, 44-52

Song, H., Matsubayashi, O., Kuramoto S. 2001. Seismic data processing of western Nankai Trough and the character of its BSR. Chinese Journal of Geophysics 44, 785-790

Song, H., Matsubayashi, O., Yang, S., et al. 2001. Geophysical researches on marine gas hydrates (II): Seismic methods, Progress in Geophysics 16(3), 110-118.

Song, H., Matsubayashi, O., Yang, S., et al. 2002. Physical property models of gas hydrate-bearing sediments and AVA character of bottom simulating reflector, Chinese Journal of Geophysics 45, 569-579.

Sorensen, H., Pedersen, K.S. & Christensen, P.L. 2002. Modeling of gas solubility in brine. Organic Geochemistry 33, 635–642.

Sorkhabi, R. 2004. Deepwater, petroleum's third wave. Geotimes 49(12), 30-31)

Spence, G.D., Chapman, N.R., Hyndman, R.D. & Cleary, C. 2001. Fishing trawler nets massive "catch" of methane hydrates. Eos 82, 621.

Spence, G.D., Hyndman, R.D., Chapman, N.R., Riedel, M., Edwards, N. & Juan. 2003. Cascadia Margin, Northeast Pacific Ocean: Hydrate Distribution from Geophysical Investigations. In: Max, M.D. (Ed.), 2000, 2003. Natural Gas Hydrate in Oceanic and Permafrost Environments, Kluwer Academic Publishers, The Netherlands, 183-198.

Spiess, V., Kasten, S., Schneider, R., Zuehlsdorff, L., Bohrmann, G., Sahling, H., M. Breitzke, M., J. Bialas, J., M. Ivanov, M. & Meteor M56 Shipboard Scientific Party. 2003. Gas hydrates and fluid venting in ultradeep large scale pockmarks at the southwest African margin off Congo. Geophysical Research Abstracts 5, 11657.

Spiess, V., Zühlsdorff, L., Seifert, L & Hirsceh, K. 2004. 3D-Seismic and Acoustic Imaging of Gas Migration and Gas Hydrate Accumulations Beneath Pockmarks in Hemipelagic Sediments off Congo, SW Africa. In: Proceedings 85(47) AGU Fall Meeting, San Francisco, Ocean Sciences General Contributions: Gas Hydrates II, OS41C-0497.

Spina, F., Max, M.D. & Bovio, E. 1998. GIS and E-Maps as the basis for data fusion and near Real-Time information distribution: present capabilities, REA specific requirements and directions for development. In: Pouliquen, E., Kirwan, A.D. Jr. & Pearson, R.T. (Eds.), Rapid Environmental Assessment, SACLANTCEN Conference Proceedings CP-44, March 10-13 1997, Lerici (La Spezia), Italy, 279-283.

Stagg, H.M.J., Symonds, P.A., Exon, N.F., Auzende, J.M., Dickens, G. & van de Beuque, S., 2000. Potential gas hydrates in Australia's marine zones. 15th Australian Geological Convention Abstracts, Sydney, 477.

SC. 2003. Statistics Canada Report on Energy Supply-Demand in Canada, www.statcan.ca.

Stauffer, D. & Aharony. A. 1994. Introduction to percolation theory. Revised second edition. Taylor and Francis, Bristol, Pennsylvania, USA

Stauffer, K.E., Ahmet, A., Kuzela, R.C. & Smith, M.A. 1999. Revised MMS (Minerals Management Service) regulations on shallow geohazards in the Gulf of Mexico. 31[st]5 Annual Society of Petroleum Engineers, et al., Offshore Technology Conference. Houston TX May 3-6, 1999 OTC-10728), 79-81.

Stern, L., Circone, S., Kirby, S. & Durham, W. 2001. Anomalous Preservation of Pure Methane Hydrate at 1 atm. Journal of Physical Chemistry(B) 105, 1756- 1762.

Stern, L., Circone, S., Kirby, S. & Durham, W. 2002. Reply to "Comments on anomalous preservation of pure methane hydrate at 1 atm." Journal Physical Chemistry B 106(1), 228-230.

Stern, L., Kirby, S. & Durham, W. 1996. Peculiarities of Methane Clathrate Hydrate Formation and Solid-State Deformation, Including Possible Superheating of Water Ice. Science 273, 1843.

Stern, L., Kirby, S. Durham, W., Circone, S. & Waite, W. 2000. Chapter 25: Laboratory Synthesis of Pure Methane Hydrate Suitable for Measurement of Physical Properties and Decomposition Behavior. In: Max, M.D. (Ed.), Natural Gas Hydrate in Oceanic and Permafrost Environments, Kluwer Academic Publishers, 323-348.

Stevenson, J. 2004. Mackenzie Valley pipeline Costs Jump, CNEWS, http://cnews.canoe.ca/CNEWS/Canada/2004/10/07/660036-cp.html

Stone, H.L. 1970. Probability Model for Estimating Three-Phase Relative Permeability, Transactions of the SPE of AIME 249, 214-218.

Subramanian, S. & Sloan, E.D. 2002. Solubility effects on growth and dissolution of methane hydrate needles, Proceedings from the Fourth International Conference on Gas Hydrates, Yokohama, Japan. 856-861.

Suess, E., et al. 2001. Sea floor methane hydrate at hydrate ridge, Cascadia margin. In: Natural Gas Hydrates: Occurrence, Distribution, and Detection, Paull. C.K & Dillon, W.P. (Eds.), AGU, Washington, DC, 87-98.

Suess, E., et al. 2001. Sea floor methane hydrates at Hydrate Ridge, Cascadia Margin. In: Paull, C.K. & Dillon, W.P. (Eds.), Natural gas hydrates occurrence, distribution, and detection. American Geophysical Union Geophysical Monograph 124, 87-98.

Suess, E., et al., 1999. Gas hydrate destabilization: Enhanced dewatering, benthic material turnover and large methane plumes at the Cascadia convergent margin. Earth Planetary Science Letters 170, 1-15.

Surdam, R.C. 1995. Wyoming and the 21st Century: The Age of Natural Gas, from Proceedings of a Workshop on the Future of Natural Gas in Wyoming: Laramie, Wyoming, Institute for Energy Research.

Svensen, H., Planke, S., Malthe-Sørenssen, A., Jamtveit, B., Myklebust, R., Torfinn Rasmussen Eidem, T.R. & Rey, S.S. 2004. Release of methane from a volcanic basin as a mechanism for initial Eocene global warming. Nature 429, 542-545.

Takeya, S., Hori, A., Takeo, H. & Uchida, T. 2000. Freezing-memory effect of water on nucleation of CO_2 hydrate crystals, Journal of Physical Chemistry (B)104, 4164-4168.

Takahashi, M., Kawamura, T., Yamamoto, Y., Ohnari, H., Himuro, S. & Shakutsui, H. 2003. Effect of Shrinking Microbubble on Gas Hydrate Formation" Journal of Physical Chemistry (B)107(10), 2171-2173.

Takahashi, H. & Tsuji, 2005. Multi-well exploration program in 2004 for natural hydrate in the Nankai-trough. Procedings Offshore Technology Conference (CD-ROM), Houston Texas, 2-5 May, 2005, (OTC 17162), 10pp.

Takahashi, H., Yonezawa, T. & Takedomi, Y. 2001. Exploration for natural hydrate in Nankai Trough wells offshore Japan. Proceedings SPE Offshore Technology Conference (CD-ROM), Houston TX, 30 April to 3 May. (OTC-13040), 12pp.

Tanaka, S. 2003. Introduction of Research Consortium for Methane Hydrate Resources in Japan, presented at Mallik International Symposium, 8-10 December 2003, Chiba, Japan.

Tang, G.Q. & Morrow, N.R. 1999. Influence of brine composition and fines migration on crude oil/brine/rock interactions and oil recovery. Journal of Petroleum Science and Engineering 24, 99-111.

Tarso, P.de, Menezes, L. & Morais, E.R. 2003. Shallow electromagnetic exploration for hyrocarbons at Sergipe-Alagoas Basin, Brazil. The Leading Edge 22, 1127-1129; DOI: 10.1190/1.1634917.

Taylor, A.E., Dallimore, S.R., Hyndmann, R.D. & Wright, F. 2003. Comparing the response of permafrost and marine gas hydrates to the end of the last Ice Age, and their sensitivity to climate warming. In: Procedings of the Mllik International Symposium, 78-79.

Taylor, M., Dillon, W.P. & Pecher, I.A. 2000. Trapping and migration of methane associated with the gas hydrate stability zone at the Blake Ridge Diapir: new insights from seismic data. Marine Geology 164, 79–89.

Thomas, D.J., Zachos, J.C., Bralower, T.J., Thomas, E. & Bohaty, S. 2002. Warming the fuel for the fire: evidence for the thermal dissociation of methane hydrate during the Paleocene-Eocene thermal maximum. Geology, 30, 1067-1070, 2002.

Tinoco, Jr., I., Sauer, K. & Wang, C.J. 1995. Physical Chemistry: Principles and Applications in Biological Sciences, Prentice Hall, New Jersey, 271-273.

Tohidi, B., Danesh, A., Burgass, R.W. & Todd, A.C. 1994. Phase equilibria in the presence of saline water systems and its application to the hydrate inhibition effect of produced water. SPE 28884, Proceedings SPE European Petroleum Conference, London, U.K., October, 25-27.

Tohidi, B., Ostergaard, K.K., Burgass, R.W. & Llamedo, M. 2002. Measuring hydrate phase boundary in porous media. In: Proceedings of the VI International Conference on "Gas in Marine Sediments". Special Issue, Geologiya I Geofizika 43(7), Siberian Brancch, Russian Academy of Sciences, 685-688.

Tomutsa, L., Freifeld, B., Kneafsey, T.J. & Stern, L.A. 2002. X-ray computed tomography observation of methane hydrate dissociation. Proceedings SPE Gas Technology Symposium, Calgary, Alberta CA, 30 April – 2 May, 2002, SPE 75533, 8pp.

Torres, M.E., et al. 1999. Geochemical observations on Hydrate Ridge, Cascadia margin during R/V-ATLANTIS cruise AT 3-35b, July 1999, COAS-Data Report174, 87 pp., Oregon State Univ., Corvallis, Oregon, 1999.

Torres, M.E., et al. 2002. Fluid and chemical fluxes in and out of sediments hosting methane hydrate deposits on Hydrate Ridge, OR: I. Hydrological provinces. Earth Planetary Science Letters 201, 525-540.

Torres, M.E., Wallmann, K., Tre´hu, A.M., Bohrmann, G., Borowski, W.S. & Tomaru, H. 2004. Gas hydrate growth, methane transport, and chloride enrichment at the southern summit of Hydrate Ridge, Cascadia margin off Oregon. Earth and Planetary Science Letters 226, 225-241

Townend, J. 1997. Estimates of conductive heat flow through bottom-simulating reflectors on the Hikurangi margin and southwest Fiordland, New Zealand, Marine Geology 141, 209-220.

Tréhu, et. al. 2004. Three-dimensional distribution of gas hydrate beneath southern Hydrate Ridge: constraints from ODP Leg 204. Earth and Planetary Science Letters 222, 845-862.

Tréhu, A.M. & Bangs, N. 2001. 3-D seismic imaging of an active margin hydrate system, Oregon continental margin, report of cruise TTN112, Oregon State University Data Report 182, ref. 2001-2, Corvallis, Oregon.

Tréhu, A.M., Bohrmann, G., Rack, F.R., Torres, M.E. & Leg 204 Scientific Party. 2003. Gas hydrate distribution and dynamics beneath Southern Hydrate Ridge: ODP Leg 204 JOIDES Journal 29(2), 5-8, (also CD-ROM).

Tréhu, A.M., et al. 1999. Temporal and spatial evolution of a gas hydrate-bearing ridge on the Oregon continental margin. Geology 27(10), 939-942.

Tréhu, A.M. & Flueh, E.R. 2001. Estimating the thickness of the free gas zone beneath Hydrate Ridge, Oregon continental margin, from seismic velocities and attenuation. Journal of Geophysical Research 106, 2035-2045.

Tréhu, A.M., Glemings, P.B., Bangs, N.L., Chevallier, J., Gracia, E., Johnson, J.E., Liu, C.-S., Liu, X., Riedel, M. & Torres, M.E. 2004. Feeding methane vents and gas hydrate deposits at south Hydrate Ridge. Geophysical Research Letters 31, L23310, doi:10.1029/2004GL021286.

Tréhu, A. M., Lin, G. Maxwell, E. & Goldfinger, C. 1995. A seismic reflection profile across the Cascadia subduction zone offshore central Oregon: New constraints on methane distribution and crustal structure. Journal of Geophysical Research 100(B8), 15,101– 15,116.

Tréhu, A.M., Long, P.E., Torres, M.E., Bohrmann, G., Rack, F.R., Collett, T.S., Goldberg, D.S., Milkov, A.V., Riedel, M., Schultheiss, P., Bangs, N.L., Barr, S.R., Borowski, W.S., Claypool, G.E., Delwiche, M.E.,

Dickens, G.R., Gracia, E., Guerin, G., Holland, M., Johnson, J.E., Lee, Y.-J., Liu, C.-S., Su, X., Teichert, B., Tomaru, H., Vanneste, M., Watanabe, M. & Weinberger, J.L. 2004. Three-dimensional distribution of gas hydrate beneath southern Hydrate Ridge: Constraints from ODP Leg 204. Earth and Planetary Science Letters, 222(3-4), 845-862.

Tréhu, A.M., Stakes, D.S., Bartlett, C.D., Chevallier, J., Duncan, R.A., Goffredi, S.K., Potter, S.M. & Salamy, K.A. 2003. Seismic and seafloor evidence for free gas, gas hydrates, and fluid seeps on the transform margin offshore Cape Mendocino. Journal Geophysical Research. 108(B5), 2263, doi:10.1029/2001JB001679

Tréhu, A.M., Torres, M.E., Moore, G.F., Suess, E. & Bohrmann, G. 1999. Temporal and spatial evolution of a gas-hydrate-bearing accretionary ridge on the Oregon continental margin. Geology 27, 939-942.

Treude, T. Ziebis, W., Boetius, A. & Jorgensen, B.B. 2003. Anaerobic oxidation of methane above gas hydrates at Hydrate Ridge, Cascadia Margin (SO165, July/August 2002). Geophysical Research Abstracts 5, 01729.

Tryon, M. D., Brown, K.M & Torres, M.E. 2002. Fluid and chemical flux in and out of the sediments, hosting methane hydrate deposits on Hydrate Ridge, OR, II: Hydrological processes. Earth and Planetary Science Letters 201, 541-557, 2002.

Tsouris, C., Brewer, P., Peltzer, E., Walz, P., Riestenberg, D., Liang, L. & West, O.R. 2004. Hydrate Composite Particles for Ocean Carbon Sequstration: Field Verification. Environmental Science Technology 38, 2470-2475.

Tsuji, Y. & Emmermann, R. 2003. Preface: Mallik International Symposium "From Mallik to the Future" Program Abstracts, From Mallik to the Future. Mallik International Symposium in Makuhari, Japan. December 8-10, (iii).

Tsuji, Y., Furutani, A., Matsuura, S. & Kanamori, K. 1998. Exploratory surveys for evaluation of methane hydrates in the Nankai Trough Area, offshore central Japan. In: Proceedings of the International Symposium on Methane Hydrates resources in the near future? October 20-22. JNOC-TRC, Chiba City, Japan, 15-26.

Tsuji, Y. & Nonaka, Y. 2004. Multi-Well Drilling Campaign METI "Tokai-oki to Kumano-nada" in 2004 for Methane Hydrate Evaluation in the Nankai Trough area, off Central Japan, AAPG Hedberg Conference, Vancouver, BC, September 12-16, 2004.

Tsuji, Y., Ochai, K., Namikawa, T., Shimizu, S., Hayashi, M., Oikawa, N., Nakamiizu, M., Matsushima, J., Simizu, T. & Shibata, T. 2003. Occurrence of Methane Hydrate and BSRs in the Nankai Trough, Offshore Japan (Abs.), Geological Society of America 35(6), September 2003, 533.

Uchida, T., Dallimore, S.R. & Collett, T.S. 2001. Comparison of natural gas hydrate coccurrence observed in the JAPEX/JNOC/GSC Mallik 2L-38 well, Mackenzie Delta, N.W.T. CA, with recently obtained natural gas hydrates, Candada Geological Survey. Annual Canadian Society of Petroleum Geologists convention (CD-ROM), Calgary, Alberta 18-22 June, 2001. (Abs.), 035-1 - 035-3,

Uchida, T., Ebinuma, T. & Narita, H. 2000. Observations of CO_2-hydrate decomposition and reformation processes, Journal of Crystal Growth 217, 189-200.

Uchida, T., Mizukoshi, I., Lu, H., Ishida, H. & Matsumoto, R. 2001. Occurrence of subsurface gas hydrate at the Nankai Trough: Implication for the geological and geochemical characteristics. Annual AAPG-SEPM convention (Denver, CO, 3-6 June 2001. (Abs A204), 1pp.

Uchida, T., Moriwaki, M., Takeya, S., Ikeda, I.Y., Ohmura, R., Nagao, J., Minagawa, H., Ebinuma, T., Narita, H., Kazutoshi Gohara, K. & Mae, S. 2004. Two-Step Formation of Methane–Propane Mixed Gas Hydrates in a Batch-Type Reactor. Journal American Institute of Chemical Engineers 50, 518–523.

Uchida, T., Ochqiai, K., Uchida, T., Kato, A., Sasaki, H., Waseda, A., Kono, F., Takeya, S. & Sawamura, F. 2003. Occurrence and physical properties of natural gas hydrate and associated gas hydrte-bearing sediments. From Mallik to the Future. Mallik International Symposium in Makuhari, Japan. December 8-10, 82.

Uchida, T., Takeya, S., Chuvilin, E.M., Ohmura, R., Nagao, J., Yakushev, V.S., Istomin, V.A., Minagawa, H., Ebinuma, T. & Narita, H. 2004. Decompostion of methane hydrates in sand, sandstone, clays, and glass beads. Jorunal Geophysical Research 109, B05206 doi:10.1029/ 2003JB002771, 12pp.

U.S. Bureau of Mines. 1946. Report on the Investigation of the Fire at the Liquefaction, Storage, and Regasification Plant of the East Ohio Gas Co., Cleveland, Ohio, October 20, 1944.

U.S. Science Support Program (USSSP). 2004. 311-Cascadia Margin Hydrates http://www.usssp-iodp.org/ Science_Support/Sailing_Information/cascadia.html

USGS. 2004. United States Geological Survey National Assessment of Oil and Gas Resources Update (October, 2004) http://energy.cr.usgs.gov/oilgas/ noga/index.htm.

USGS. 2000. Mineral Commodity Summaries, February 2000, 119.

Ussler III, W., & Paull, K. 2001. Ion exclusion associated with marine gas hydrate deposits. In: Paull. C.K & Dillon, W.P. (Eds.), Natural Gas Hydrates: Occurrence, Distribution, and Detection, AGU, Washington, DC, 41-51.

Valentine, D.L. & Reeburgh, W.S. 2000. New perspectives on anaerobic methane oxidation. Environmental Microbiology 2(5), 477-484.

van de Beuque, S., Auzende, J-M., Lafoy, Y., Bernardel, G., NERCISSIAN, A., Regnier, M., Symonds, P. & Exon, N., 1998B. Transect sismique continu entre l'arc des nouvelles-hébrides et la marge orientale de l'Australie: program FAUST (French Australian Seismic Transect). Comptes Rendu Academie Science Paris, Sciences de la terre et des planetes, 327 serie II, 761-768.

van de Beuque, S., Stagg, H.M.J., Ssyers, J., Willcox, J.B. & Symonds, P.A. 2003. Geological framework of the northern Lord Howe Rise and adjacent areas. Geoscience Australia Record 2003/01, 108 pp.

van der Meer, R. 1996. Grading in mud volcanic breccia from the Mediterranean Ridge. Marine Geology 132, 165-173.

Vanniste, M., De Batist, M., Versteeg, W., Golmshtok, A & Kremlev, A. 2001. Multi-frequency seismic study of gas hydrate – bearing sediments in Lake Baikal. Marine Geology 172, 1-21.

Vanneste, M., Mienert, J., Guidard, S., & HYDRATECH-INGGAS partners. 2002. "Arctic" gas hydrates offshore Western Svalbard, Norway. Proceedings of the 4th International Conference on Gas Hydrates. Yokohama, May 19-23, 222-227.

Vanneste, M., Poort, J., De Batist, M., & Klerkx, J. 2002. Atypical heat flow near gas hydrate irregularities and cold seeps in the Baikal Rift Zone; Marine and Petroleum Geology, 19(10), 1257-1274.

Van Rensbergen, P., Poort, J., Kipfer, R., De Batist, M., Vanneste, M., Klerkx, J., Granin, N., Khlystov, O., Krinitsky, P. 2003. Evidence of near-surface sediment mobilization and methane venting in relation to hydrate dissociation in Southern Lake Baikal, Siberia; In: Van Rensbergen, P., Hillis, R., Maltman, A., Morley, C. (Eds.), Subsurface Sediment Mobilization, Geological Society of London Special Publication, 216, 207-221.

Vassilev, A. & Dimitrov, L. 2002. Spatial and quantitative evaluation of the Glack Sea gas hydrates. In: Proceedings of the VI International Conference on "Gas in Marine Sediments". Special Issue, Geologiya I Geofizika 43(7), Siberian Brancch, Russian Academy of Sciences, 672-684.

Vincelette, R.R., Beaumont, E.A. & Foster, N.H. 1999. Classification of exploration traps. AAPG Treatise, Petroleum Geology Handbook (ISBN 0-89181-602-X), 2-1 to 2-42.

Volgin, A.V. & Woodside, J.M. 1996. Sidescan sonar images of mud volcanoes from the Mediterranean Ridge; possible causes of variations in backscatter intensity. Marine Geology 132, 39-53.

Waite, W.F., deMartin, B.J., Kirby, S.H., Pinkston, J. & Ruppel, C.D. 2002. Thermal conductivity measurements in poous mixtures of methane hydrate and quartz sand. Geophysical Research Letters 29(24), 2229, doi:10.1029/2002GL015988.

Waite, W.R., Helgerud, M.B., Nur, A., Pinkston, J.C., Stern, L.A., Kirby, S.H. & Durham, W.B. 2000. Laboratory measurements of compressional and shear wave speeds through methane hydrate. In: Gas Hydrates: Challenges for the Fugure, Holder, G. & Bishnoi, P. (Eds.), Annals of the New York Academy of Sciences 912, 1003-1011.

Waite, W.F., Winters, W.J. & Mason, D.H. 2004. Methane hydrate formation in partially water-saturated Ottawa sand. American Mineralogist 89, 1202-1207.

Wakefield, S.J. & O'Sullivan, G.M. 1996. The inorganic geochemistry of a Mediterranean Ridge mud breccia. Marine Geology 132, 203-214.

Wehner, H., Faber, E. & Hufnagel, H. 2000. Characterization of low and high molecular-weight hydrrocarbons in sediments from the Blake Ridge, Sites 994, 995, and 997 (ODP Leg 164). Proceedings of the Ocean Drilling Program (ISSN 0884-5891). Scientific Results 164, (Also CD-ROM), 47-58.

Weinberger, J.L. & Brown, K.M. 2004. Structural controls on hydrate distriubution and morphology at Hydrate Ridge, Oregon. In: Proceedings 85(47) AGU Fall Meeting, San Francisco, Ocean Sciences General Contributions, Gas Hydrates I OS34B-02.

Wellsbury, P.K., Goodman, W., Cragg, B.A. & Parkes, R.J. 2000. In: Paull, C.K., Matsumoto, R., Wallace, P.J., Dillon, W.P. (Eds.), The geomicrobiology of deep marine sediments from Blake Ridge containing methane hydrate (Sites 994, 995 and 997). Procedings of the Ocean Drilling Program, Scientific Results 164. Ocean Drilling Program, College Station, TX, 379-391.

Wellsbury, P. & Parkes, R.J. 2003. Deep biosphere: Source of mehtane for oceanic hydrate. In: Max, M.D. (Ed.) 2003. Natural Gas Hydrate in Oceanic and Permafrost Environments. Kluwer Academic Publishers, The Netherlands, 91-104.

White, R.S. & Louden, K.E. 1982. The Makran continental margin: Structure of a thickly sedimented convergent plate boundary, In: Watkins, J.S. & Drake, C.L. (Eds.), Studies in continental margin geology, American Association of Petroleum Geologists Memoir 34, 499-518.

Weitemeyer, K., Constable, S., Key, K. & Behrens, J. 2005. The use of marine EM methods for mapping gas hydrates. Proceedings Offshore Technology Conference (CD-ROM), Houston, TX, 2-5 May 2005, 5pp.

Willcox, J.B., Sayers, J., Stagg, H.M.J. & van de Beuque, S. 2001. Geological framework of the Lord Howe Rise and adjacent ocean basins. In: Hill, K.C. & Bernecker, T. (Eds.), Eastern Australasian Basins Symposium, A Refocused Energy Perspective for the Future. Petroleum Exploration Society of Australia, Special Publication, 211-225.

Willcox, J.B., Symonds, P.A., Bennett, D. & Hinz, K., 1981. Lord Howe Rise area, offshore Australia: preliminary results of a co-operative Federal Republic of Germany / Australia geophysical survey. Bureau of Mineral Resources, Geology & Geophysics, Report 228.

Willcox, J.B., Symonds, P.A., Hinz, K. & Bennett, D., 1980. Lord Howe Rise, Tasman Sea. Preliminary geophysical results and petroleum prospects. BMR Journal of Australian Geology & Geophysics 5, 225-236.

Williams, T., Deskins, G., Ward, S.L. & Hightower, M. 2001. Sound coiled-tube drilling practices. Final Report. U.S. Department of Energy DOE/NETL-2002/1170.

Willoughby, E. C. & Edwards, N. 2000. Shear velocities in Cascadia from seafloor compliance measurements. Geophysical Research Letters 27, 1021– 1024.

Willoughby, E.C., Schwalenberg, K., Edwards, R.N., Mir, R., Spence, G.D. & Hyndman. 2005. Assessment of marine gas hydrate deposits: A comparative study of seismic, electromagnetic and seafloor compliance methods. (Poster). International Gas Hydrate Workshop, Victoria, BC, Canada, 8-11 May, 2005.

Winters, B. 2005. Physical properties of sediment containing methane gas hydrate (Abs). Procedings, American Chemical Society National Meeting, Symposium on Gas Hydrates and Clathrates, San Diego, California, March 14.

Winters, W.J., Dallimore, S.R., Collett, T.S., Medioli, B.E., Matsumoto, R., Brennan-Alpert, P. & Katsube, J.T. 2003. Physical Property Relationships of Sediments from the JAPEX/JNOC/GSC et al. Mallik 5L-38 Gas Hydrate Production Research Well, Mackenzie Delta, NWT Canada. From Mallik to the Future. Mallik International Symposium in Makuhari, Japan. December 8-10, 88-89.

Winters, W.J., Dillon, W.P., Pecher, I.A. & Mason, D.H. 2000. GHASTLI; determining physical properties of sediment containing natural and laboratory-formed gas hdyrate. In: Max, M.D. (Ed.) Natural gas hydrate in oceanic and permafronst environments. Kluwer Academic Publishers. Dordrecht, Netherlands, 311-322.

Winters, W.J., Pecher, I.A., Waite, W.F. & Mason, D.H. 2004a. Physical properties and rock physics models of sediment samples containing natural and laboratory-formed methane gas hydrate, American Mineralogist 89, 1221-1227.

Winters, W.J., Waite, W.F. & Mason, D.H. 2004b. Strength and acoustic properties of Ottawa Sand containing laboratory-formed methane gas hydrate. In: The study of Gas Hydrates. Taylor, C.E. & Kwan, J.T. (Eds.), Springer, 213-236.

Winters, W.J., Waite, W.F., Mason, D.H., Dillon, W.P. & Pecher, I.A. 2002. Sediment properties associated with gas hydrate formation: Gas Hydrates Conference, 4th International, Yokohama, Japan, 19-23 May 2002, 722-727.

Wood, WT., Gardner, J., Hagen, R.A., Coffin, R.B., Pohlman, J.W., Hart, P.E. & Hutchinson, D.R. 2004. Modeling Heat and Fluid Flux of Seafloor Mounds in the Gulf of Mexico (Abs.), In: Proceedings 85(47) AGU Fall Meeting, San Francisco, Ocean Sciences General Contributions: Gas Hydrates I OS34B-08.

Wood, W.T. & Ruppel, C. 2000. Seismic and thermal investigations of the Blake Ridge hydrate area: A synthesis, In: Paull, C.K., Matsumoto, R., Wallace, P.J., Dillon, W.P. (Eds.), Proceedings ODP Scientific Results, College Station, TX, (Ocean Drilling Program) Leg 164, 253-264, 2000.

Wood, W.T ., Gettrust, J.F. & Spychalski, S.E. 2003. A new deep-towed, multi-channel seismic system. Sea Technology, Septermber, 6pp.

Wood, W.T, Gettrust, J.F., Chapman, N.R., Spence, G.D. & Hyndman, R.D. 2002. Decreased Stability of methane hydrates in marine sediments owing to phase boundary roughness, Nature 420, 656-660.

Wood, W.T. & Gettrust, J.F. 2001. Deep-towed seismic investigations of methane hydrates, In: Paull, C.K. & Dillon, W.P. (Eds.), Natural Gas Hydrates: Occurrence, Distribution, and Dynamics, AGU Monograph 124, 165-178.

Wood, W.T., Holbrook, W.S. & Hoskins, H. 2000, In situ measurements of P-wave attenuation in the methane hydrate and gas-bearing sediments of the Blake Ridge, In: Paull, C.K., R. Matsutmoto, P.J. Wallace, and W.P. Dillon (Eds.), Proceedings ODP, Scientific Results, 164, College Station, TX, 265–272.

Woodside, J.M. & Volgin, A.V. 1996. Brine pools associated with Mediterranean Ridge mud diapirs; an interpretation of echo-free patches in deep tow sidescan sonar data. Marine Geology 132, 55-61.

Woodside, J.M., Ivanov, M.K., Limonov, A.F., & Shipboard Scientists of the ANAXIPROBE Expedition. 1998. Shallow gas and gas hydrates in the Anaximander Mountains region, Eastern Mediterranean Sea.

In: Henriet, J.-P. & Mienert, J. (Eds.), Gas Hydrates: Relevance to World Margin Stability and Climatic Change. Geological Society London, Special Publication 137, 177-193.

WP. 2004. Internet site: http://walrus.wr.usgs.gov/globalhydrate/ U.S. Geological Survey. Same reference as Kvenvolden & Lorenson, 2000. Downloadable PDF poster.

Wu, N. 2004. Prospect and indicative characteristics of gas hydrate in the northern continental slope of the South China Sea. . Proceedings 32nd International Geological Congress, Florence, Italy, 1250

Wu, S., Zhang, G., Huang, Y., Liang, J. & Wong, H.K. 2005. Gas hydrate occurrence on the continental slope of the northern South China Sea. Marine and Petroleum Geology 22(3), 403-412.

Xu, W. & Ruppel, C. 1999. Predicting the occurrence, distribution, and evolution of methane gas hydrate in porous marine sediments. Jorunal Geophys. Research 104, 5081–5096.

Xu, W. 2002. Phase Balance and Dynamic Equilibrium During Formation and Dissociation of Methane Gas Hydrate. Proceedings of the Fourth International Conference on Gas Hydrates 1, 195-200.

Xu, W. 2004. Modeling dynamic marine gas hydrate systems. American Mineralogist 89, 1271-1279.

Xu, W., Lowell, R.P. & Peltzer, E.T. 2001. Effect of seafloor temperature and pressure variations on methane flux from a gas hydrate layer: Comparison between current and late Paleocene climate conditions. Journal of Geophysical Research 106, 26,413-26,423.

Wu., S., Zhang, G., Huang, Y., Liang, J. & Wong, H.K. 2005. Gas hydrate occurrence on the continental slope of the northern South China Sea. Marine and Petroleum Geology 22(3), 403-412.

Xun, H., Bangs, N.L. & Stoffa, P.L. 2004. Estimation of Free Gas Saturation Using AVO Analysis on 3D Seismic Data at South Hydrate Ridge, Cascadia Accretionary Complex. In: Proceedings 85(47) AGU Fall Meeting, San Francisco, Ocean Sciences General Contributions: Gas Hydrates II OS41C-0492.

Yang, J., Zhang, X., Chen, J., He, Q., Ye, Y. & Gong, J. 2003. Gas hydrate hunting in China seas. Geophysical Research Abstracts 5, 00854.

Yang, S.O., Cho, S.H., Lee, H. & Lee, C.S. 2001. Measurement and prediction of phase equilibria for water + methane in hydrate forming conditions. Fluid Phase Equilibria 185, 53-63.

Yakushev, V.S. & Chuvilin, E.M. 2000. Natural gas and gas hydrate accumulations within permafrost in Russia. Cold Regions and Technology 31, 189-197.

Yao, B. 1998. Preliminary survey of gas hydrates in north of South China Sea. Marine Geology and Quaternary Geology 18(4), 11-18.

Yoshikawa, K., Kondo, Y., Kimura, T. & Fujimoto, T. 2003. Production method for hydrate and device for proceeding the same. U.S. Patent 6,653,516. (November 25, 2003).

Yuan, J. & Edwards, R.N. 2000. The assessment of marine gas hydrates through electrical remote sounding: Hydrate without a BSR? Geophysical Research Letters 27, 2397–2400.

Yuan, J., & Edwards, R.N. 2000. The assessment of marine gas hydrates through electrical remote sounding: Hydrate without a BSR? Geophysical Research Letters 27, 2909-2922.

Yuan, T, Hyndman, R.D., Spence, G.D. & Desmos, B. 1996. Seismic velocity increase and deep-sea hydrate concentration above a bottom simulating reflector on the northern Cascadian slope. Journal of Geophysical Research 101, 13655-13671.

Yuan, T., Spence, G.D. Hyndman, R.D., Minshull, T.A. & Singh, S.C. 1999. Seismic velocity studies of a gas hydrate bottom-simulating reflector on the northern Cascadia continental margin: Amplitude modeling and full waveform inversion. Journal of Geophysical Research 104, 1179–1191.

Yuan, T., Spence, G.D., Hyndman, R.D., Minshull, T.A. & Singh, S.C. 1999. Seismic velocity studies of a gas hydrate bottom-simulating reflector on the northern Cascadia continental margin: Amplitude modeling and full waveform inversion. Journal of Geophysical Research 104, 1179~1191.

Zatsepina, O. & Buffett, B.A. 2001. Experimental study of the solubility of CO_2 hydrate in a porous medium. Fluid Phase Equilibria 192, 85–102.

Zhang, C.L., Li, Y., Wall, J.D., Larsen, L., Sassen, R., Huang, Y. Wang, Y., Peacock, A., White, D., Horita, J. & Cole, D.R. 2002. Lipid and carbon isotopic evidence of methane-oxidizing and sulfate reducing bacteria in association with gas hydrates from the Gulf of Mexico. Geology 30, 239-242.

Zhang, C.L., Pancost, R.D., Sassen, R., Qian, Y. & Macko, S.A. 2003. Archeal lipid biomarkers and isotopic evidence of anaerobic methane oxidation associated with gas hydrates in the Gulf of Mexico. Organic Geochemistry 34, 827-836.

Zhang, G., Huang, Y., Zhu, Y. et al. 2000. Prospect of gas hydrate resources in the South China Sea. Marine Geology & Quaternary Geology 22(1), 75~81.

Zhang, Y. 2003. Methane escape from gas hydrate systems in marine environment, and methane-driven oceanic eruptions. Geophysical Research Letters 30, doi:10.1029/2002GL016658.

Zhang, Y. & Xu, Z. 2003. Kinetics of convective crystal dissolution and melting, with applications to methane hydrate dissolution and dissociation in seawater. Earth and Planetary Science Letters 213, 133-148.

Ziff, P. 2004, LNG and Texas Natural Gas, Texas Independent Producers and Royalty Owners Association, Speech to Summer Policy Meeting, San Antonio. http://www.tipro.org/TIPRO-Web/pdf/meetingpres/sumpol2004/Paul%20Ziff.pdf.

Zühlsdorff, L., Spiess, V., Hübscher, C. & Breitzke, M. 1999. Seismic reflectivity in sediments at the eastern flank of the Juan de Fuca Ridge: Evidence for fluid migration? Journal of Geophysical Research 104, 15,351– 15,364.

Zykov, M & Chapman, R. 2005. 3-D velocity model of hydrocarbon vent site in Cascadia region offshore Vancouver Island. (Poster) 4th International Gas Hydrate Workshop, Victoria, BC, Canada, 8-11 May, 2005.

Miscellaneous Information

Author Address list

William P. Dillon, Ph.D.
U.S. Geological Survey
(also HEI)
Quissett Campus
Woods Hole, MA 02543
[508-457-2224 / 508 457-2310]
bdillon@.usgs.gov

Arthur H. Johnson
Hydrate Energy International (HEI)
Chief of Exploration
Hydrate Energy International
612 Petit Berdot Drive
Kenner, LA 70065-1126
tel & Fx: 504-464-6208, cell: 504-220-6208
art_johnson@hydrate-energy.com
artjohnson51@hotmail.com

Michael D. Max, Ph.D.
MDS Research
(also HEI)
1601 3rd Street South
St. Petersburg, FL 33701
[727-821-3993 / 727-821-3340]
mmax@mdswater.com.

With:

George J. Moridis, Ph.D.
(& Michael Kowalsky)
Research Area Leader, Transport and Thermodynamics
Lawrence Berkeley National Laboratory, University of California
Earth Sciences Division, 1 Cyclotron Rd., MS 90-1116
Berkeley, CA 94720
GJMoridis@lbl.gov

George Taft, Esq.
US Department of State
Office of the Legal Adviser
Washington, D.C. 20520
[202-647-7036 /].
TaftG@state.gov

John P. Osegovic, Ph.D.
Shelli R. Tatro, Ph.D.
Sarah A. Holman, Chem. Eng.
Marine Desalination Systems, L.L.C.
1601 3rd St. South
St. Petersburg, FL 33701

And:
Peter G. Brewer
Senior Scientist
MBARI
P.O. Box 628
7700 Sandholt Road
Moss Landing CA 95039
[831-775-1706 / 831-775-1620]
e-mail: brpe@mbari.org

GAS HYDRATE FRESH WATER RESERVOIRS (Peter G. Brewer)

The quantity of gas hydrates on earth has been estimated as about 1 x 10^{19} grams of carbon (Kvenvolden, 1993) as methane, and therefore 1/12x10^{19} moles of carbon. For every gas molecule there are 6 molecules of water, and therefore we have 6/12x10^{19} moles of water tied up as hydrate, or 0.5x10^{19} moles of fresh water, for brine is excluded on freezing. This is about 9x10^{16} liters of water, and thus the amount of water tied up as gas hydrate is a very large number. It is approximately equal to the volume of all the worlds lakes, or twice the amount of sea ice in the Arctic and Antarctic combined (Table 1), although there are clearly large uncertainties in the hydrate estimate. A volume of water this large would cover the surface of the ocean with a fresh water layer about 3.24 m deep; small compared with the fresh water fluxes associated with ice age changes, but still an impressive number.

Water Reservoir	Volume (liters x 10^{16})
World Lakes	12
Gas Hydrates	~ 9
Arctic Sea Ice	2-5
World Rivers	1.2
Antarctic Sea Ice	0.5-3
Antarctic land ice	28x10^6 km^3
Greeland land ice	2.7x10^6 km^3

Table 1. Fresh water reservoirs of Earth.

EARLIEST RECORD OF ARTIFICIALLY PRODUCED GAS HYDRATE

The first record of production of gas hydrate has been regarded as Michael Faraday's 1823 reproduction of the chlorine hydrate fabrication experiment reported by Sir Humphrey Davy in 1810 (Max, 2000, 2003). 30 Years earlier, however, Joseph Priestley (1790) recorded an experiment of his that took place on January 6-7, 1779 using 'vitriolic acid vapors', SO_3 or SO_2. All the observations were made with reference to pure water over the course of two days of unseasonably cold weather. The temperatures recorded by Priestley appear to be Farenheit, but also could have referred to a home-made temperature scale. In his experiment Preistly filled a container with fresh water and exposed it to an excess of SO_3 or SO_2 vapor. The water was found to include SO_3 upon freezing, and to 'freeze' at temperatures above that at which normal water remained liquid. The water-SO_3 solid (hydrate) also had a density greater than the ground water, as it sank in the water after it formed, in contrast to pure water ice, which floats.

Although not knowing precisely what he had done and certainly knowing nothing of clathrates as a mineral species, Priestley clearly identified and described the characteristiic behavior of a negatively buoyant gas hdyrate. He also observed that whereas normal water excludes air upon freezing, the strange ice held the gas and exsolved it when it melted, releasing a strong smell of sulfur trioxide. He also noted that the absorbtion of the gas by a volume of water resulted in a lower 'freezing' temperature for the water than water which had not absorbed this gas.

"…. water impregnated with viriolic acid air (SO_3) that may be converted into ice, whereas water impregnated with fluor acid air will not freeze. I had observed with respect to marine acid air and alkaline air, that they dissolve ice, and that water impregnated with them is incapable of freezing, at least in such `a degree of cold as I had exposed them to. The same, I find, is the case with fluor acid air, but it is not so at all with vitriolic acid air, which entirely contrary to my expectation, I find to be altogether different from marine acid air in this respect and to resemble fixed air. But whereas water impregnated with fixed air discharges it when it is converted into ice, water impregnated with vitriolic acid air, and then frozen, retains it as strongly as ever."

Coastal Systems and Continental Margins

1. B.U. Haq (ed.): *Sequence Stratigraphy and Depositional Response to Eustatic, Tectonic and Climatic Forcing*. 1995 ISBN 0-7923-3780-8
2. J.D. Milliman and B.U. Haq (eds.): *Sea-Level Rise and Coastal Subsidence*. Causes, Consequences, and Strategies. 1996. ISBN 0-7923-3933-9
3. B.U. Haq, S.M. Haq, G. Kullenberg and J.H. Stel (eds.): *Coastal Zone Management Imperative for Maritime Developing Nations*. 1997 ISBN 0-7923-4765-X
4. D.R. Green and S.D. King (eds.): *Coastal and Marine GeoInformation Systems: Applying the Technology to the Environment*. 2001 ISBN 0-7923-5686-1
5. M.D. Max (ed.): *Natural Gas Hydrate in Oceanic and Permafrost Environments*. 2000 ISBN 0-7923-6606-9; Pb 1-4020-1362-0
6. J. Chen, D. Eisma, K. Hotta and H.J. Walker (eds.): *Engineered Coasts*. 2002 ISBN 1-4020-0521-0
7. C. Goudas, G. Katsiaris, V. May and T. Karambas (eds.): *Soft Shore Protection*. An Environmental Innovation in Coastal Engineering. 2003 ISBN 1-4020-1153-9
8. D.M. FitzGerald and J. Knight (eds.): *High Resolution Morphodynamics and Sedimentary Evolution of Estuaries*. 2005 ISBN 1-4020-3295-1
9. M.D. Max, A.H. Johnson and W.P. Dillon: *Economic Geology of Natural Gas Hydrate*. 2006 ISBN 1-4020-3971-9